INFRARED AND MILLIMETER WAVES

VOLUME 8 ELECTROMAGNETIC WAVES
IN MATTER, PART I

CONTRIBUTORS

G. W. Chantry

W. F. X. Frank

F. Gervais

P. Goy

A. Hadni

B. Jensen

U. Leute

J. C. Maan

T. Ohyama

E. Otsuka

S. Perkowitz

INFRARED AND MILLIMETER WAVES

VOLUME 8 ELECTROMAGNETIC WAVES IN MATTER, PART I

Edited by **KENNETH J. BUTTON**
NATIONAL MAGNET LABORATORY
MASSACHUSETTS INSTITUTE OF TECHNOLOGY
CAMBRIDGE, MASSACHUSETTS

1983

ACADEMIC PRESS
A Subsidiary of Harcourt Brace Jovanovich, Publishers

New York   London
Paris   San Diego   San Francisco   São Paulo   Sydney   Tokyo   Toronto

ACADEMIC PRESS, INC.
111 Fifth Avenue, New York, New York 10003

United Kingdom Edition published by
ACADEMIC PRESS, INC. (LONDON) LTD.
24/28 Oval Road, London NW1 7DX

Library of Congress Cataloging in Publication Data

Main entry under title:

Infrared and millimeter waves.

  Vol. 8-   edited by K.J. Button.
  Includes bibliographies and indexes.
  Contents: v.1. Sources of radiation.--v.2. Instrumenta-
tion.--v.3. Submillimeter techniques.--[etc.]--v.8. Electromagnetic waves in matter.
  1. Infra-red apparatus and appliances. 2. Millimeter
wave devices. I. Button, Kenneth J.
TA1570.152   621.36'2   79-6949
ISBN 0-12-147708-8 (v. 8)

PRINTED IN THE UNITED STATES OF AMERICA

83 84 85 86   9 8 7 6 5 4 3 2 1

621.362
INF

CONTENTS

LIST OF CONTRIBUTORS

Numbers in parentheses indicate the pages on which the authors' contributions begin.

G. W. CHANTRY[1] (1), *National Physical Laboratory, Teddington, Middlesex, United Kingdom*

W. F. X. FRANK (51), *AEG-Telefunken, Geschäftsbereich Hochfrequenztechnik, Ulm, Federal Republic of Germany*

F. GERVAIS (279), *Centre de Recherches sur la Physique des Hautes Températures, Centre National de la Recherche Scientifique, 45045 Orléans, France*

P. GOY (341), *Laboratoire de Physique de l'Ecole Normale Supérieure, Paris, France*

A. HADNI (173), *Laboratoire Infrarogue Lointain, Université de Nancy I, 54506 Vandoeuvre, Nancy, France*

B. JENSEN (127), *Department of Physics, Boston University, Boston, Massachusetts 02215*

U. LEUTE (51), *Universität Ulm, Abteilung Angewandte Physik, Ulm, Federal Republic of Germany*

J. C. MAAN (387), *Max-Planck-Institut für Festkörperforschung, Hochfeld Magnetlabor, 166X, F38042 Grenoble, France*

T. OHYAMA (213), *Department of Physics, College of General Education, Osaka University, Toyonaka, Osaka 560, Japan*

E. OTSUKA (213), *Department of Physics, College of General Education, Osaka University, Toyonaka, Osaka 560, Japan*

S. PERKOWITZ (71), *Department of Physics, Emory University, Atlanta, Georgia 30322*

[1]Present address: British Embassy, 77-Friedrich Ebert Allee, D-5300 Bonn, Federal Republic of Germany.

PREFACE

This is the first volume in this treatise to deal exclusively with the millimeter and submillimeter wave properties of materials and the methods of measuring and interpreting these properties. A second volume on this topic is now in preparation, and plans for a third are in progress.

The table of contents speaks eloquently for the theme of this subseries on electromagnetic wave interaction in matter. G. W. Chantry's dielectric materials studies are the foundation of a branch of modern classical physics that he inspires with his semiannual Gordon conferences. This general treatment of dielectrics is naturally followed by the work of W. F. X. Frank and U. Leute on the far-infrared spectroscopy of high polymers and then by several semiconductor chapters. S. Perkowitz gives a general treatment of the spectroscopy of semiconductors; B. Jensen provides a most thorough treatment of free carrier behavior in semiconductors (which is most important in the far infrared); A. Hadni then reviews pyroelectric detectors.

We were most fortunate to obtain a contribution on cyclotron resonance by T. Ohyama and E. Otsuka, who have done a great deal of work in that area. F. Gervais contributed a chapter on his very innovative spectroscopic work. The chapter by Phillipe Goy appeared to me to be so timely that I designated it as the opening invited paper at the Miami Beach conference in December 1981. Finally, J. C. Maan was invited to prepare the chapter on semiconductor layered structures because he has the ability to make this specific example of a larger, emerging topic sound simple.

A second volume devoted to electromagnetic waves in matter is being readied for press. Volumes 9, 10, and 11 begin a subseries on millimeter wave components and techniques; all three of these volumes are in production and scheduled for publication very soon.

CONTENTS OF OTHER VOLUMES

CHAPTER 1

Properties of Dielectric Materials

G. W. Chantry

National Physical Laboratory
Teddington, Middlesex
United Kingdom

I. Introduction

The study of liquids and polymers by means of far-infrared and submillimeter spectroscopy forms part of the much larger topic of dielectric physics. A dielectric medium is one in which there are no free charges so the dc conductivity is zero, but the medium can sustain displacement currents and these may have lossy components. Thus dielectrics are all materials that are not metallic, semiconducting, or ionized. If an external field E is applied to a dielectric, the field inside the dielectric is given by

$$D = (\varepsilon/\varepsilon_0)E, \qquad (1)$$

where ε is the permittivity of the dielectric and ε_0 the absolute permittivity of free space (8.85418×10^{-12} F/m). In nearly all dielectric work, however, it is customary to write $\varepsilon = \varepsilon/\varepsilon_0$ to avoid the constant apperance of the ε_0 factor; then ε so defined is the *relative permittivity of the medium,* i.e., the permit-

Copyright © 1983 by Academic Press, Inc.
ISBN 0-12-147708-8

tivity relative to that of the vacuum. The relative permittivity is complex because it will have a lossy component, and one therefore usually writes

$$\hat{\varepsilon} = \varepsilon' - i\varepsilon'', \tag{2}$$

where the caret is used to signify an explicitly complex quantity. The use of complex relative permittivities goes very well with the use of the complex representation to describe alternating fields. The dielectric medium will thus sustain two currents in response to an ac field of circular frequency ω, a quadrature or displacement current determined by ε' and an in-phase or lossy current determined by ε''. The complex relative dielectric conductivity $\hat{\sigma}$ is therefore also complex with components

$$\hat{\sigma} = j\omega\varepsilon' + \omega\varepsilon'', \tag{3}$$

and the current, in the complex plane, is not perpendicular to the voltage but departs from this position by an angle δ whose tangent is given by

$$\tan \delta = \varepsilon''/\varepsilon'. \tag{4}$$

In most practical cases, δ will always be small and it is permissible to write

$$\delta = \varepsilon''/\varepsilon', \tag{5}$$

which is the justification for the common practice of quoting the dimensionless ratio $\varepsilon''/\varepsilon'$ in microradians.

The presence of a slab of dielectric in a volume that would otherwise be vacuum leads to a polarization P whose magnitude is given by the field in the dielectric minus the applied field; in other words,

$$P = (\hat{\varepsilon} - 1)E. \tag{6}$$

This equation is not very useful as it stands because the frequencies at which P will fall to zero, and at which consequentially $\hat{\varepsilon}$ will equal unity, lie in the x-ray region, whereas all dielectric work is restricted to the region of millimeter and longer wavelengths. It is more usual, therefore, to write

$$P = (\hat{\varepsilon} - \varepsilon_\infty)E, \tag{7}$$

where ε_∞ is a high-frequency limiting permittivity. This is unfortunately a rather amorphous and poorly defined quantity because it is not possible, in fact, to find any measurable frequency at which dispersion does not exist. It is rather to be considered as a parameter: the permittivity in frequency regions so high that the particular dispersion mechanism under consideration will no longer have any effect. It is also helpful to normalize the polarization by writing

$$P_n = \frac{\hat{\varepsilon} - \varepsilon_\infty}{\varepsilon_s - \varepsilon_\infty} E, \tag{8}$$

where ε_s is the static relative permittivity. Then P_n will go from one at $\omega = 0$ to zero at $\omega = \infty$.

At extra-high frequencies and beyond ($\nu > 30$ GHz) it is not possible, with the currently available techniques, to measure field parameters such as the vectorial electric field strength and its phase. Rather, one measures energy flow; and in this situation one is led naturally to introduce a refractive index n which is the ratio of phase velocity in free space (i.e., c) to that in the medium. The space dependence of the field is then given by

$$E = E_0 \exp(-i\omega nx/c), \tag{9}$$

but the measurable quantity, the intensity, is given by

$$I = \tfrac{1}{2}\varepsilon_0 cE_0^2. \tag{10}$$

The propagating medium will usually be lossy, so one has a progressive attenuation of the field given by Lambert's law

$$E = E_0 \exp(-\tfrac{1}{2}\alpha x), \tag{11}$$

which also can be written as

$$I = I_0 \exp(-\alpha x). \tag{12}$$

Here α is the power absorption coefficient (usually measured in nepers per centimeter). Equations (9) and (11) can be combined to give

$$E = E_0 \exp(-i\omega \hat{n}x/c), \tag{13}$$

where the complex refractive index $\hat{n}$ is defined by

$$\hat{n} = n - i(\alpha/4\pi\bar{\nu}), \tag{14}$$

in which the wave number

$$\bar{\nu} = \nu/c = \omega/2\pi c \tag{15}$$

is introduced. The two formalisms, in terms of either $\hat{\varepsilon}$ or $\hat{n}$, are readily connected by means of Maxwell's celebrated relationship

$$\hat{\varepsilon} = \hat{n}^2. \tag{16}$$

Identifying real and imaginary components in Eq. (16) gives the simple relations

$$\varepsilon' = n^2 - (\alpha/4\pi\bar{\nu})^2 \tag{17a}$$

and

$$\varepsilon'' = 2n(\alpha/4\pi\bar{\nu}), \tag{17b}$$

which have the not-so-simple converses

$$n = (\varepsilon'/2)^{1/2}\{[1 + (\varepsilon''/\varepsilon')^2]^{1/2} + 1\}^{1/2} \qquad (18a)$$

and

$$\alpha/4\pi\bar{v} = (\varepsilon'/2)^{1/2}\{[1 + (\varepsilon''/\varepsilon')^2]^{1/2} - 1\}^{1/2}. \qquad (18b)$$

It is a fairly widespread practice in dielectric work to introduce the absorption index k and to write

$$\hat{n} = n - ik, \qquad (19)$$

from which it immediately follows that

$$k = \alpha/4\pi\bar{v}. \qquad (20)$$

This is a rather unsatisfactory practice because confusion with the wave vector, also designated by k, arises; nevertheless it is firmly established, especially in solid-state work. Another convention, fortunately now tending to become obsolete, is the use of the extinction index κ defined by

$$\hat{n} = n(1 - i\kappa). \qquad (21)$$

It is also worth mentioning that there is no universally agreed convention for the sign of the imaginary components in these complex quantities. One can find $\hat{\varepsilon}$, for example, defined with either sign. There is no physical difference; the two signs merely represent the sense of the conventional rotation of the vectors in the Argand diagram: negative going with clockwise and positive with anticlockwise. However, as soon as one has defined any one of these quantities, all the rest become fixed. Thus, because it seems more natural to consider the vector $\hat{E}(t)$ rotating anticlockwise into the first quadrant, one writes $\hat{E}(t) = E_0 \exp(i\omega t)$; then the sign for the imaginary component of $\hat{\varepsilon}$ will be negative because the charge will lag behind the driving field.

Dielectric spectroscopy is the determination of the variation of $\hat{\varepsilon}$ or $\hat{n}$ with frequency. In principle there is no upper frequency limit to dielectric spectroscopy; but to keep the concept useful, it is usual to introduce a division that separates phenomena involving long-range cooperative motion from phenomena involving highly localized, essentially independent motion. The former is the class to be treated by dielectric concepts, whereas the latter is treated by the methods of, for example, molecular spectroscopy. This division is helpful, but it is arbitrary, and physical phenomena may well transgress its boundaries. Thus the positions and intensities of lines in molecular spectra can be calculated by purely local models; but calculating their shapes demands a more global treatment involving long-range interactions. The division does make sense spectroscopically, because at frequencies below 6 THz (200 cm^{-1}), one is often talking of dielectric phenomena whereas at higher frequencies one seldom is. So, for the purposes of this review, the upper limit to dielectric spectroscopy will be taken to be 200 cm^{-1}.

One can approach dielectric physics from two directions: one can consider the macroscopic theory, which essentially stems from the application of universally valid physical laws such as Maxwell's equations; or one can consider the microscopic theory, which shows how the motion of the constituent entities leads to the observed dielectric properties. The macroscopic theory gives results that are exact or nearly so but of somewhat limited information content. The microscopic theory is much more interesting because it can, in principle, give a wealth of information; however, the computational difficulties involved in its full elaboration are formidable and can be side-stepped only by means of rather drastic approximations. Therefore the microscopic theory does not give quite as much insight as one might hope; nevertheless considerable progress has been made.

II. The Macroscopic Theory

The complex permittivity $\hat{\varepsilon}$ is the natural quantity for those who have coherent sources and detectors and can measure field parameters. Typically this means those who work below 30 GHz, but the rapid development of techniques for measuring the higher-frequency regions is tending to extend this limit upward. The complex refractive index $\hat{n}$ is the natural quantity for those limited to optical techniques with incoherent black body sources and thermal detectors. Rather interestingly, the closing up of the millimeter–submillimeter gap has enabled direct tests of Maxwell's relation Eq. (16) to be carried out by comparing $\hat{\varepsilon}$ with $\hat{n}^2$ at a fixed frequency; but the results, as expected, entirely support the theory. Therefore one can use either formalism; but because that involving $\hat{\varepsilon}$ is simpler, it is usually preferred. We now consider those properties of $\hat{\varepsilon}$ that are independent of particular microscopic models, calling this branch of the topic the *macroscopic theory*.

A. THE RESPONSE FUNCTION

The behavior of dielectric media can be modeled quite well by simple RC circuits. There are many possible combinations, but they all contain a common element, a resistor R and a capacitor C in series. It is therefore important to analyze the response of this simple circuit combination to various time dependences of the applied field. The situation is illustrated in Fig. 1. A driving voltage $V(t)$ is applied across the resistor and capacitor in series, and we are interested in the resulting voltage across the capacitor $V_c(t)$. The basic differential equation is

$$V(t) - [dV_c(t)/dt]RC = V_c(t), \tag{22}$$

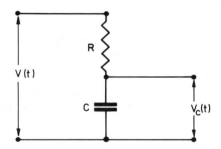

FIG. 1   A simple *RC* circuit used to model a lossy dielectric.

and we seek solutions valid in the long time limit (i.e., $t \gg RC$) when starting transients will have died out. Taking Laplace transforms throughout gives

$$V(s) = V_c(s) + RCsV_c(s) \tag{23}$$

from which

$$V_c(s) = V(s)/(1 + RCs). \tag{24}$$

Inverting the Laplace transform and invoking the convolution theorem gives

$$V_c(t) = \int_0^t V(t')\tau^{-1} \exp[-(t - t')/\tau] \, dt', \tag{25}$$

which is a special case of the more general relation

$$V_c(t) = \int_0^t V(t')\dot{R}(t - t') \, dt. \tag{26}$$

The function $R(t)$ whose time derivative appears in this equation is the response function. For the simple RC circuit it takes the elementary form

$$R(t) = \exp(-t/\tau), \tag{27}$$

where $\tau = RC$. It is called the response function because, if $V(t)$ changes abruptly, $R(t)$ measures the subsequent response. Thus suppose that $V(t)$ is constant for a time t_0 ($\gg \tau$) but then drops to zero. What will be the voltage across the capacitor at a subsequent time t ($> t_0$)? Substituting in Eq. (26) and noting that $\exp(t_0/\tau)$ is much greater than unity gives the answer

$$V_c(t) = V_0 \exp(-(t - t_0)/\tau). \tag{28}$$

It t_0 is allowed to decrease toward zero, this becomes

$$V_c(t) = V_0 \exp(-t/\tau) = V_0 R(t). \tag{29}$$

In this limit, we are considering an impulse, or delta, function, so the response function may be defined as the result at a subsequent time $t \, (\gg \tau)$ of the application of a unit area impulse at time zero.

Equation (26) is general, but in most dielectric cases, and certainly for most high-frequency dielectric cases, one is considering the response to a cosinusoidal (or what is the same thing, a complex exponential) driving voltage. One has thus

$$\hat{V}(t) = V_0 \exp i\omega_0 t, \tag{30}$$

and hence has

$$\hat{V}_c(t) = V_0 \exp i\omega_0 t/(1 + i\omega_0 \tau). \tag{31}$$

On rationalizing, this gives the in-phase and quadrature components mentioned previously. From Eq. (18), it follows that

$$[\hat{\varepsilon}(\omega) - \varepsilon_\infty]/(\varepsilon_s - \varepsilon_\infty) = 1/(1 + i\omega_0 \tau), \tag{32}$$

which is a special case of the general relation

$$[\hat{\varepsilon}(\omega) - \varepsilon_\infty]/(\varepsilon_s - \varepsilon_\infty) = \int_0^\infty \dot{R}(t) \exp(-i\omega t) \, dt. \tag{33}$$

In other words the scaled complex relative permittivity is given by the complex Fourier transform of the time derivative of the response function. This is a most valuable result because it permits the frequency dependence of the complex relative permittivity to be calculated once the response function is known. Of course we have carried out all the previous analyses in terms of a simple model, but the results are general. Equation (33) applies universally, and there are many dielectrics that obey Eq. (32). Equation (26), which is a form of Duhamel's integral, is likewise universal; and in fact, Eqs. (33) and (26) may be transformed into one another by the use of the Fourier convolution theorem. Thus one has

$$V_c(t) = V(t) * \dot{R}(t) = \int \mathscr{F}(V(t)) \cdot \mathscr{F}(\dot{R}(t)) \exp i\omega t \, d\omega, \tag{34}$$

where the asterisk denotes convolution and the symbol $\mathscr{F}$ means "the Fourier transform of." The Fourier transform of $V_0 \exp i\omega_0 t$ is the impulse, or delta, function $\delta(\omega - \omega_0)$, so one has

$$V_c(t) = V_0 \int \delta(\omega - \omega_0)\{\mathscr{F}[\dot{R}(t)] \exp(-i\omega t)\} \, d\omega, \tag{35}$$

which, because the delta function used in integration is a shift operator, becomes

$$V_c(t)/V_0 = \mathscr{F}[\dot{R}(t)] \exp i\omega_0 t, \tag{36}$$

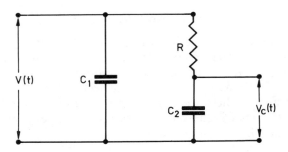

FIG. 2 Simplest equivalent circuit for the response of a dielectric material.

and Eq. (33) follows at once. The proof of Eq. (26) itself is also best tackled by means of the delta function formalism. The necessary physical assumptions are that the system (i.e., the dielectric or its model) be linear and that the principle of superposition be valid. With these assumptions, one can take the polarization at a time t to be the result of the decaying voltages from all previous times. The simplest model for a dielectric that has this characteristic is shown in Fig. 2.

A pulse applied as $V(t)$ will instantly charge C_1, but then afterward the charge on C_1 will be shared with C_2 according to a characteristic relaxation time

$$\tau = C_1 C_2 R/(C_1 + C_2). \tag{37}$$

When the next pulse is applied, the current through R will be determined by the difference between the voltage thus applied and that already present across the terminals of C_1. The voltage on C_2 would thus be

$$V_2(\Delta) = V_1(0) - [V_1(0) - V_2(0)]R(\Delta). \tag{38}$$

where Δ is the interval between the pulses and $R(\Delta)$ the value of the response function at time Δ. After $(n + 1)$ pulses, one would have

$$V_2[(n + 1)\Delta] = V_1(n\Delta) - [V_1(n\Delta) - V_2(n\Delta)]R(\Delta). \tag{39}$$

By continually expanding the square bracket in Eq. (39) and letting $\Delta \rightarrow 0$ so that the sequence of delta pulses approximates $V(t)$, one can obtain the summation

$$V_2[(n + 1)\Delta] = [1 - R(\Delta)] \sum_0^n V_1(i\Delta)[R(\Delta)]^{(n-i)}. \tag{40}$$

In the limit as Δ approaches zero, this expression turns into the integral

$$V_2(t) = \int_0^t V_1(t')\dot{R}(t - t') \, dt', \tag{41}$$

as quoted previously. Integrating this by parts gives an equivalent result which is sometimes used:

$$V_2(t) = V_1(t) - \int_0^t \frac{dV(t)}{dt} \cdot R(t - t') \, dt'. \tag{42}$$

B. THE CORRELATION FUNCTION AND THE COLE–COLE PLOT

In the microscopic theory one looks for a particular physical form for the interaction between the molecules making up the dielectric and then deduces the macroscopic properties. However, it is possible to make meaningful statements about the time dependence of the polarization, and hence about $\hat{\varepsilon}(\omega)$, in purely phenomenological terms. One notes that by definition there are no free charges in a dielectric, so the motion under analysis is that of dipoles, quadrupoles, etc. But because the interaction of the higher multipoles with the driving electromagnetic field is so much weaker, one can usually ignore them and consider only dipoles. If one has a dipole $\mu(t)$ in a dielectric medium, a correlation function $f(t)$ can be defined by

$$f(t) = \frac{\langle \mu(t) \cdot \mu(0) \rangle}{\mu(0) \cdot \mu(0)}, \tag{43}$$

where $\mu(0)$ was the dipole at time $t = 0$. There are many reasons why the average value of the dot product of $\mu(t)$ and $\mu(0)$ should tend to zero as time advances; thus one might invoke rotational diffusion brought on by Brownian motion, but it is a general result that if the medium is dissipative, as all dielectric media must be to a greater or lesser extent, then $f(t)$ will approach zero after a sufficiently long time. The importance of the correlation function is that it is a microscopic response function, and one has from Eq. (33) the immediate result

$$[\hat{\varepsilon}(\omega) - \varepsilon_\infty]/(\varepsilon_s - \varepsilon_\infty) = \int_0^\infty \dot{f}(t) \exp(-i\omega t) \, dt. \tag{44}$$

The most discussed form for the correlation function is the simple exponential decay

$$f(t) = \exp(-t/\tau), \tag{45}$$

which, from the analysis just given leads immediately to the famous Debye equation (Debye, 1929)

$$[\hat{\varepsilon}(\omega) - \varepsilon_\infty]/(\varepsilon_s - \varepsilon_\infty) = 1/(1 + i\omega\tau), \tag{46}$$

whose real and imaginary parts are

$$\varepsilon'(\omega) = \varepsilon_\infty + (\varepsilon_s - \varepsilon_\infty)/(1 + \omega^2\tau^2) \tag{47a}$$

and

$$\varepsilon''(\omega) = (\varepsilon_s - \varepsilon_\infty)\omega\tau/(1 + \omega^2\tau^2). \tag{47b}$$

Debye derived his equation originally in terms of a specific model: a dipole attempting to rotate synchronously with the field but prevented from doing so by the viscous drag of the surrounding medium. He was even able to relate the one parameter τ that appears to the macroscopic viscosity η via the equation

$$\tau = 4\pi\eta a^3/kT. \tag{48}$$

This equation was for a particular model, essentially a very dilute solution of large dipolar spheres in a nonpolar solvent whose molecules were much smaller than the spheres; but the answers that it gives, $\tau \simeq 10^{-11}$s, do match within a factor of two or so the observed relaxation times. Debye, in deriving his equation, did have to make some drastic assumptions, one of which was that the only time-dependent effect was the viscous drag. In other words there were no inertial effects and an isolated molecule could undergo instantaneous rotational acceleration. This is clearly not so; and because the short-time behavior of $f(t)$ will be wrong, the high-frequency behavior of $\hat{\varepsilon}(\omega)$ will be incorrect. To illustrate this, one may calculate α as a function of ω from Eq. (18b) in its approximate form

$$\alpha = \omega\varepsilon''/c\varepsilon_\infty^{1/2}, \tag{49}$$

which, using Eq. (47b), gives the result

$$\alpha = (\varepsilon_s - \varepsilon_\infty)\omega^2\tau/\varepsilon_\infty^{1/2}c(1 + \omega^2\tau^2). \tag{50}$$

As $\omega \rightarrow \infty$, this gives the limiting behavior

$$\alpha_\infty = (\varepsilon_s - \varepsilon_\infty)/\varepsilon_\infty^{1/2}c\tau, \tag{51}$$

that is, the absorption coefficient reaches a limiting "Debye plateau" and all polar liquids would be expected to be virtually opaque throughout the optical regions in complete contradiction to the observed behavior. The fault can be rectified by making a suitable modification to $f(t)$; this modification can be deduced by applying a constraint to the possible mathematical forms for $f(t)$. So far we have only required $f(t)$ or $R(t)$ to be decay functions, either monotonic or damped oscillatory; but for a physical system, time-reversal symmetry must apply, and this demands that $R(t) = R(-t)$. The only way this can be satisfied near $t = 0$ is for the response function to have a zero slope as shown in Fig. 3.

The response function should ideally be an even function that behaves

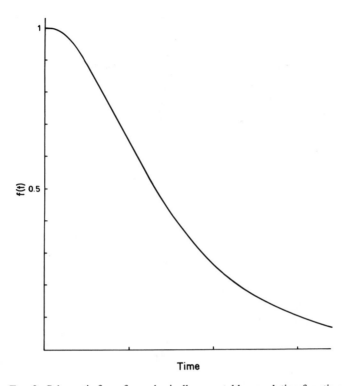

FIG. 3 Schematic form for a physically acceptable correlation function.

like $\exp(-|t|/\tau)$ at long times. There have been several improved correlation functions suggested over the years. One of the earliest was that of Rocard and Powles, (Rocard, 1933; Powles, 1948).

$$f(t) = (\tau_1 - \tau_2)^{-1}[\tau_1 \exp(-t/\tau_1) - \tau_2 \exp(-t/\tau_2)], \qquad (52)$$

which leads to a complex permittivity

$$(\hat{\varepsilon} - \varepsilon_\infty)/(\varepsilon_s - \varepsilon_\infty) = [(1 + i\omega\tau_1)(1 + i\omega\tau_2)]^{-1}. \qquad (53)$$

The short-time behavior of Eq. (52) can be illustrated by a series expansion

$$f(t) = 1 - \frac{t^2}{2}\left(\frac{1}{\tau_1^2} - \frac{1}{\tau_2^2}\right); \qquad (54)$$

this correct behavior is reflected in $\varepsilon''(\omega)$ which goes as ω^{-2} for large ω, and therefore $\alpha_\infty \to 0$. The desired "recovery of transparency at high frequencies" is therefore achieved. Other examples are available to the interested reader in the literature (Birnbaum and Cohen, 1970; Davies et al., 1968;

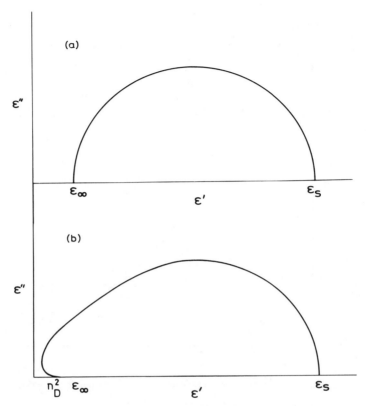

FIG. 4   (a) Debye and (b) "short-time-correct" forms of the Cole–Cole plot.

Kestement, *et al.,* 1978; Kubo, 1958a,b; Lassier *et al.,* 1969; Leroy *et al.,* 1967; McConnell, 1980; Sack, 1957).

Despite the incorrect high-frequency behavior, the Debye equation has been extensively used in the analysis of experimental data on dielectric relaxation because it is valid, other things being equal, at frequencies less than the reciprocal of the relaxation time. Relaxation times, as previously mentioned, tend to be of the order 10^{-11}s, so one is talking of time frequencies of 10^{11} Hz and below, that is, frequencies up to the microwave band. This in fact is the band where most measurements have been carried out and where the workers have found Debye's formalism helpful, but in practical use it is common to subject Eq. (46) to some preliminary algebraic manipulation. Thus it was pointed out by Robert and Kenneth Cole (1941) that Eq. (46) is really the equation of a circle in the complex plane. Simple manipulation leads to

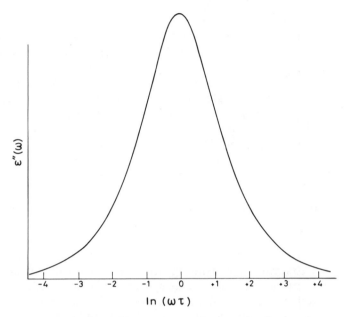

FIG. 5 Spectral variation of ε'' for an ideal Debye dielectric.

$$[\varepsilon' - \tfrac{1}{2}(\varepsilon_s + \varepsilon_\infty)]^2 + \varepsilon''^2 = \tfrac{1}{4}(\varepsilon_s - \varepsilon_\infty)^2, \tag{55}$$

i.e., a circle centered on the ε' axis at $\tfrac{1}{2}(\varepsilon_s + \varepsilon_\infty)$ and with radius $\tfrac{1}{2}(\varepsilon_s - \varepsilon_\infty)$. For passive media both ε' and ε'' are constrained to be positive, so Eq. (55) represents a semicircle in the right upper quarter with intercepts on the ε' axis at ε_s and ε_∞. This is illustrated in Fig. 4.

Plots of ε'' versus ε' are widely used by dielectricians and are universally known as Cole–Cole plots. It is readily shown that any function whose corresponding correlation function behaves correctly at $t = 0$ will not cut the ε' axis perpendicularly at its left intercept (as does the Debye plot) but will instead approach the axis asymptotically as shown in Fig. 4b.

The Cole–Cole plot has the virtue that it immediately gives the parameters ε_s and ε_∞, and with a bit of ingenuity, one can even infer τ because the maximum on the arc corresponds to $\omega\tau = 1$. However, it does have the disadvantage that the independent variable ω does not appear explicitly. Thus an alternative approach, in which one writes Eq. (47b) in the equivalent form

$$\varepsilon''(\omega) = (\varepsilon_s - \varepsilon_\infty)/[2 \cosh(\ln \omega\tau)], \tag{56}$$

is often used. So a plot of $\varepsilon''(\omega)$ versus $\ln \omega$ will be a symmetrical bell-shaped curve (Fig. 5) whose maximum occurs at $\omega = 1/\tau$ and whose width at half-maximum will be 2.634 Napierian logarithmic units, i.e., $2 \ln_e(2 + \sqrt{3})$, or else 1.144 decimal logarithmic units. Another method is to plot either $\omega\varepsilon''$ or ε''/ω against ε' which in both cases leads to a straight-line plot whose slope gives τ.

The fundamental assumption in the modern derivations of the Debye equation is that there is a single relaxation time τ and that the correlation function does behave like an exponential for sufficiently long values of time. It would not be surprising if real systems had a spread of relaxation times described by a function $\phi(\tau)$. Then one would have

$$f(t) = \int_0^\infty \phi(\tau) \exp(-t/\tau) \, d\tau, \tag{57a}$$

which gives for the complex permittivity

$$\frac{\hat{\varepsilon}(\omega) - \varepsilon_\infty}{\varepsilon_s - \varepsilon_\infty} = \int_0^\infty \frac{\phi(\tau) \, d\tau}{1 + i\omega\tau}. \tag{57b}$$

The complex permittivity resulting from Eq. (57b) will not usually be such that a plot of its imaginary versus its real components will be semicircular. One could have depressed arcs, that is, semicircular arcs with a center lying below the real axis, or else skew arcs. This is an enormous topic with vast amounts of experimental data having been reported; but because these departures are primarily of interest to the low-frequency practitioners, only the briefest outline of the theoretical approach will be attempted here. The theoreticians usually have chosen to work in the frequency domain rather than in the time domain and have chosen the simple Debye equation as a starting point for their elaborations. Some examples are

$$(\hat{\varepsilon} - \varepsilon_\infty)/\varepsilon_s - \varepsilon_\infty) = 1/[1 + (i\omega\tau)^{1-\alpha}], \qquad 0 < \alpha < 1 \tag{58}$$

from Cole and Cole (1941),

$$(\hat{\varepsilon} - \varepsilon_\infty)/(\varepsilon_s - \varepsilon_\infty) = 1/(1 + i\omega\tau)^\beta, \qquad 0 < \beta < 1 \tag{59}$$

from Davidson and Cole (1950, 1951),

$$(\hat{\varepsilon} - \varepsilon_\infty)/(\varepsilon_s - \varepsilon_\infty) = 1/[1 + (i\omega\tau)^{1-\alpha}]^\beta, \qquad 0 < \alpha < 1, \quad 0 < \beta < 1 \tag{60}$$

from Havriliak and Negami (1966, 1967). Starting with the Debye equation in the form Eq. (55), one can obtain the generalization (Fuoss and Kirkwood, 1941)

$$\varepsilon''(\omega) = \tfrac{1}{2}(\varepsilon_s - \varepsilon_\infty) \, \mathrm{sech}[\alpha \ln(\omega\tau)], \tag{61}$$

which can also be written as

$$\varepsilon''(\omega) = \frac{\varepsilon_s - \varepsilon_\infty}{2} \cdot \frac{1}{(\omega/\omega_0)^\alpha + (\omega/\omega_0)^{-\alpha}} \tag{62}$$

with, of course, $\omega_0 = 1/\tau$. Jonscher (1977) has suggested a "universal" dielectric response which is a generalization of this, namely,

$$\varepsilon''(\omega) = \frac{A}{(\omega/\omega_1)^{-m} + (\omega/\omega_2)^{1-n}}, \qquad 0 < m \le 1, \quad 0 \le n < 1. \tag{63}$$

Williams and Watts (1970) are of the smaller group who have chosen to work in the time domain. Their correlation function

$$f(t) = \exp[(-t/\tau)^\beta], \qquad 0 < \beta \le 1, \tag{64}$$

does correspond with the behavior observed for some systems, but it has the severe disadvantage that when it is substituted into Eq. (44), the integral cannot be evaluated in closed form for arbitrary values of β. There is not, of course, any guarantee that the correlation functions shown by real physical systems will take on simple analytical forms or that even if they do they will have analytical Fourier transforms. However, the availability of high-speed digital computers has lessened this difficulty. It is now possible to obtain $(\hat{\varepsilon} - \varepsilon_\infty)/(\varepsilon_s - \varepsilon_\infty)$ by numerical transformation of an arbitrary form for $f(t)$; and conversely, given experimental results for the permittivity, one can derive $f(t)$ by a numerical inverse Fourier transformation followed by numerical integration. Nevertheless, whenever an analytical transform exists for one of these Debye generalizations, it interesting to examine it. Some illuminating examples have been discussed in the comprehensive textbook of Böttcher and Bordewijk (1978). One should remark that none of these generalizations obeys the requirements of time-reversal symmetry at the high-frequency end, but in the context in which they would be used this is only a minor point because they are all expressly limited to frequencies less than ω_0. What happens at frequencies beyond ω_0 will be taken up later.

C. CORRECTIONS FOR THE INTERNAL FIELD

The treatment given so far is formally open to the objection that we are trying to relate directly a microscopic quantity, the decay of molecular correlation, with a macroscopic quantity, the dielectric permittivity, without allowing for the fact that the relaxing molecular dipole is surrounded by a polarizable medium with which it can react. In other words we are ignoring the effects of the internal field. This topic is an old one with contributions from Debye, Clausius and Mosotti, Onsager, Glarum, Cole, etc.; but the most satisfactory treatment of the dynamic case came from the work of Fatuzzo and Mason (1967) in the late 1960s. This work was extended by

Klug, Kranbuehl, and Vaughan (1969) and by Rivail (1969) to give the result

$$\frac{[\hat{\varepsilon}(\omega) - \varepsilon_\infty][2\hat{\varepsilon}(\omega) + \varepsilon_\infty]\varepsilon_s}{(\varepsilon_s - \varepsilon_\infty)(2\varepsilon_s + \varepsilon_\infty)\hat{\varepsilon}(\omega)} = \int_0^\infty \dot{f}(t) \exp(-i\omega t) \, dt, \qquad (65)$$

which is now the accepted form. This equation is rather more complicated than the simple Eq. (44), but careful work (Titulaer and Deutch, 1974; Finsy and VanLoon, 1975) has shown it to give excellent agreement with observation. In this light one should think of a macroscopic relaxation time that is not necessarily the same as the microscopic one. Equation (65) is undoubtedly the correct form to use in the analysis of dielectric relaxation, but it is readily seen that for $\omega \rightarrow 0$ it goes over to the simpler Eq. (44); and it can be shown, although less readily, that even for high frequencies the difference between the two is not great. This explains why the simple Debye equation has been used so successfully for so many years to analyze experimental results. However, the conditions under which Eq. (65) will give substantially the same results as Eq. (44) cannot always be assumed to exist. Thus analysis via Eq. (65) is advisable before one concludes that new phenomena have been observed.

D. CAUSALITY AND THE KRAMERS–KRONIG RELATIONS

The putting together of a propagative quantity (ε' or n) and an attenuative quantity (ε'' or n) into a single complex quantity has more justification than mere convenience. So far, in all cases, the two quantities have been related to one another; and although Eq. (2) is set up to be deliberately noncommital, it is an unavoidable fact that for any physical system these two quantities will *always* be related. This follows at once from Eq. (33), where ε' and ε'' are given, in essence, by the cosine and sine transforms of a common function. However, it is illuminating to pursue the matter further because a great deal of physical insight results and equations emerge that give either ε' or ε'' in terms of the other without the need to evaluate the response function as an intermediary. The necessary connection between, say, ε' and ε'' arises through the operation of the Principle of Causality, an absolute law of physics that can be stated crudely as "causes must always precede their effects." This principle restricts the possible forms that complex response functions such as $\hat{\varepsilon}$ or $\hat{n}$ may take and in fact requires that the real and imaginary parts of a causal response function be related by integral transforms.

The necessity of a connection can be derived by a thought experiment. Imagine that a brief impulse is applied to a system that has an input (to which the impulse is applied) and an output (at which the effects are monitored). Now the input pulse (of the form $I = 0$ for $t < 0$, $I = I_{max}$ for

$0 < t < \Delta$, $I = 0$ for $t > \Delta$) can be resolved into its Fourier components which stretch from $t = -\infty$ to $t = +\infty$, but these cancel everywhere except for $0 < t < \Delta$. The output of the system will be the input modified by the transfer function of the system. One can think of a particular case in which the system is a narrow-band filter that does not pass frequencies lying between v_0 and $v_0 + \delta v$ but that passes all other frequencies unattenuated. The Fourier input components that have frequencies lying within the reject band will be removed or, what is the same thing, their complements will be added to the output. One will therefore expect to observe ripples at the output of the system in anticipation of a pulse that is yet to be applied. The White Queen would not find this remarkable and mathematically there is, in fact, nothing remarkable about it; but in the non-Wonderland in which we actually live, the principle of causality applies and forbids the appearance of *any* signal at the output for times less than zero. By definition, the amplitudes of the components that are passed are not affected, so the only parameter left is the *phase*. It follows that all the transmitted waves will undergo phaseshifts appropriate to ensure that there is once more perfect cancellation for times less than zero. The situation is illustrated schematically in Fig. 6.

The result is that the application of an arbitrary (in the mathematical sense) boundary condition, the principle of causality, leads to the inevitable appearance of dispersion (frequency-dependent phase shifts) whenever there is absorption. This is a general result that applies to all physical systems and, in particular, to dielectrics propagating electromagnetic waves. One can press the argument to derive the actual form for the variation of the dispersion ($\varepsilon'(\omega)$) given the form of the absorption ($\varepsilon''(\omega)$), but it is quicker to use another approach in which the powerful theorems that arise in the theory of the functions of a complex variable are used.

There are two strands to this approach. We begin by defining the very important Hilbert transform

$$\mathcal{H}\{f(t)\} = \frac{P}{\pi} \int_{-\infty}^{+\infty} \frac{f(t')}{t' - t} \, dt', \tag{66}$$

where the P means *Cauchy principal value*. The singularity at $t = t'$ is dealt with by integrating from $-\infty$ to $t - \Delta$ and from $t + \Delta$ to $+\infty$, adding the two results, and letting Δ tend to zero. From its definition it follows that

$$\mathcal{H}\{f(t)\} = -(1/\pi t') * f(t'), \tag{67}$$

where the asterisk signifies convolution. The Hilbert transformation is therefore equivalent to convolving a function with the reciprocal function. The next step is to note that, if the response $R(t)$ of a system is described by a function $F(t)$ for $t > 0$, then the principle of causality demands that

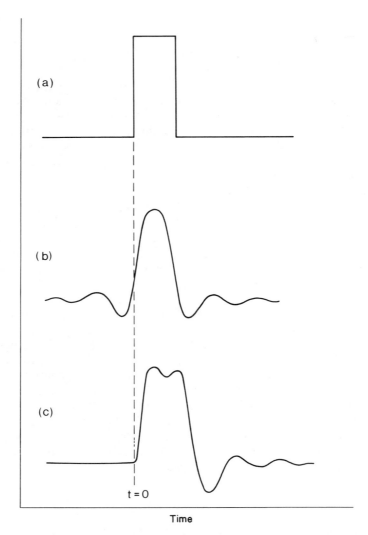

(a)

(b)

(c)

t = 0

Time

FIG. 6　A rectangular pulse (a), when input to a narrow-band filter, might be expected to give the output (b); but the principle of causality demands that the output be of the form (c).

$$R(t) = H(t) \cdot F(t), \tag{68}$$

where $H(t)$ is the Heaviside step function equal to zero for $t < 0$ and equal to unity for $t > 0$. The transfer function of the system $\hat{T}(v)$ is given by the Fourier transform of $R(t)$ and hence

$$\hat{T}(\omega) = \int_{-\infty}^{+\infty} H(t) \cdot F(t) \exp(-i\omega t) \, dt. \tag{69}$$

Then $R(t)$, a one-sided function, can be expressed as the sum of two perfectly well behaved double-sided functions $R_e(t)$ and $R_o(t)$, the even and odd components, where $R_e(t) = R_o(t) = \frac{1}{2}R(t)$ for $t > 0$ and $R_e(t) = -R_o(t)$ for $t < 0$. Introducing the signum function $\text{sgn}(t) = +1$ for $t > 0$ and -1 for $t < 0$, one may write

$$R(t) = H(t) \cdot F(t) = [1 + \text{sgn}(t)]R_e(t), \tag{70}$$

and substituting this in Eq. (69) gives

$$\hat{T}(\omega) = \phi(\omega) + \mathcal{F}\{\text{sgn}(t) \cdot R_e(t)\}, \tag{71}$$

where $\phi(\omega)$ is the (real) Fourier transform of $R_e(t)$. The Fourier transform of the signum function $\text{sgn}(t)$ is the imaginary reciprocal function $+i/\pi\omega$, so by the convolution theorem the second term on the right-hand side of Eq. (71) may be expressed as the convolution

$$\mathcal{F}\{\text{sgn}(t) \cdot R_e(t)\} = +(i/\pi t) * R_e(t), \tag{72}$$

and we finish with

$$\hat{T}(\omega) = \phi(\omega) - i\mathcal{H}\{\phi(\omega)\}. \tag{73}$$

Thus for any causal transfer function, the real and imaginary components are Hilbert transforms of one another. In the particular case of the complex permittivity, one would have

$$\varepsilon'(\omega_0) = \varepsilon_\infty + \frac{P}{\pi} \int_{-\infty}^{+\infty} \frac{\varepsilon''(\omega)}{\omega - \omega_0} \, d\omega, \tag{74a}$$

and

$$\varepsilon''(\omega_0) = -\frac{P}{\pi} \int_{-\infty}^{+\infty} \frac{\varepsilon'(\omega)}{\omega - \omega_0} \, d\omega. \tag{74b}$$

The second and related strand starts by noting that the restriction of ω to positive values has been lifted in the first strand and that one might just as well go further and allow ω to have complex values. One can now translate the ideas of the thought experiment into mathematical terms. This produces the results that the mathematically arbitrary imposition of causality has produced an asymmetry in the time plane [$R(t)$ defined only for $t > 0$] and that there will be a corresponding asymmetry in the Fourier conjugate plane defined by the complex variable $\hat{\omega}$. This asymmetry takes the form that all the poles of the complex response function, that is, all the values of $\hat{\omega}$ for which $\hat{\varepsilon}$ is infinite, will lie in the upper half-plane. The complex response function is therefore analytic throughout the lower half-plane. We can now use the Cauchy integral formula

$$f(a) = \frac{1}{2\pi i} \oint \frac{f(z)}{(z-a)} \, dz, \tag{75a}$$

which applies for any closed contour around $z = a$, because we will only be considering regions within which $f(z)$ is analytic. The only difficulty is that our path of integration will be along the real axis and this includes the singularity because we need to know the value of $f(a)$ with a on the axis. The principal value convention gets around this; and noting that $\hat{\varepsilon}(\hat{\omega})$ will go to zero as $\hat{\omega}$ goes to infinity, we may write

$$\hat{\varepsilon}(\omega_0) = -\frac{1}{i\pi} P \int_{-\infty}^{+\infty} \frac{\hat{\varepsilon}(\omega)}{\omega - \omega_0} \, d\omega. \tag{75b}$$

Substituting Eq. (2) into Eq. (75b) and identifying real and imaginary parts gives Eqs. (74) immediately.

This treatment is very beautiful — in the case of the Cauchy integral formula almost breathtakingly so; after all, it is almost unbelievable that, if one merely knows the values of an analytical function around an arbitrary closed path, one can calculate its value at *any* point within the contour. Unfortunately, as they stand, the Kramers–Kronig equations, Eqs. (74), have the fatal experimental flaw that one cannot know the values of $\varepsilon(\omega)$ for negative ω. Fortunately, the natures of the Hilbert transform and of the dielectric response function compel $\varepsilon'(\nu)$ to be an even function and $\varepsilon''(\nu)$ to be an odd function. Using this fact, one may readily show that one can write equivalent Kramers–Kronig equations

$$\varepsilon'(\nu_0) = \varepsilon_\infty + \frac{2}{\pi} P \int_0^\infty \frac{\nu \varepsilon''(\nu)}{\nu^2 - \nu_0^2} \, d\nu \tag{76a}$$

and

$$\varepsilon''(\nu_0) = -\frac{2\nu_0}{\pi} P \int_0^\infty \frac{\varepsilon'(\nu)}{\nu^2 - \nu_0^2} \, d\nu, \tag{76b}$$

which involve only positive frequencies. These equations are widely used because it is often difficult to measure one component at a desired frequency, whereas it may be much easier to measure the other over an extended frequency range.

Corresponding Kramers–Kronig equations may be written for any other conjugate pair, for example, n and k, but the most useful are those involving the reflectivity, because frequently a specimen may be so heavily absorbing that there is no possibility of measuring its optical properties in transmission. Faced with this situation, one can either measure the modulus and phase of the complex reflectivity [see Birch and Parker (1980) for details] or else measure just the power reflectivity ρ and use the Kramers–Kronig equation

$$\theta(v_0) = \frac{1}{\pi} P \int_0^\infty \left[\frac{d}{dv} (\ln \rho)^{1/2} \right] \left[\ln \left| \frac{v_0 + v}{v_0 - v} \right| \right] dv \qquad (77)$$

to calculate the phase. Knowing ρ and θ, one can calculate the complex reflectivity $\hat{r}$ from

$$\hat{r} = \rho \exp i\theta, \qquad (78)$$

and the complex refractive index from

$$\hat{n} = (1 + \hat{r})/(1 - \hat{r}). \qquad (79)$$

III.   The Microscopic Theory

When one attempts to get down to the molecular level and to relate the observed dielectric properties to the motions of the dipolar molecules in the fluid being studied, one is immediately overwhelmed by the immense complexity of the problem. The situation is that there is an enormous number ($\sim 10^{23}$) of dipoles spaced randomly with respect to one another and free to move in all directions. The problem of setting up the coupled differential equations for this system, based on Newton's laws, and then solving them is too difficult to contemplate. What can we do? One approach is to consider various limiting situations and to derive analytical solutions for them. The second approach is to use these simple solutions as the bases for rather heuristic extensions into regions beyond their strict applicability. The third approach is very different: one attempts to deal with the randomness of the situation by using Monte Carlo simulations in a large digital computer. This is the method of molecular dynamics. All of these approaches are currently being pursued with considerable diligence, but there is still a long way to go before we have a satisfactory microscopic theory.

Debye was the first to tackle the problem (Debye, 1929). He derived a solution by making fairly draconian approximations. These were

(a)   that the dipole in question was embedded in a sphere much larger than the molecules randomly bombarding it;

(b)   that the effects of the Brownian motion could be replaced by a torque force equivalent to white noise;

(c)   that dipole–dipole interactions could be ignored (in other words that the system was in the limit of virtually zero concentration);

(d)   that there were no inertial effects (in other words that the molecule could accelerate instantaneously). For most cases this is tantamount to insisting that ω be less than $1/\tau$.

Debye then went on to derive an explicit functional form for ε by rather lengthy arguments. The modern version, however, notes that this treatment gives a correlation function of the form of Eq. (45) and this gives immediately the relation for the permittivity, Eq. (46). Apart from simplifying the derivation, this approach shows at once that, if a system obeys the Debye equation, one cannot deduce thereby that one is necessarily dealing with the relaxation of rigid dipoles brought about by viscous drag. In fact, any model (and there is an infinity of them) that leads to Eq. (45) will perforce give a permittivity obeying the Debye equations. This is a disappointing conclusion because one of the objectives of science is to derive information about the microscopic world from observations on the macroscopic world, and it is helpful in this if there is a one-to-one correspondence between the two. For dielectric measurements this is clearly not the case.

Attempts to relax *any* of the approximations that Debye found necessary lead to great difficulties; nevertheless these attempts have been made. In fact, it is not too fanciful to imagine that a great deal of the dielectric theory of the last 50 years has been concerned with trying to move a few steps further than Debye did. The literature is very extensive and the reader is referred to the classic paper by Nee and Zwanzig (1970) as a means of getting an impression of this branch of physics. Here we will concentrate on the attempts to get around the restriction to low frequencies because clearly this is a *sine qua non* for a theory valid at submillimeter frequencies. The approach of Rocard and Powles (Rocard, 1933; Powles, 1948) has already been mentioned, but it can be shown that this particular solution is a member of an infinite family of solutions. The case of the dipolar sphere has been worked out by several people, the names Kubo (1958a,b), Sack (1957), and Mori (1965) being prominent. Their work leads to an iterative solution that is most elegantly displayed in the form of a continued fraction:

$$\frac{\hat{\varepsilon} - \varepsilon_\infty}{\varepsilon_s - \varepsilon_\infty} = 1 - \cfrac{i\omega t\gamma}{i\omega t\gamma + \gamma \cfrac{}{1 + i\omega t\gamma + 2\gamma \cfrac{}{2 + i\omega t\gamma + 3\gamma \cfrac{}{3 + \cdots}}}}, \qquad (80)$$

where $\gamma = I/\tau^2 kT$ with I being the moment of inertia. The theory of continued fractions is not as well known as the corresponding theory of infinite series, but similer concepts arise. Thus, just as one may approximate an infinite series by discontinuing it after a finite number of terms, so one

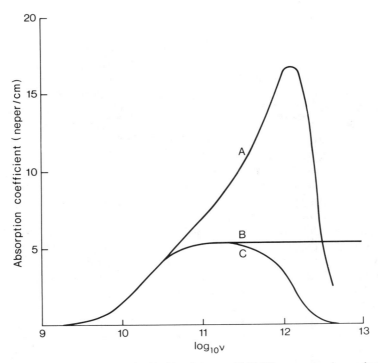

FIG. 7 Dielectric spectra for liquid chlorobenzene (C_6H_5Cl): curve A, observed spectral profile; B, Debye theoretical profile; and C, theoretical profile when inertial effects are taken into account.

may approximate a continued fraction by stopping it after a finite number of divisions. The terminations are usually known as convergents and the various convergents of Eq. (80) are

$$1, \quad \frac{1}{1 + i\omega t}, \quad \frac{1}{(1 + i\omega t)(1 + i\omega t\gamma)}, \quad \text{etc.} \tag{81}$$

One obtains a frequency-independent result, the Debye equation, the Rocard–Powles, etc. This is quite an advance, but it is possible to generalize Eq. (80) so that it involves more than two parameters. Then it becomes possible to fit almost any observed profile. McConnell at the Dublin Institute for Advanced Studies [see, for example, McConnell (1980)] has done a lot of work in this field.

The difficulty that arises in this study is that the observed absorption spectra do not look anything like the predictions of Eq. (80). For example, Fig. 7 shows the predictions of the Debye theory for chlorobenzene, the corresponding improved version with roll-off according to Eq. (80), and the actual observed result. It will be seen that far from rolling-off, the absorption coefficient rises very rapidly in the submillimeter region; and by the peak, near 35 cm^{-1}, it is twice the Debye plateau value. Thus the absorption in the submillimeter region is far higher than might be expected. This additional absorption is often referred to as the Poley absorption after J. Ph. Poley (1955) who predicted its existence, very bravely for the time, from extrapolations of slight indications that he found in his microwave data. It is now established that all polar liquids show this Poley absorption with an integrated intensity roughly proportional to the square of the molecular dipole moment. Nonpolar liquids also show a submillimeter absorption band, but usually this is an order of magnitude weaker than that of the polar liquids.

The task facing the theoretician is to give some satisfactory account of the origin of these Poley absorption bands. The first thing to do is to clear away explanations that are certainly not correct. Thus one can dismiss the possibility that they arise from the superposition of difference bands because the phenomenon is general and the bands, if anything, intensify as the temperature falls whereas difference bands would weaken. One also can dismiss the suggestion that they arise from a second rolled-off Debye process. This is for two reasons — because one cannot imagine what would be the nature of this second process and, more cogently, because Chamberlain (1967) has shown that the dispersion characteristics are not correct. The remaining explanations can be collected into two rather different categories: one can assume (a) that totally new phenomena are manifesting themselves in the submillimeter region and that the observed profile is the sum of independent processes, or (b) that the whole profile is an entity and that the low-frequency dielectric dispersion is merely the tail of the strong submillimeter process. The main theory in category (a) is the liquid-lattice theory of Chantry and Gebbie (1965). This assumes that the Poley absorption is the degenerate remains of the discrete lattice spectrum which would have been observed were the specimen crystalline. The structure is lost because of the loss of the long-range order in the liquid. The main evidence in favor of this theory is that the integrated absorption of the Poley band and the crystalline lattice spectrum are nearly equal. In Fig. 8 are shown some typical results for the spectrum of a liquid and of the same specimen when polycrystalline. This theory has been detailed for the particular case of chlorobenzene by Chantry (1977) and its consequences for the refraction spectrum have been investigated by Hill (1978). Even if this theory proves too simplistic, one significant element, namely, the importance of attractive Hooke's law forces in determining the microdynamics of the liquid, will remain; and this will have to be incorporated into any acceptable theory.

The earliest theory of type (b) was that of Kroon and Van der Elsken

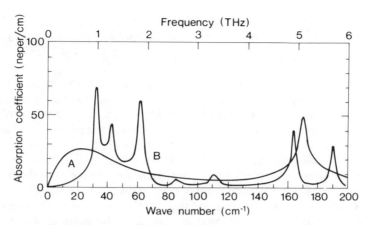

FIG. 8   Absorption spectra of pentadeuterobromobenzene (C_6D_5Br): (A) the liquid at 300K and (B) the crystalline solid at $\sim$ 100 K. The features at 33, 43, 62, 86, and 110 cm^{-1} are lattice bands. The two bands near 160 and 190 cm^{-1} are the crystal-field-split components of the band at 170 cm^{-1}, which is observed as a singlet for the liquid. (Courtesy J. W. Fleming.)

(1967). They took the opposite view; instead of visualizing the liquid phase as a melted crystal, they assumed it to be a condensed gas. The Poley absorption was then the smeared out pure-rotation spectrum. This theory is attractive because the generally accepted line-shape function for pure rotation, that of Van Vleck and Weisskopf, does go over into the Debye equation at very low frequencies. The main objection to the theory is that the band is in the wrong place. Thus for chlorobenzene, the peak in the pure-rotation envelope should be nearer 3 cm^{-1} than 30 cm^{-1}! The reply to this criticism, supported by the elegant work of Birnbaum (personal communication, 1968), is that in the liquid the simple $\Delta J = \pm 1$ selection rules will no longer apply and that transitions with $\Delta J = \pm 2, \pm 3$, etc., may become strongly allowed. This will have the effect of moving the peak out to higher frequencies. Extensions of the approach of Kroon and Van der Elsken have been discussed by several authors, and Evans and his colleagues have done much to weld together this microscopic approach with a more general one based on the Kubo–Mori formalism (Rowlinson and Evans, 1975).

The liquid-lattice theory of Chantry and Gebbie (1965) was in fact anticipated in a very significant early paper by Hill, who was interested in discrepancies between ε_∞ and n_D^2. It follows from the Kramers–Kronig arguments given earlier that a step in refractive index must arise from the presence of absorption bands, and Hill in fact predicted the existence of an intense submillimeter absorption process some time before it was observed. She developed a model that would give such an absorption: this was a molecule vibrating in a surrounding cage of molecules, each of which was subject to diffusion because of the Brownian motion. Her model was later tidied up and put into tight mathematical form by Wyllie (1971). His version is usually called the *itinerant oscillator* model. It has achieved a fair degree of success. Lassier and Brot (1969) studied a model in which a molecule would oscillate with simple harmonic motion at small amplitude in a deep potential well but would then suddenly undergo a large amplitude jump into a different potential well. Their model has the correct form for the correlation function near $t = 0$ and the shapes they calculate do look remarkably like those observed. The Lassier–Brot model starts from the liquid-lattice ideas, but at the end it has changed into a global theory that encompasses both low-frequency relaxation and high-frequency resonance. More recently (1977) Hill has again taken up the question of the bearing of the refraction spectrum on this matter, especially the point of whether the diffuse absorption that is observed could be explained by the superposition of a finite number of much broadened lattice bands.

The theoretical position is still very active, but it would be idle to pretend that we are close to a generally accepted theory. Probably more carefully considered experiments will be necessary to provide the theoreticians with relatively simple and easily characterised situations to analyze. Even so, what has emerged so far has shown submillimeter dielectrics to be a new and very fruitful branch of chemical physics.

IV. Experimental Methods

The dielectric spectra of materials at submillimeter wavelengths are obtained by two distinct methods. Broad-band continuous spectra are determined by means of Fourier transform spectrometry (FTS) whereas discontinuous, point-by-point, spectra are determined by the use of a Mach–Zehnder interferometer and a laser source. Both of these techniques are described in detail in other chapters in this series. Birch and Parker (1980) describe FTS and Afsar (1983) describes laser techniques. Fourier transform spectrometry comes in two variants: straightforward transmission FTS, which is exactly equivalent to normal infrared spectroscopy though very much more efficient, and dispersive Fourier transform spectrometry (DFTS), which has no analog in conventional spectroscopy. Normal FTS has the specimen in one of the two passive arms of the interferometer (source-beam divider or beam divider–detector) whereas DFTS features the specimen in one of the active arms (beam divider to either mirror). The two common positions are illustrated schematically in Fig. 9. DFTS has the intriguing feature that any structure introduced into the interferogram because of discrete absorptions in the specimen is found only

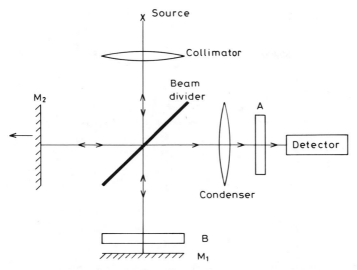

FIG. 9 The usual positions of the specimen for (A) conventional Fourier transform spectrometry and (B) dispersive Fourier transform spectrometry.

to one side of the displaced brightest fringe, whereas in conventional FTS such structure is distributed symmetrically about the now unique zero-path-difference fringe. The reasons for this remarkable behavior, an example of which is shown in Fig. 10 lie in the principle of causality and have been discussed by Chamberlain (1979). The connections with the arguments given in Section IID are very close. DFTS is experimentally more complicated than conventional FTS and requires more sophisticated computational techniques (Afsar *et al.*, 1976; Honijk *et al.*, 1973a, 1973b, 1977), but it does give both n and α in one run; moreover, the severe difficulty with the interface problems (Chamberlain, 1972) can be overcome naturally in the data reduction process (Afsar *et al.*, 1976; Honijk *et al.*, 1973a, 1973b, 1977).

The interface problems arise because there will always be losses by reflection and phase changes at the interface of the specimen with the outside world. In the case of liquid in a windowed cell, studied by conventional FTS, these effects can be particularly serious (Kilp *et al.*, 1978). An ingenious way of getting around this difficulty was developed by Kilp (1977) who invented a variable-path cell in which the distance between the two windows is fixed. The reflection effects are therefore constant; and if the absorption spectrum is measured as a function of liquid thickness, absolute absorption coefficients can be deduced. A diagram of Kilp's cell is shown in Fig. 11.

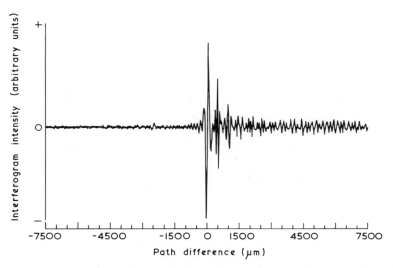

FIG. 10 Example of a dispersive interferogram. The sample was gaseous ammonia and the structure is due to pure rotational absorption. (Courtesy J. R. Birch.)

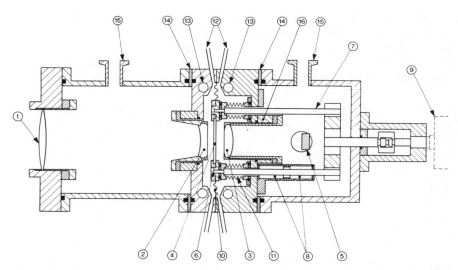

FIG. 11 Schematic diagram of Kilp's cell for determining accurate infrared spectra of liquids: (1), (2), (3), lenses; (4) dividing polytetrafluorethylene (PTFE) septum; (5) 45° mirror assembly; (6) septum mount; (7) guide rail; (8) ball-bush; (9) barrel micrometer; (10) flexible PTFE diaphragm; (11) bellows; (12) inlet and outlet tubes; (13) water cooling and thermostating channel; (14) insulating ring; (15) pipe connector to vacuum line; (16) spacer for adjusting overall cell length. (Kilp *et al.*, 1978.)

In the early days of DFTS there were problems with obtaining the correct absolute level in the refraction spectrum. It was usual to take one measurement at a spot frequency, provided by a laser, so that the relative measurements could be put on an absolute scale. Nowadays DFTS has advanced to the stage where its measurements can claim absolute status just as well as the laser measurements, so the two sets of data are combined together in the usual way to reduce the degree of experimental imprecision in the overall plot. The broad-band methods are currently limited at the low frequency end by the rapid fall off in source power, and measurements below 3 cm^{-1} are very difficult and are imprecise. Few measurements seem to have been taken above 200 cm^{-1} by DFTS although there is little reason to prevent such measurements being taken. The lack of interest certainly stems from the point made earlier that 200 cm^{-1} marks the dividing line between dielectric-type processes and purely resonant ones. The coverage from essentially dc up to 10^{12} Hz is now complete and attempts have been made (Afsar *et al.*, 1977) to quantify the experimental precision attainable over this range. The difficult region remains 30–150 GHz; therefore the

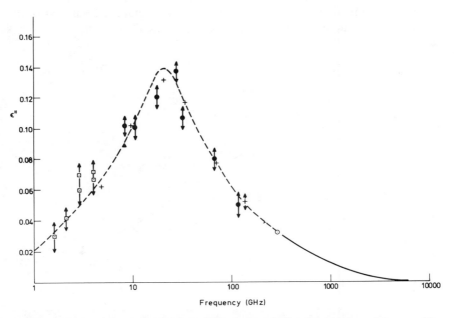

FIG. 12 Dielectric loss spectrum for a 10% solution of chlorobenzene in cyclohexane. The estimated random experimental imprecision is denoted by ↕. [The collaborative program was sponsored by the Community Reference Bureau of the EEC, the groups involved being Leiden (□), Brussels (+), Göttingen (▲), Nancy (●), Mainz (⊙), and NPL 1 (—).]

recent development (Kuno, 1979) of IMPATT solid-state sources for this region is particularly welcome. The current state of the high-frequency dielectric art is shown in Fig. 12, where ε' and ε'' for a 10% solution of chlorobenzene in cyclohexane are shown. The error bars show the presently attainable precision.

V.   Some Illustrative Examples of Submillimeter Dielectric Measurements

A.   POLAR LIQUIDS

The Poley absorption was first discovered for a polar liquid, chloroben-zene. Since then a wide range of polar liquids has been investigated and in every case a Poley band has been observed. As a further example, the spectrum of liquid chloroform is shown in Fig. 13. The general characteris-tics of the Poley band are that its maximum lies in the $20-70$ cm^{-1} region, its half-width (FWHM) is of the order 30 cm^{-1}, its integrated area is roughly proportional to μ^2, and in no case is there any resolvable structure. The diffuseness cannot be due to thermal broadening because cooling the liquid to just above its freezing point produces merely a marginal narrowing and certainly no structure. There is, however, a marked blue-shift in the peak

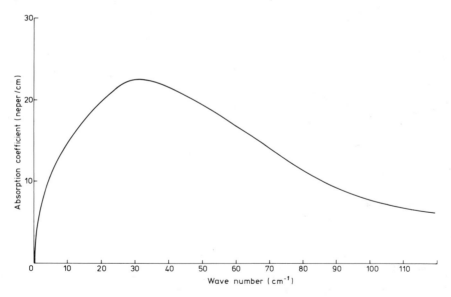

FIG. 13   Far-infrared absorption spectrum of liquid chloroform. [From Goulon et al., 1943.]

absorption frequency. In the case of chlorobenzene, the liquid can be supercooled but there is nothing observed except the familiar diffuse band. But the moment the liquid crystallizes, the sharp lattice spectrum promptly appears. Chantry's explanation in terms of very much broadened Lorentzian liquid-lattice bands is fairly plausible for chlorobenzene; but presumably it would be a general theory, in which case one might have thought that for some cases a particular combination of band intensities and band positions would lead to some observable discrete structure. So far this has not been observed. Chantry's theory, while taking full account of the short-range quantum-mechanical exchange forces, takes no account of the long-range electrostatic forces. A theory stemming from the Nee and Zwanzig approach in which the liquid is treated as a dipolar plasma has been developed by Lobo et al. (1973). They predict modes of the plasma to lie in the submillimeter region; and Ascarelli has claimed to have observed (Ascarelli, 1976, 1977) the longitudinal mode for the very polar and intensely absorbing liquid, nitromethane, by working in reflection. Afsar et al. have tried similar experiments but do not report the observation of the plasma (Afsar et al., 1978). The next stage in this type of approach will be to combine the short-range exchange forces and the long-range electrostatic forces in a unified theory. This has some formidable mathematical hurdles to overcome. From the experimental side the next step will be careful studies of solutions of polar liquids in nonpolar solvents, a topic taken up again in Section VB.

B. NONPOLAR LIQUIDS

The simple dielectric theories that relate the loss to the magnitude of the relaxing dipole would indicate that nonpolar liquids should be transparent up to frequencies where vibrational absorption starts to be manifest. Indeed, nonpolar liquids are much more transparent than are polar liquids but they are not perfectly so. The first microwave indications of loss in nonpolar liquids came from the careful work of Whiffen (1949), which was followed by strong suggestions from early far-infrared work (Wyss et al., 1964), and finally to measurements spanning the whole region of the absorption. Some examples of loss in nonpolar liquids are shown in Fig. 14. It will be seen that the absorption is about an order of magnitude weaker than that for a typical polar liquid but that it varies widely from one nonpolar molecule to another. The form of the absorption is quite similar to that of the polar liquids in the far infrared but very different in the microwave. It is almost as though one were studying a pure Poley absorption with no Debye process to give a characteristic microwave loss. The line shape to be expected for this "pure Poley" absorption process has been studied by Davies and Evans (1976). They use a truncation of the Mori continued fraction solution of the

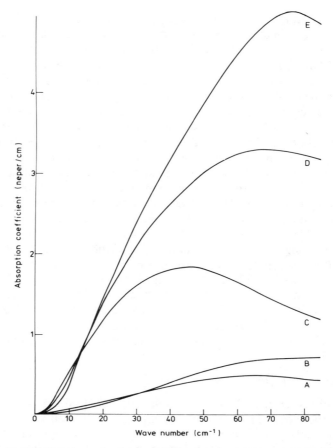

FIG. 14   Far-infrared absorption spectra of some nonpolar liquids: curve A, *trans*-decalin; B, cyclohexane; C, carbon tetrachloride; D, carbon disulfide; and E, benzene.

relevant Volterra integro-differential equation (Berne and Harp, 1970) to derive the expression

$$\alpha(\omega) = \frac{AK_0(0)K_1(0)\gamma\omega^2}{\gamma^2[K_0(0) - \omega^2] + \omega^2\{\omega^2 - [K_0(0) + K_1(0)]\}^2}. \qquad (82)$$

In this equation, the two K's are parameters related to the mean square torques on the molecules and γ is a line-width parameter. A is, strictly speaking, not a parameter because it is related to the total dispersion $\varepsilon_s - \varepsilon_\infty$ over the band by the equation

$$A = (\varepsilon_s - \varepsilon_\infty)/(\varepsilon_\infty^{1/2}c); \qquad (83)$$

but because of the rather indeterminate nature of ε_∞ it is best treated as a parameter in practice. One can thus rewrite Eq. (82) in the form

$$\alpha(\bar{v}) = \frac{A'\bar{v}^2}{\Delta\bar{v}^2(\bar{v}_1^2 - \bar{v}^2)^2 + \bar{v}^2(\bar{v}_2^2 - \bar{v}^2)^2}. \tag{84}$$

For CCl_4 at room temperature the values of the four quantities appearing in Eq. (84) are $A' = 73.37 \times 10^6$, $\Delta\bar{v} = 115\ \text{cm}^{-1}$, $\bar{v}_1 = 30\ \text{cm}^{-1}$, and $\bar{v}_2 = 87.56$ cm^{-1}. With these values the form of the absorption coefficient variation is as shown in Fig. 15. The agreement with the experimental data given in Fig. 14 is very good.

If one enquires into the nature of the processes responsible for the finite absorption of nonpolar liquids in the far infrared, three explanations might occur.

(a) The absorption might be due to higher multipole interactions.
(b) It might be due to transient dipoles induced by collision.

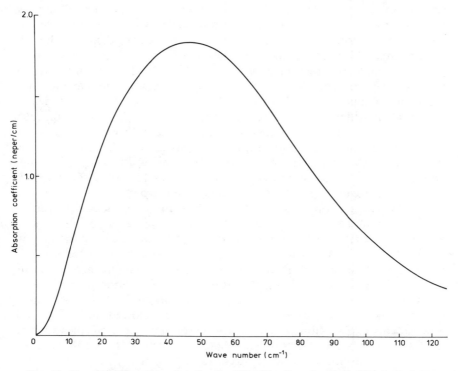

FIG. 15 Theoretical line-shape function for CCl_4. [Based on the work of Davies and Evans (1976).]

(c)   It might be due to "borrowing" of intensity from strong intermolecular modes.

The first of these is rejected at once in its simple form because the observed intensity is far too high (only one order of magnitude down on the dipolar case) and because there is no obvious correlation with, for example, the quadrupole moment. Thus CCl_4 whose lowest finite multipole moment is an octapole absorbs more strongly than does cyclohexane.

The second explanation is far more plausible and was in fact Whiffen's suggested interpretation of his pioneering observation of microwave loss in nonpolar liquids (Whiffen, 1949). Explanations along this line would agree well with several other phenomena, for example, the induction of otherwise forbidden infrared bands in gases subjected to high pressure. The difficulty is that lattice bands also appear in the spectrum of crystalline nonpolar substances, and in the crystal there can be no question of intermolecular collisions. The integrated intensity of the crystalline spectrum is closely similar to that of the liquid spectrum, so it seems likely that the absorption process must be common to both. One can make progress in resolving this difficulty by noting that the idea of a multipole expansion of an electric charge distribution is useful only when one is considering distances that are large compared with those separating the charges. It clearly is not relevant to the case of molecules separated by, at most, the sum of their van der Waals radii. If one considers CCl_4 as an example, one has a nonpolar molecule made up of highly polar bonds and containing four highly polarizable atoms. There will thus be strong electrical interactions when CCl_4 molecules are in contact, and the interaction will vary rapidly with the distance separating the molecules. This mimics very closely a "collision" situation. As expected, the absorption bands that are observed for nonpolar liquids and crystals are very sensitive to hydrostatic pressure: an order of magnitude intensification can be produced by quite moderate pressure.

The third suggestion rests on the notion that the separation of the modes of vibration of an ensemble into internal and external modes is only an approximation. It is a good approximation in most practice, but in the limit one has to realise that all modes of vibration that actually occur are mixtures of the pure basic types. The lattice modes can therefore acquire finite intensity because the motion is not quite 100% the motion of rigid molecules but involves a small admixture of internal motion. This small admixture will contain contributions from strongly allowed intermolecular modes. This explanation would clearly be relevant for molecules such as CCl_4 and CO_2; but for homonuclear diatomics such as N_2 and Br_2, a more thorough quantum-mechanical analysis of the electrical changes associated with the unit cell modes of vibration would be required. This borrowing mechanism is couched in language very different from that of the collision mechanism; nevertheless, at the fundamental level they may well be alternative descriptions of the same physical phenomenon. Some experiments that throw light on this point have been carried out by Gerschel and his colleagues (Gerschel *et al.*, 1976). These workers have studied the far-infrared spectrum of, say, CS_2 while the originally gaseous specimen is taken via the triple point into the liquid phase.

C. SOLUTIONS OF POLAR MOLECULES IN NONPOLAR SOLVENTS

The early studies of Poley absorption were carried out on pure liquids such as chlorobenzene. They served to reveal all the new phenomena and to found a new branch of chemical physics. However, from a theoretical point of view they were unsatisfactory because there was no way of avoiding a many-body formalism if the system was to be analyzed properly. The obvious approach was to study polar molecules dissolved in nonpolar solvents. At sufficient dilution one should be able to use essentially single-body theory, the only problem being to obtain the spectra in the first case. Submillimeter interferometric techniques were rapidly improved in the late 1960s and early 1970s, and it is now possible to produce spectra of such precision that one can readily study solutions of polar molecules at the 1-in-10 molecular ratio and below. The first result that emerged was that the absorption appeared to be simply additive. This was so startling a result that it stimulated several further careful investigations (Chamberlain *et al.*, 1962; Davies *et al.*, 1968). These did indeed show that there are effects that can be attributed to interactions between the polar molecule and its hosts (Pardoe, 1970); but nevertheless these are small and one still has the very surprising result that the various molecules appear to absorb essentially independently. Some illustrative results due to Kilp and Afsar (Kilp *et al.*, 1978) are shown in Figs. 16 and 17.

Clearly some explanation is required for this remarkable observation, especially because all the theories so far ennumerated have been in terms of strong interactions between the molecules. One fairly obvious line of explanation would be that the solutions so far studied have been of polar molecules dissolved in solvents which, apart from being nonpolar, are otherwise very similar, i.e., chlorobenzene dissolved in cyclohexane. One might expect the vibrational frequencies here to be roughly the same; and because the major contribution to the intensity comes from the oscillation of the molecular dipole, one would expect to obtain a roughly similar spectrum. Experimental data are unfortunately sparse, but it would seem desirable to study systems in which the host molecules are as different as possible from the solute molecules. The problem here is that substances tend to dissolve in substances not too different from themselves. Another line

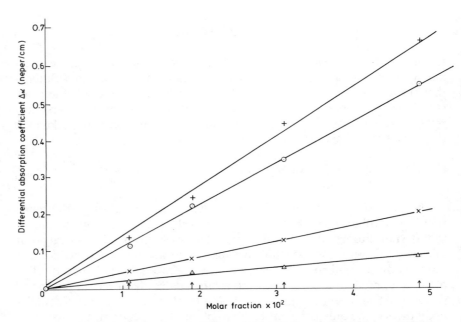

FIG. 16 Differential absorption of solutions of chlorobenzene in cyclohexane at 20°C as a function of molar fraction (x) at a series of wave numbers spanning the Poley absorption band. 30 cm^{-1} (+), 50 cm^{-1} (⊙), 90 cm^{-1} (✕), and 110 cm^{-1} (△). (Kilp *et al.*, 1978.)

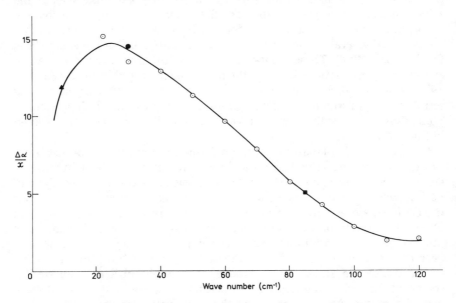

FIG. 17 Plot of the extrapolated absorption coefficient for pure chlorobenzene in cyclohexane at 20°C, derived from dilute solution data as in Fig. 16. Data-point symbols are frequency-multiplied klystron, $\bar{\nu} = 9.51$ cm^{-1} (▲); HCN laser, $\bar{\nu} = 29.712$ cm^{-1} (●); and H$_2$O laser, $\bar{\nu} = 84.323$ cm^{-1} (■). This curve differs only slightly from that actually observed for pure chlorobenzene. (Kilp *et al.*, 1978.)

would be to choose pairs for which strong interactions are known to occur. Chloroform and benzene would be a good combination. This system was studied early on by Chantry and Gebbie (1967), but these authors were more concerned with the remarkable lattice spectra produced (see Fig. 18) when the liquid was frozen and did not report detailed results on the liquid mixture. Another interesting approach is to try (as much as possible) to satisfy the first assumption made by Debye and choose solute molecules much larger than the solvent molecules. Whiffen reported some microwave observations on camphor dissolved in cyclohexane (Whiffen, 1966 and Afsar *et al.,* 1978) have recently contributed some millimeter and sub-millimeter observations on the same system. The combined results are shown in Fig. 19. Camphor is a large but essentially spherical molecule, and it will be

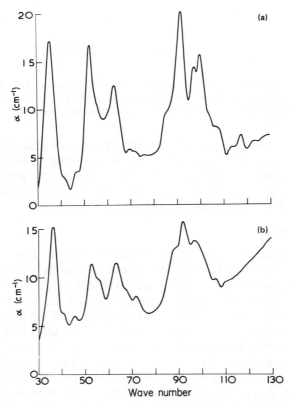

FIG. 18 Far-infrared spectra of the polycrystalline solids produced by freezing two different mixtures of chloroform and benzene. These spectra establish the existence of the stoichiometric compounds (a) $CHCl_3 : C_6H_6$ and (b) 2 $CHCl_3 : C_6H_6$ at $\approx$ 130 K.

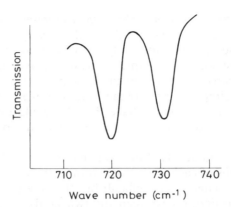

Wave number (cm⁻¹)

FIG. 20　Factor group splitting of the 725 cm⁻¹ band of polyethylene.

of small molecules and one made up of macromolecules is that, in the latter case, the unit-cell analysis needs to be done twice because the macromolecule is essentially infinite in extent along its length. Thus if one considers an isolated perfect polyethylene macromolecule (Fig. 21), one sees that there is a unit cell made up of two consecutive CH_2 groups. To derive the spectrum one needs only to analyze the possible vibrations of this six-atom grouping (Zbinden, 1964). The forces linking the two CH_2 groups to one another and to the symmetrically equivalent CH_2 groups in the adjoining cells are comparable to those within the CH_2 groups, so a division of the modes into internal and external types would not be appropriate. Thus there will be no low frequency lattice spectrum; in fact, the lowest frequency intersection of a branch with the $k = 0$ axis is that at 725 cm⁻¹ mentioned previously. By the same token, the modes corresponding to pure "internal motion of the CH_2 groups will be widely split. The net result, bearing in mind the D_{2h} factor group symmetry, is that the permitted modes and frequencies will be A_g: 2848, 1640, 1061 cm⁻¹; A_u: 1306 cm⁻¹; B_{1g}: 2883, 1168 cm⁻¹; B_{1u}: 1176 cm⁻¹; B_{2g} : 1415, 1131 cm⁻¹; B_{2u}: 2919, 725 cm⁻¹; B_{2g}: 1295 cm⁻¹; B_{2u}: 2851, 1468 cm⁻¹.

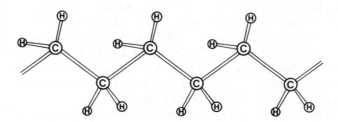

FIG. 21　Configuration of the perfect linear polyethylene macromolecule.

Four of the eighteen (3 × 6) modes will have zero frequency, the three translations and the rotation about the z axis. This contrasts with the normal molecular crystal case in which there are six modes of zero frequency. These zero frequency modes are the intersections at $k = 0$ of the acoustic branches.

When one considers the macromolecules packed into a crystal, one is dealing with the situation of packing rods; then the only degrees of freedom available are the rotation of a rod about the z axis relative to its neighbours (the so-called setting angle) and the displacement of a rod along z relative to its neighbours. Depending on the values chosen for these quantities, one can have one or more macromolecular segments passing through the unit cell. In the case of polyethylene there are two segments per unit cell and the factor group is also found to be D_{2h}. The results are the splitting of the intramolecular fundamentals, as mentioned above and illustrated in Fig. 20, and the appearance of a lattice spectrum. Far-infrared spectra of highly crystalline polyethylene at low temperatures are shown in Fig. 22. The lattice spectrum can be thought of as arising from combining the acoustic branches

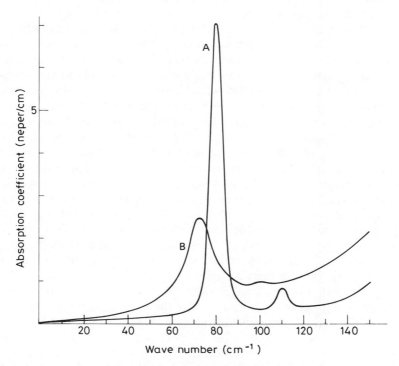

FIG. 22 Far-infrared spectra of highly crystalline polyethylene at (A) liquid helium temperature and (B) room temperature.

of the two macromolecules in-phase and out-of-phase. The in-phase combination of the x, y, and z translations gives, again, zero frequency $k = 0$ modes; but the out-of-phase combination will have a finite frequency. This is equivalent to saying that the presence of two types of macromolecule in the lattice, different by symmetry, splits the acoustic branches just as it does the optical branches. In the case of polyethylene one gets three out-of-phase modes, B_{1u} at 80 cm^{-1}, B_{2u} at 110 cm^{-1}, and a forbidden A_u mode thought to lie near 40 cm^{-1}. The two allowed modes are shown in Fig. 22. Their frequencies are temperature sensitive because the crystal will expand on warming and the van der Waals forces will fall in magnitude. Thus the 78 cm^{-1} B_{1u} mode is observed at 73 cm^{-1} for room-temperature specimens.

In real polymer crystals, it is very uncommon to observe long stretches of linear macromolecules, that is, extended chain crystallites. More often one gets chain-folding and the crystals form as thin lamellae, often no more than 200-Å thick. The chain-folding, forming a sort of concertina, is equivalent to disorder, which will be manifest in the infrared spectrum by line broadening and perhaps by the appearance of forbidden lines. Also, frequencies corresponding to maxima of branches, normally forbidden by the $k = 0$ selection rule, may appear because they correspond to peaks in the density of states function. In polyethylene, for example, a broad continuum near 200 cm^{-1} is thought to arise from disorder induced absorption. Chain-folding is not the only cause of disorder, and polymeric materials can contain large regions that are completely amorphous. This leads to the simplest model of a real polymer, the two-phase theory, which sees the material as made up to tiny crystalline regions dispersed like plums in a pudding throughout an amorphous matrix. It is important to realize that one and the same macromolecule may wander from an amorphous region into a crystalline region and then out again into another part of the amorphous matrix. Studies of partly crystalline polymers have provided some of the best evidence for the validity of the liquid-lattice theory. Thus as the crystallinity falls, the lattice bands broaden; eventually the discrete structure is lost and one has a broad continuous band very similar to those exhibited by liquids. A good example is provided by the random copolymers of ethylene and vinyl acetate. As the amount of vinyl acetate monomer is increased, the crystallinity of the resulting material is reduced. Some of the observed (Chantry et al., 1973) spectra are shown in Fig. 23.

Apart from the fundamental interest of dielectric studies on polymers, there are very good practical reasons for the investigations. Polymers are very useful materials for transmitting extra-high frequency and submillimeter radiation, and it is desirable to try to make them as transparent as possible so that only the very minimum of precious signal will be lost. From what has been said above, the more crystalline the material the better it will

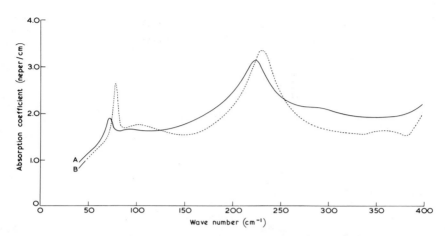

FIG. 23(a) Far-infrared absorption spectra of an ethylene (99.8%)–vinyl acetate (0.2%) random co-polymer: curve A, room temperature, and B, 130 K. (Chantry *et al.*, 1973.)

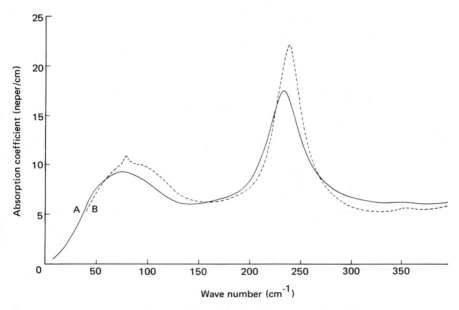

FIG. 23(b) Far-infrared absorption spectra of an ethylene (96.5%)–vinyl acetate (3.5%) random co-polymer: curve A, room temperature, and B, 130 K. (Chantry *et al.*, 1973.)

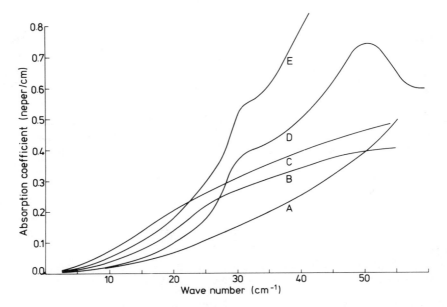

FIG. 24 Far-infrared absorption spectra of some nonpolar polymers: curve A, polyethylene; B, polypropylene; C, TPX, a polymer based on poly-4-methyl pentene-1; D, unsintered polytetrafluorethylene; and E, sintered polytetrafluorethylene.

be because the frequencies of interest lie well below those of the lattice modes. The best materials for this purpose are polyethylene, polypropylene, TPX (poly-4-methyl pentene-1) and polytetra-fluorethylene. Their far-infrared absorption spectra are shown in Fig. 24. In the connection of highest possible transparency, the constitution of the polymer is all important. Thus the sintered forms of polytetrafluorethylene are much more absorbing than the unsintered form. Unfortunately unsintered polytetrafluorethylene has zero mechanical strength, so some form of compromise is necessary. For polyethylene, which can be readily characterized, it is meaningful to attempt analytical expressions for the variation of the dielectric parameters with frequency. Thus for the best grade of polyethylene it has been suggested (Chantry *et al.,* 1971) that the absorption coefficient can be represented by

$$\alpha(\bar{v}) = 8 \times 10^{-4}\bar{v} + 1.4 \times 10^{-4}\bar{v}^2. \tag{84}$$

Because ε' is virtually constant over the whole range at $\varepsilon' = 2.135$, one can also write

$$\varepsilon''(v) = (1.81 + 0.32\bar{v})10^{-4}. \tag{85}$$

E. Plastic Crystals

In most molecular crystals, the van der Waals forces resisting rotation of a molecule are of the same order as those resisting displacement, so rotational motion is restricted to small amplitude libration in a fairly deep potential well. In some cases, however, the structure of the molecule may give it an effectively spherical or cylindrical shape; then rotation may be opposed only by very weak forces. The best known example is found for all the molecules of the series $CCl_x(CH_3)_{4-x}$ ranging from carbon tetrachloride (x = 4) to neopentane (x = 0). These substances show several phases in the solid form, and only the lowest temperature phase is a crystal with true long-range order. The other phases show various degrees of rotational disorder. This is manifest in several ways. Thus the entropy change on freezing is very small and is likewise small at all the subsequent phase changes, apart from the final one where a large entropy change is found. The higher temperature phases are mechanically soft and will often deform merely under gravity. It was this property that gained them the sobriquet of "plastic crystals," a name which is rather unfortunate because confusion with "polymer crystals" is a real

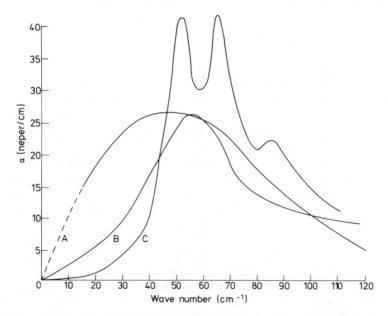

FIG. 25 Far-infrared spectra of three solid phases of methyl chloroform (Cl_3CCH_3): curve A, phase 1, 233 K; B, phase 2, 213 K; and C, phase 3, 148 K. Only the lowest-temperature phase C shows a discrete lattice spectrum.

possibility. In any of the higher temperature phases of a plastic crystal, one has positional order but rotational disorder. Thus the crystal is a sort of half-way-house between the liquid and the truly crystalline state. It is therefore of considerable interest to examine the far-infrared spectra of these higher temperature phases because such studies might be expected to throw some more light on the origin of the liquid-state Poley bands.

The spectra of methyl chloroform (Lassier *et al.*, 1969) (x = 3), or to give its modern name 1,1,1,-trichloroethane, are shown in Fig. 25. It will be seen that the spectrum of the highest temperature phases is almost indistinguishable from that of a typical polar liquid. The spectra of the lowest temperature phases show discrete, if rather broad, bands. It can be hypothesized that the residual broadness would be reduced still further at liquid helium temperature because it presumably arises from rotational freedom of the methyl group, which is frozen-out at a sufficiently low temperature. It seems reasonable to suggest that a valid interpretation of these observations is that crystalline disorder of a sufficient extent can broaden the lattice bands into a continuum. This is precisely the basic assumption of the liquid-lattice theory, and it can be presumed therefore that the plastic crystal studies carried out so far afford strong support for this interpretation of the Poley bands.

F. GLASSES

A glass is basically a supercooled liquid and examples are known from most types of physical system. The most technically important category of glass is that based on mixtures of various silicates; indeed, the word *glass* in ordinary usage is restricted to just this meaning. However, in infrared physics several other glass-forming systems, for example that of arsenic and sulphur, are of interest; and amorphous semiconductors, which are really glasses, are quite important. The spectra of glasses in the far-infrared region consist usually of featureless continua which at low frequencies have the absorption coefficient varying as the square of the frequency (Strom *et al.*, 1974, 1977). Theoretical analysis of the modes of vibration of the completely disordered lattice have been carried out by refined mathematical methods (Bell *et al.*, 1968, 1970, 1971) and the results have been shown to agree well with observation. The most recent studies of silicate glasses are by Birch and Pardoe (1975, 1979) who have used DFTS to show that the continuum is not quite featureless. They have found the interesting result that there appear to be features at very low frequencies indeed (< 10 cm^{-1}) that affect the dispersion in the observable region ($\bar{v} > 10$ cm^{-1}) sufficiently to be noted. The main conclusion however is the same as that emerging from the other studies, namely, that it is merely disorder of *any* kind that broadens the lattice bands into a continuum. One concludes therefore by stating that all the dielectric studies of materials in the far infrared support the liquid-lattice theory which in its simplest form states that the Poley absorption is due to a lattice mode spectrum broadened into a featureless continuum by the spatial and temporal disorder characteristic of the liquid state.

REFERENCES

Afsar, M. N. (1983). *In* "Infrared and Millimeter Waves" (K. J. Button, ed.), Vol. 7, Academic Press, New York.
Afsar, M. N., Chamberlain, J., Chantry, G. W., Finsy, R., and Van Loon, R. (1977). *Proc. IEE* **124**, 575.
Afsar, M. N., Chamberlain, J., and Hasted, J. B. (1976). *Infrared Phys.* **16**, 587.
Afsar, M. N., Chantry, G. W., Birch, J. R., and Kilp, H. (1978). *Infrared Phys.* **18**, 843.
Afsar, M. N., Chantry, G. W., Ferry, R., and Wilkinson, G. W. unpublished work (1978).
Ascarelli, G. (1976). *Chem. Phys. Lett.* **39**, 23.
Ascarelli, G. (1977). *Can. J. Chem.* **55**, 1916.
Bell, R. J., Bird, N. F., and Dean, P. (1968). *J. Phys. C* **1**, 299.
Bell, R. J., Dean, P., and D. C. Hibbins-Butler (1970). *J. Phys. C* **3**, 2111.
Bell, R. J., Dean, P., and D. C. Hibbins-Butler (1971). *J. Phys. C.* **4**, 1214.
Berne, B. J., and Harp, J. D. (1970). *Chem. Phys.* **17**, 63.
Birch, J. R., and Parker, T. J. (1980). "Dispersive Fourier Transform Spectroscopy." *In* "Infrared and Millimeter Waves" (K. J. Button, ed.), Vol. 2. Academic Press, New York.
Birch, J. R., Cook, R. J., Harding, A. F., Jones, R. G., and Prince, G. D. (1975). *J. Phys. D.* **8**, 1353.
Birch, J. R., Cook, R. J., and Pardoe, G. W. F. (1979). *Solid-State Commun.* **30**, 693.
Birnbaum, G., and Rosenberg, A. (1968a). *Phys. Lett.* **27A**, 272.
Birnbaum, G., and Rosenberg, A. (1968b). *J. Chem. Phys.* **48**, 1396.
Birnbaum, G., and Cohen, E. R. (1970). *J. Chem. Phys.* **53**, 2885.
Born, M., and Huang, K. (1954). "Dynamical Theory of Crystal Lattices." Oxford Univ. Press, London and New York.
Böttcher, C. J. F., and Bordewijk, P. (1978). "Theory of Electric Polarization: Dielectrics in Time-Dependent Fields," Vol. 2. Elsevier, Amsterdam.
Chamberlain, John (1967). *Chem. Phys. Lett.* **2**(7), 464.
Chamberlain, J. (1972). *Infrared Phys.* **12**, 145.
Chamberlain, J. (1979). "Principles of Interferometric Spectroscopy" (G. W. Chantry and N. W. B. Stone, eds.). Wiley, New York.
Chamberlain, J., Werner, E. B. C., Gebbie, H. A., and Slough W. (1967). *Trans. Faraday Soc.* **63**, 2605.
Chantry, G. W. (1977). *IEEE Trans.* **MTT-25**(6), 496.
Chantry, G. W., and Gebbie, H. A. (1965). *Nature* **208**, 378.
Chantry, G. W., Gebbie, H. A., and Mirza, H. N. (1967). *Spectrochim. Acta* **23A**, 2749.
Chantry, G. W. *et al.* (1971). *Chem. Phys. Lett.* **10**(4), 473.
Chantry, G. W. *et al.* (1973). *Infrared Phys.* **13**, 157.
Cole, K. S., and Cole, R. H. (1941). *J. Chem. Phys.* **9**, 341.
Davidson, D. W., and Cole, R. H. (1950). *J. Chem. Phys.* **18**, 1417.
Davidson, D. W., and Cole, R. H. (1951). *J. Chem. Phys.* **19**, 1484.

Davies, G. J., and Evans, M. W. (1976). *J. Chem. Soc. Faraday Trans. 2* **72**, 1194.

Davies, M., Pardoe, G. W. F., Chamberlain, J., and Gebbie, H. A. (1968). *Trans. Faraday Soc.* **64**, 847.

Debye, P. (1929). "Polar Molecules." Chemical Catalog Co. (Tudor), New York.

Fatuzzo, E., and Mason, P. R. (1967). *Proc. Phys. Soc. London* **90**, 741.

Finsy, R., and Van Loon, R. (1975). *J. Chem. Phys.* **63**, 4831.

Fuoss, R. M., and Kirkwood, J. G. (1941). *J. Am. Chem. Soc.* **63**, 385.

Gerschel, A., Dimicoli, I., Jaffre, J., and Riou, A. (1976). *Mol. Phys.* **32**, 679.

Goulon, J., Rivail, J. L., Fleming, J. W., Chamberlain, J., and Chantry, G. W., (1973). *Chem. Phys. Lett.* **18**, 211.

Havriliak, S., and Negami, S. (1966). *J. Polym. Sci. Part C* **14**, 99.

Havriliak, S., and Negami, S. (1967). *Polymer* **8**, 161.

Honijk, D. D., Passchier, W. F., and Mandel, M. (1973a). *Physica* **64**, 171.

Honijk, D. D., Passchier, W. F., and Mandel, M. (1973b). *Physica* **68**, 457.

Honijk, D. D., Passchier, W. F., and Mandel, M., and Afsar, M. N. (1977). *Infrared Phys.* **17**, 9.

Hill, N. E. (1963). *Proc. Phys. Soc. London* **82**, 723.

Hill, N. E. (1977). *J. Phys. C.* **10**, 459.

Hill, N. E. (1978). *J. Phys. C* **11**, 815.

Jonscher, A. K. (1977). *Nature* **267** (5613), 673.

Kestemont, E., Hermans, F., Finsy, R., and Van Loon, R. (1978). Infrared Phys. **18**, 855.

Kilp, H. (1977). *J. Phys. E.* **10**, 985.

Kilp, H., Barnes, D. C., Clutterbuck, F. W. J., Afsar, M. N., and Chantry, G. W. (1978). *Infrared Phys.* **18**, 11.

Klug, D. D., Kranbuehl, D. E., and Vaughan, W. E. (1969). *J. Chem. Phys.* **50**, 3904.

Kroon, S. G., and Van der Elsken, J. (1967). *J. Chem. Phys. Lett.* **1**, 285.

Kubo, R. (1958a). *J. Phys. Soc. Japan* **12**, 570.

Kubo, R. (1958b). "Lectures in Theoretical Physics, Vol. 1, Chap. 4. Wiley (Interscience), New York.

Kuno, H. J. (1979). *In* "Infrared and Millimeter Waves (K. J. Button, ed.), Vol. 1. Academic Press, New York.

Lassier, B., and Brot, C. (1969). Motions in Molecular Crystals, *Discuss. Faraday Soc.* **48**, 39.

Lassier, B., Brot, C., Chantry, G. W., and Gebbie, H. A. (1969). *Chem. Phys. Lett.* **3**(2), 96.

Leroy, Y., Constant, E., Abbar, C., and Desplanques, P. (1967). *Adv. Mol. Relaxation Processes* **1**, 273.

Lobo, R., Robinson, J. E., and Rodriguez, S. (1973). *J. Chem. Phys.* **59**, 5992.

McConnell, J. R. (1980). "Rotational Brownian Motion." Academic Press, New York.

Mitra, S. S., and Gielisse, P. J. (1964). "Progress in Infrared Spectroscopy" (H. A. Szymanski, ed.), Vol. 2, pp. 47. Plenum Press, New York.

Mori, H. (1965). *Prog. Theor. Phys.* **33**, 423.

Nee, T. W., and Zwanzig, R. (1970). *J. Chem. Phys.* **52**, 6253.

Pardoe, G. W. F. (1970). *Trans. Faraday Soc.* **66**, 2699.

Poley, J. Ph. (1955). *J. Appl. Sci. B* **4**, 337.

Powles, J. G. (1948). *Trans. Faraday Soc.* **44**, 802.

Rivail, J. L. (1969). *J. Chem. Phys.* **66**, 981.

Rocard, Y. (1933). *J. Phys. Radium* **4**, 247.

Rowlinson, J. S., and Evans, M. (1975). Motion of Simple Molecules in Liquids. *Ann. Rep. Chem. Soc. A*, p. 1.

Sack, R. A. (1957). *Proc. Phys. Soc. London Sec. B* **70**, 402, 414.

Stom, U., Hendrickson, J. R., Wagner, R. J., and Taylor, P. C. (1974). *Solid-State Commun.* **15**, 1871.

Strom, U., and Taylor, P. C. (1977). *Phys. Rev. B* **16,** 5512.

Titulaer, U. M., and Deutch, J. M. (1974). *J. Chem. Phys.* **60**(4), 1502.

Whiffen, D. H. (1966). Spectroscopy, Longmans, Green, London.

Whiffen, D. H. (1949). *Trans. Faraday Soc.* **45,** 124.

Williams, G., and Watts, D. C. (1970). *Trans. Faraday Soc.* **66,** 80.

Wyllie, G. (1971). *J. Phys. C* **4,** 564.

Wyss, H. R., Werder, R. D., and Gunthard, Hs. H. (1964). *Spectrochim. Acta* **20,** 573.

Zbinden, R. (1964). "Infrared Spectroscopy of High Polymers." Academic Press, New York.

CHAPTER 2

Far-Infrared Spectroscopy on High Polymers

W. F. X. Frank

AEG-Telefunken
Geschäftsbereich Hochfrequenztechnik
Ulm, Federal Republic of Germany

U. Leute

Universität Ulm
Abteilung Angewandte Physik
Ulm, Federal Republic of Germany

I. Introduction

Spectroscopy on high polymers in the low-frequency infrared region (far infrared, FIR) has been neglected during the last few decades because most spectroscopic work done by polymer scientists has concentrated on the high-frequency range above about 400 cm^{-1} where conventional infrared spectrometers as well as laser-Raman instruments are powerful analytical tools. Interpretation of the spectroscopic features in this range is guided by the concept of group frequencies, i.e., the concept that single atoms bound in a molecule undergo resonance vibrations that are governed solely by the masses involved and the corresponding force constants. Influences from other chemical bonds are assumed to be neglegible or, at least, easily calculated.

However, in the low-frequency far-infrared range, this simple concept needs expanding to take into account intermolecular as well as intramolecular coupling because intermolecular interactions lead to collective excitations, called *phonons,* especially in crystals. Furthermore, because of the lack of broad-band sources of reasonable energy, spectroscopy in the far infrared requires multiplexing instruments such as interferometers, which are more complicated than the convenient grating instruments that have been favored by chemists. Finally, low-frequency spectroscopy of high polymers has been neglected because solid-state physicists are experienced in making interferometer measurements of phonon absorptions for single crystals but are not used to handling polymers. So a situation has existed in which a whole group of technically and biologically important substances has been completely disregarded in terms of their large-scale molecular motions. Thus an important source of information has been ignored.

In this paper we point out the usefulness of studying of the low-frequency vibrational spectra of polymers. Although we present almost entirely FIR data, it should be noted that Raman spectroscopy is the necessary complementary method. Because the spectra in this region are mainly due to collective excitations in a more or less disturbed lattice, we first give a short survey of the structure of polymers and the morphological features relevant to vibrational spectra. We then outline the experimental parameters specific to polymers that influence the spectra, and we explain the theoretical methods necessary for interpretation of the spectra. Finally, we review the FIR results for the important planar molecule, polyethylene, and for polytetrafluoroethylene as an example of a helical molecule. We do not claim completeness, but we do give some instructive examples to demonstrate the basic principles of the technique.

II. High Polymers

A. THE STRUCTURE OF POLYMERIC MATERIALS

High polymers are virtually ubiquitous. They occur naturally as minerals (sulfur, selenium), as biological components (cellulose, DNA), and, in tremendously increasing quantities over the last several decades, as synthetic materials. The feature common to high polymers is that a single chemical unit (a *monomer,* or repeating unit) is repeated from hundreds to millions of times in a single molecule. The process of building up such a large molecule (a *macromolecule*) is called *polymerization.* The simplest macromolecule is a linear construction of monomers, we call a *chain molecule;* but macromolecules also may be made up of a small number of different chemical units instead of just one and may have branches or even a network instead of a simple linear chain. Nevertheless, a polymer molecule is a rather linear system with strong covalent bonds in the chain direction and only weak coupling laterally.

The chemical *constitution* of the monomeric unit is not sufficient to characterize a polymer, e.g., poly(vinyl chloride), or PVC, which is shown in the figure.

$$\left[\begin{array}{cc} H & H \\ | & | \\ -C & -C- \\ | & | \\ H & Cl \end{array} \right]_n$$

The reason is that other factors are involved. First, the degree of polymerization n is not a constant. Although nature succeeds in producing polymers of uniform length, synthetic chemistry usually yields a rather broad distribution of molecular weights, which is most often characterized by the *number average $\overline{M}_n$* and *weight averages $\overline{M}_w$*. Second, if the molecule has asymmetric side groups, they may be located on only one side (giving an *isotactic* molecule) or on both sides, alternating either regularly (*syndiotactic*) or randomly (*atactic*); this establishes the *configuration* of a molecule. (For further reading see Wunderlich, 1973–1980.)

Constitution, molecular weight, and configuration together strongly influence the conformation of macromolecules. This important concept, not present in the physics and chemistry of small molecules, refers to the set of all arrangements of a chain that can be formed by rotation around chemical bonds, taking into consideration rotational energy barriers. The barriers must not be too high; the criterion is whether or not thermal energy is sufficient to overcome the barrier with reasonable probability. For instance, rotation around a C—C single bond in polyethylene yields a new conformation, whereas rotation around the C=C double bond in poly(*trans*-1, 4-butadiene) leads to the different configuration poly(*cis*-1, 4-butadiene).

$$\left[\begin{array}{c} \qquad CH_2 \\ CH=CH \nearrow \\ \diagdown CH_2 \diagup \end{array} \right] \qquad \left[\begin{array}{c} CH=CH \\ \diagdown CH_2 \diagup \qquad \diagdown CH_2 \end{array} \right]$$

poly-*trans*-butadiene   poly-*cis*-butadiene

Chemists call macromolecules in different conformations *rotational isomers* (Flory, 1969).

The actual conformations determine the packing of the material in the condensed state, which will be either amorphous, melt or glass, purely crystalline, a state rarely achieved, or partially crystalline, which is the state of most materials of interest for FIR spectroscopy (we shall not deal with materials in solution). The partially crystalline state is usually described in terms of a two-phase model in which crystallites are embedded in amorphous material. The volume fraction of crystallites, the *crystallinity* of the sample, can be varied by several preparation methods, e.g., drawing, quenching, or annealing. Thus the morphology of a polymeric specimen — the number, size, and shape of crystallites — is the result of both molecular and experimental parameters. Therefore, the comparison of data by different workers on the "same" polymer can be very difficult.

Only the high-frequency group of vibrations, situated in a spectral region where only the main valences are deformed, are purely intramolecular and therefore uninfluenced by changes of morphology. Most of the low-frequency far-infrared vibrations are intermolecular, which means they consist of collective excitations or lattice vibrations and are determined by the packing of chains in the lattice. Although we are interested mainly in vibrations within crystals where the lattice is sufficiently ordered, the size and defects of these crystals influence line shape and position because of finite boundary conditions. Later we will illustrate this point with spectra of polyethylenes of different crystallinities.

B. Interpretation of the Spectra

The influence upon spectra of varying the morphology of a polymeric system, if the variation is well controlled, can be a valuable experimental tool for establishing correlations between spectra and morphology; i.e., examining the different responses of absorption bands to annealing or to the orientation of a sample may give valuable help in spectrum assignment. In some polymeric systems the material also can be varied chemically, for instance, by increasing the length of a side group or the length of an aliphatic link between heavier groups in the chain backbone. The study of such "homologous series" may further help in the understanding of the spectral features. Moreover, as in all other systems, one can vary temperature or pressure and determine the influence of these external variables on the frequency of a lattice band v_ℓ, i.e., $(\partial v_\ell/\partial T)_p$ and $(\partial v_\ell/\partial p)_T$. This is useful for both the assignment of bands and the determination of anharmonicity data.

Another interesting complication in polymeric systems is that a crystal is extremely anisotropic. In polyethylene, for example, thermal expansion occurs only perpendicular to the chain axis; along the chain axis there is a negative coefficient of expansion (Kobayashi and Keller, 1971; Swan, 1962). The average perpendicular compressibility exceeds that along the chain direction by a factor of a about 60 (Ito, 1980) at room termperature.

We now sketch theoretical methods for relating spectra and morphology (see Fig. 1). The most important is *group theory,* applied to polyethylene by Tobin (Tobin, 1955) as early as 1955, well before the advent of the computerized interferometers or laser Raman spectrometers necessary to measure the predicted vibrations. The principal procedure in applying group theory to macromolecules is the same as that for alkali – halogenide single crystals,

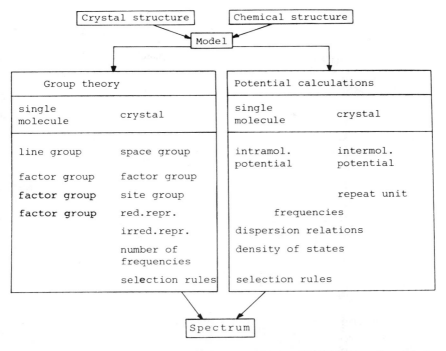

FIG. 1 Outline of methods for using group theory and potential calculations for predicting spectra.

for example; but the number of atoms in a repeating unit of a macromolecule is usually so large that special assumptions are needed to simplify the work. We will explain this using the examples of two representative polymers.

The second theoretical approach used in the polymer field is the relatively well developed method of atomic potential calculations, which yield not only conformations and their energies but also vibrational frequencies, densities of states, etc. (see, e.g., Grossman and Frank, 1977; Kobayashi and Tadokoro, 1977; Leute and Grossmann, 1981; Tasumi and Krimm, 1967; Tasumi et al., 1962, 1963).

III. Dispersion Relations

The functional correlation of a frequency with a molecular vibration and its spatial periodicity (its wavelength) is given by a *dispersion relation*. However, polymer spectroscopists and solid-state physicists use different languages to express these relations, a fact that sometimes leads to misun-

(a)   **(b)**

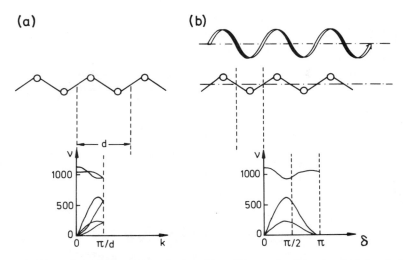

FIG. 2 Dispersion relations: (a) method used by solid-state physicists, in which $k = 2\pi/\lambda = \pi/d$ and $d = \lambda/2$: (b) method used by polymer spectroscopists, in which $\delta = (2\pi/\lambda)x = \pi/2$ and $d/2 = x = \lambda/4$.

derstanding. In Fig. 2 we compare the two approaches by means of a set of dispersion curves for a simple zig-zag chain, an example that will be useful later when we discuss polyethylene.

The usual solid-state physics method is given in Fig. 2a where the translational identity period is shown together with the dispersion relation. We consider the repeating unit of two atoms. Translation through the distance d leaves the chain (assumed to have infinite length) unaltered. The number of vibrations is given by $3n - 4$, where n is the number of atoms in the repeating unit and the value 4 comes from 3 translations of the whole chain plus 1 rotation, namely, that around the chain axis; the other two rotations around axes perpendicular to the chain direction cannot take place. We thus have $3 \times 2 - 4 = 2$ observable proper vibrations belonging to the antisymmetric vibration of the two atoms, along the chain direction and perpendicular to it. (For low-molecular-weight substances, the number of possible vibrations is $3n - 6$.)

The selection rule $k = 0$ says that only in-phase vibrations of all elementary cells of a crystal can be optically active. It is assumed there is a transition moment that provides the interaction of the mechanical motion with the electromagnetic field. In Fig. 2a, we find these two vibrations at the intersection of the optical branches with the ordinate, at $k = 0$ (above 1000 cm^{-1}). The asymmetry of the translation unit causes all branches to have different slopes within the first Brillouin zone ending at the boundary π/d. There we find the shortest wavelength in the system, $\lambda = 2d$.

In Fig. 2b, we show the method used by polymer spectroscopists, i.e., v versus the phase difference δ. They take into account that a zig-zag chain usually contains an additional symmetry element, a screw axis. Using this idea, we obtain a simpler form of the dispersion relation: rotating one mass element around the axis and shifting it by $d/2$ where d is the distance used in Fig. 2a, leaves the chain in a state identical to initial state. What we have done is to consider the zig-zag chain to be a 2/1 helix (2 elements per 1 turn). The vertical line at $\delta = \pi/2$ in Fig. 2b corresponds to $k = \pi/d$ in Fig. 2a, a result easily obtained by considering the phase difference δ between adjacent elements.

The advantage of the point of view of the polymer spectroscopist is a large reduction in the number of branches, which will greatly help in the investigation of more complicated helix molecules (see the later discussion of polytetra-fluoroethylene.) The point is that this method still describes a dispersion relation, but not a Brillouin zone. How can we find the $k = 0$ modes in the new dispersion diagram? We consider an m/n helix; if m units are wound up in n turns, we obtain the identical position of the chain with the phase difference given by $m/n \, \delta = 0, 2\pi, 4\pi$. This means, in our simple example, that the $k = 0$ modes are at $\delta = 0$ and π. Of course, we find the same two frequencies in the region above 1000 cm^{-1} in a way similar to that in the dispersion relation in Fig. 2a.

IV. Applications

As examples, we describe the submillimeter absorption spectra of two materials well known to spectroscopists, polyethylene, PE, and polytetra-fluoroethylene, PTFE. These substances, because of their low absorption coefficients, are suitable for spectrometer lenses, transmission filters, diaphragms, etc. They will be used to show the characteristic features of the absorption spectra of polymers and to elucidate the principal methods of the interpretation of spectra. Both polymers are partially crystallized, so both crystalline and amorphous portions of the samples contribute to the spectra.

A. POLYETHYLENE

1. *Single Chain*

We first outline the procedure needed to obtain the full vibrational spectrum, beginning with a single chain. The chemical structure of PE is shown in Fig. 3a. The molecule is in its lowest energy state, the all-*trans* planar zig-zag form. The symmetry elements of the structure, depicted in Fig. 3b, are the twofold chain axis, $3C_2$, the center of inversion i, and the mirror plane 3σ. The space group can be reduced to a line group because of its translational symmetry; it is isomorphic to the point group D_{2h}.

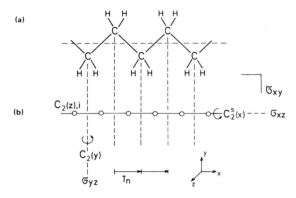

FIG. 3 Polyethylene single chain: (a) the chemical structure and (b) the symmetry elements of the structure (identity operation E not depicted here).

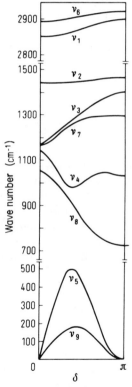

FIG. 4 Vibrational spectrum of a single-chain polyethylene (from Tasumi *et al.*, 1962, 1963).

With $3 \times 6 = 18$ normal modes and using the character table of D_{2h}, we obtain $18 - (3T + 1R) = 14$ modes, eight of which are Raman active, five IR active, and one inactive in both Raman and IR. In the Brillouin-zone with abscissa k, there are 18 branches corresponding to the 18 normal modes; in the $\delta -$ plot of Fig. 4 (from Tasumi *et al.*, 1962, 1963), we find these vibrations at $\delta = 0$ and $\delta = \pi$. The four improper vibrations are those at $v = 0$. We see from Fig. 4 that no vibrations occur in the low-frequency region because of the presence of only the strong force constants which result from main valence bonds. We do not deal with the high-frequency part of the spectrum which is well investigated.

2. *The Crystal*

We now consider polyethylene crystal and look for modifications of the single-chain spectrum, which occur in the low-frequency region. The orthorhombic crystal of PE is shown viewed in the direction parallel to the chains (Fig. 5a) and in the b-axis direction (Fig. 5b). The symmetry elements happen again to be E, $3C_2$, i, and 3σ, which leads to the same space group D_{2h} as for the single chain.

In order to find the modes of the orthohombic cell containing 2×2 CH$_2$ units, we proceed as follows. Because the main valence bonds in the single

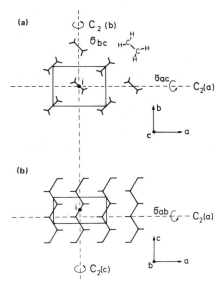

FIG. 5 Polyethylene orthorhombic crystal: (a) view parallel to the chains and (b) view in the b-axis direction.

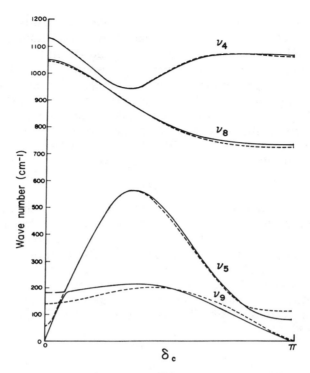

FIG. 6 Vibrational spectrum of an orthorhombic crystal of polyethylene (from Tasumi and Krimm, 1967).

chain are very strong compared with the intermolecular interaction in the crystal, we expect the type of dispersion relation to be similar to that for the single chain; however, the single chain "sees" different force constants in the a-axis and b-axis directions. This causes a frequency splitting of all nine branches shown in Fig. 4, giving 18 branches. The dispersion relations are now to be calculated in all three crystal directions: δ_a, δ_b, and δ_c. Of particular interest is δ_c, which gives the phase differences between neighbouring unit cells along the c-axis, as shown in Fig. 6 (Tasumi and Krimm, 1967).

Again, the modes at $\delta_c = 0$ and $\delta_c = \pi$ are optically active, so we obtain 36 frequencies. Only the low-frequency range is affected, essentially because of lattice interaction. There are $2 \times 4 = 8$ frequencies at the bottom of Fig. 6; three of them are acoustical and give zero frequencies and five form the lattice modes. The solid lines in the figure show vibrations parallel to the a-axis; the dotted lines indicate those along the b-axis. The symmetry species are marked in Fig. 7, and the frequencies of the two IR active modes, B_{1u} and

species	motion	transition moment	wave number
A_u	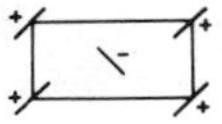	−	59 cm⁻¹ (theor.)
B_{1u}	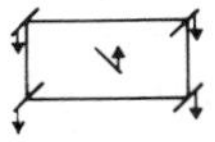	‖ a-axis	73 cm⁻¹ (exp., 20°C)
B_{2u}	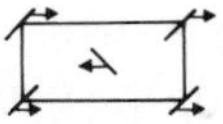	‖ b-axis	~108 cm⁻¹ (exp., −160°C)

FIG. 7   Symmetry species and frequencies of the Au, B_{1u}, and B_{2u} modes (from Tasumi and Krimm, 1967).

B_{2u}, and the inactive A_u mode are shown. The A_u value was calculated by Tasumi and Krimm (1967).

We also can obtain the polarization properties of the infrared vibrations: B_{1u} is parallel to the *a*-axis, B_{2u} is parallel to the *b*-axis; the A_u mode, inactive both in Raman and IR, is polarized along the *c*-axis.

3.   *Far Infrared Results*

We now review the experimental observations giving the confirmation of the previous theoretical predictions. The B_{1u} vibration located at 72 cm⁻¹ at room temperature was detected in 1964 by several groups (Bertie and Whalley, 1964; Franzel and Butler, 1964, McKnight and Möller, 1964). We give an example for this absorption band in Fig. 8 (from Bank and Krimm,

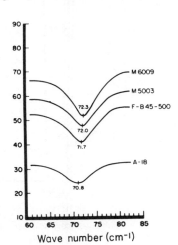

FIG. 8   The B_{1u} absorption band of crystal polyethylene at T = 30°C; curves refer to samples of different molecular weights (from Bank and Krimm, 1968).

1968), illustrating its sensitivity to the morphological differences of the various samples. All samples are polyethylene but are of different molecular weights and of different degrees of branching. The B_{2u} band predicted at 108 cm^{-1} was unobservable at room temperature; however, Dean and Martin (1967) were successful in finding it at 2 K. Their results are shown in Fig. 9.

We now illustrate the effect of the various experimental parameters depicted above. One method of orienting the chains is extrusion whereby the a-axis is forced parallel to the direction of extrusion (Fig. 10a). Another method is drawing (Fig. 10b); here, the chains (the c-axis direction) become parallel to the drawing direction. Both these treatments result in a dichroism, shown in Fig. 11 (from Bank and Krimm, 1968) and Fig. 12 (from Frank and Rabus, 1974). By measuring the dichroism spectroscopically one may deduce the degree of orientation, which is valuable information because its influence on the mechanical properties of the material.

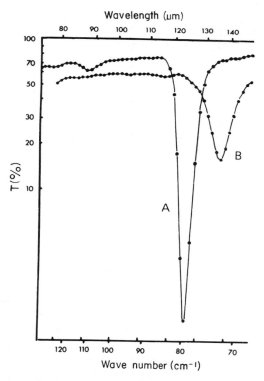

FIG. 9   The B_{2u} absorption band of crystal polyethylene, A at 2 K, B at room temperature (from Dean and Martin, 1967).

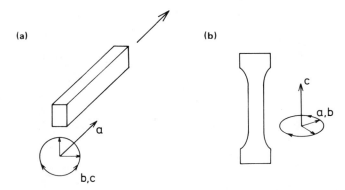

FIG. 10   Orientation of chains: (a) by extrusion and (b) by homogeneous drawing.

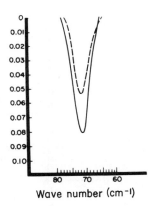

FIG. 11   Dichroism resulting from orientation by extrusion; electrical vector parallel (full line) and perpendicular (dotted line) to extrusion direction (from Bank and Krimm, 1968).

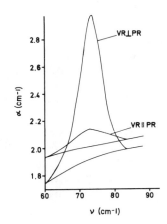

FIG. 12   Dichroism resulting from orientation by drawing: VR ∥ PR means electrical vector parallel to drawing direction; VR⊥PR means perpendicular orientation (from Frank and Rabus, 1974).

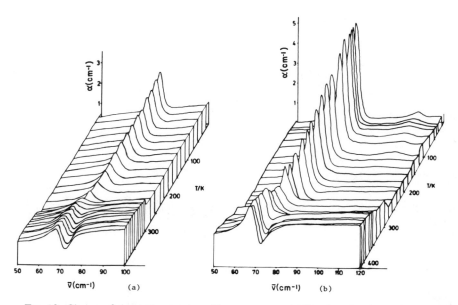

FIG. 13 Change of absorption spectra with temperature: (a) for a low-density polyethylene with many branching points (from Frank, 1979) and (b) for a linear high-density polyethylene (from Frank *et al.*, 1977).

The effects of varying both crystallinity and temperature can be seen in Fig. 13 (from Frank, 1979). The absorption spectra are plotted as a function of temperature for two typical kinds of PE. The results for a low-density polyethylene (LDPE), prepared by introducing many branching points into a C—C chain backbone, is shown in Fig. 13a. In this example we have approximately 40 branching points per 1000 carbon atoms which results in side chains of different lengths. (The material is Lupolen 1804 H from BASF, Germany; for sample preparation see Frank *et al.*, 1977.) The branching points inhibit crystallization, so this type of PE has a relatively low volume fraction of crystallinity (here, 40%). Figure 13b shows the results for a linear PE of high density (HDPE, Lupolen 6011 H by BASF). The number of branching points in this case is only about 3 per 1000 carbon atoms (from which comes the term "linear"); therefore, the degree of crystallinity is comparatively high (about 80%). For sample preparation, see Frank *et al.* (1977). The different degree of crystallinity of these samples are reflected in the different peak heights of the B_{1u} mode.

Increasing the temperature shifts the band toward lower wave numbers, from 81 cm^{-1} to 68 cm^{-1} between 12 and about 400 K. Finally, the

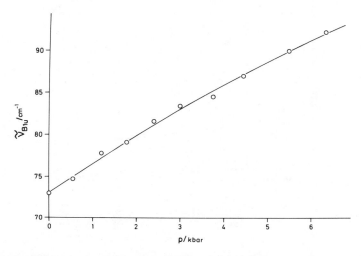

FIG. 14 Change in E_{1u} frequency with increasing pressure, for polyethylene sample Lupolen 6011L (from Leute and Grossman, 1981).

application of external pressure leads to a frequency shift in the opposite direction (Leute, 1981a, 1981b; Leute and Frank, 1980; Leute and Grossman, 1981). When measured up to several kbar, the shift exceeds that due to temperature variation, as shown in Fig. 14. This result, together with high-pressure x-ray crystallography, makes it possible to discuss lattice anharmonicity and Grueneisen parameters (Leute and Grossman, 1981).

B. POLYTETRAFLUORETHYLENE

1. *Predicted Spectrum of the Single Chain*

Polytetrafluoroethylene (PTFE) can be described as a polyethylene in which all hydrogens are replaced by fluorine atoms. The solid-state phases of PTFE near room temperature differ from those of PE in so far as the molecule is no longer a planar zig-zag chain but is twisted into a helix (because the large fluorine atoms do not allow a planar structure). This helix undergoes a phase transition at 19°C.

The conformation of the phase that exists at 1 bar, and not too far above 19°C, is a $\frac{15}{7}$ helix (15 units in 7 turns). Below 19°C we have a $\frac{13}{6}$ helix, more tightly twisted and differing more clearly from the planar $\frac{2}{1}$ helix of PE. We now discuss the absorption spectrum of this more complicated single helix in terms of the dispersion relation given previously. The dispersion curves from Hannon *et al.*, 1969, are depicted in Fig. 15. In this case, the advantage of the δ-plot becomes obvious: we calculate the phase angles along the *c*-axis,

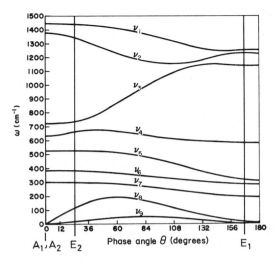

FIG. 15 Dispersion curves for polytetrafluoroethylene (from Hannon *et al.*, 1969).

where $k = 0$ modes are to be found, following Chantry *et al.* (1974). For the $\frac{15}{7}$ helix (symmetry group D_{15}) these modes occur at

$$\theta_1 = (7 \times 360°)/15 = 168°,$$
$$\theta_2 = 360° - 2 \times 168° = 24°,$$
$$\theta_3 = 0°.$$

The corresponding θ_1 for the $\frac{13}{6}$ helix would be $166°$ and θ_2 would be $28°$. The intersections of the vertical lines with the dispersion branches are assigned to the mode types A_1, A_2, E_1, and E_2. These intersections give the possible frequencies, shown in Fig. 15.

2. Experiments

In order to discuss the experimental results, we will briefly outline the procedure, because the story is rather more complicated than that of polyethylene. From the analysis of the single chain we expect a spectrum with vibrations at the intersection of E_1 and E_2 with ν_8 and ν_9 (see Fig. 15). There are only three modes because the fourth is of zero frequency. What was observed first in partially crystallized material above 19°C was one band near 50 cm^{-1} (Chantry *et al.*, 1972), shown in Fig. 16a. Cooling the samples below the phase transition gave five bands in this region, as shown in Fig. 16b. The Raman spectrum also shows band splittings (Boerio and Koenig, 1971) but the x-ray diagram reported by Krimm (1960) and Kilian (1962) indicates that the crystal unit cell, above and below the phase transition temperature, contains only one molecular segment. This contradicts the

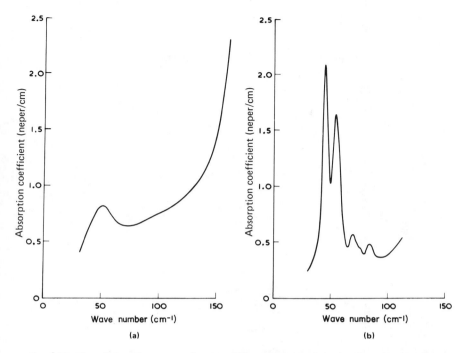

FIG. 16 Experimental spectra of polytetrafluoroethylene: (a) single band in partially crystallized material above 19°C and (b) bands from samples cooled below phase transition (from Chantry *et al.*, 1972).

spectroscopic result which requires more than one unit in the elementary cell. Several groups, including Boerio and Koenig (1971), Chantry *et al.* (1972, 1974, 1977), and Piseri *et al.* (1973), have been involved in discussion about this phenomenon. The discussion was summarized by Chantry *et al.* (1977), as follows: below 19°C the assumption of a correlation splitting of bands is necessary, which includes a unit cell containing more than one segment; above 19°C the assumption of a conformational disorder (Piseri *et al.*, 1973) that activates the vibrational modes seems to be reasonable.

3. *Experimental Determination of Dispersion Branches*

Polytetrafluoroethylene provides an excellent example for the fact that FIR spectroscopy, especially of polymers, opens the possibility of determining the complete set of dispersion curves for a substance, at least in the low-wave-number region. The basis for this procedure is the breakdown of the selection rules, not only when the crystalline order disappears but also when the chain length becomes so short that cyclic boundary conditions

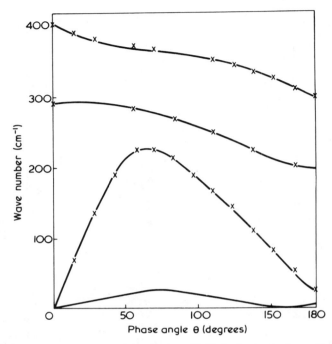

FIG. 17   Far-infrared spectrum of an oligomer ($C_{14}F_{30}$) of polytetrafluoroethylene, in the form of a dispersion diagram (Chantry *et al.*, 1977).

(implying an infinite chain length) are no longer acceptable. As an example, we refer to Chantry et al. (1977) on oligomers of PTFE ($C_{14}F_{30}$, in this case). The spectrum in the far infrared becomes enriched with bands, which are presented in Fig. 17 in the form of a dispersion diagram showing frequencies at angles of $n\left(\frac{180}{14}\right)$, $n = 0, ..., 14$. The shape of the branches is approximately the same as for the polymeric material, but the breakdown of the selection rules activates each frequency of the branch. This, especially for polymeric systems, suggests a new possibility for determining dispersion relations, other than the neutron-scattering method which is effective but usually requires very expensive equipment.

V.   Conclusion

In the preceding paragraphs we have discussed only chemically simple systems, as physicists like to do. However, the field of FIR spectroscopy of high polymers also includes more complicated molecules, the sidegroups of which involve further intermolecular interactions (hydrogen bonds, etc.).

High polymers can be found as synthetic polymers and are abundant in biology, but discussion of these requires knowledge of many individual chemical features and therefore has to be detailed and specific. In the limited treatment presented here, we have shown the theoretical and experimental tools for understanding the spectrum of lattice vibrations. The importance of these vibrations lies in the fact that they determine many technically relevant properties such as thermal conductivity, expansion, and specific heat, and that they elucidate others such as mechanical moduli and optical properties. Thus by comprehending the influence of structure on lattice vibrations, one approaches an important aim of polymer science, namely, the design of polymers for specific purposes.

REFERENCES

Bank, M. I., and Krimm, S. (1968). *J. Appl. Phys.* **39,** 4951.
Bertic, J. E., and Whalley, E. (1964). *J. Chem. Phys.* **41,** 575.
Boerio, F. J., and Koenig, J. L. (1971). *J. Chem. Phys.* **54,** 3667.
Chantry, G. W., *et al.* (1972). *Chem. Phys. Lett.* **16,** 141.
Chantry, G. W., *et al.* (1974). *Polymer* **15,** 69.
Chantry, G. W., *et al.* (1977). *Polymer* **18,** 37.
Dean, G. D., and Martin, D. H. (1967). *Chem. Phys. Lett.* **1,** 415.
Flory, P. J. (1969). "Statistical Mechanics of Chain Molecules." Wiley (Interscience), New York.
Frank, W. F. X. (1979). In "Nonmetallic Materials and Composites at Low Temperatures" (R. P. Reed and G. Hartwig, eds.), p. 51. Plenum, New York.
Frank, W., and Rabus, G. (1974). *Colloid Polym. Sci.* **252,** 1003.
Frank, W., Schmidt, H., and Wulff, W. (1977). *J. Polym. Sci. C. Polym. Symp.* **61,** 317.
Franzel, A. D., and Butler, J. P. (1964). *J. Opt. Soc. Am.* **54,** 1059.
Grossmann, H. P., and Frank, W. (1977). *Polymer* **18,** 341.
Hannon, M. J., Boerio, F. J., and Koenig, J. L. (1969). *J. Chem. Phys.* **50,** 2829.
Ito, I. (1980). "High Pressure Science and Technology" (K. D. Timmerhaus and M. S. Barber, eds.), Vol. 1, p. 482. Pergamon, Oxford.
Kilian, H. G. (1962). *Kolloid-Z* **185,** 13.
Kobayashi, Y., and Keller, A. (1971). *Polymer* **11,** 114.
Kobayashi, M., and Tadokoro, H. (1977). *J. Chem. Phys.* **66,** 1258.
Krimm, S. (1960). *Fortschr. Hochpolym. Forsch.* **2,** 51.
Leute, U. (1981a). *Polym. Bull.* **4,** 89.
Leute, U. (1981b). *Proc. Eight Int. AJRAPT Conf., Uppsala* 1981, in press.
Leute, U., and Frank, W. F. X. (1980). *Infrared Phys.* **28,** 327.
Leute, U., and Grossman, H. P. (1981). *Polymer* **22,** 1335.
McKnight, R. V., and Möller, K. D. (1964). *J. Opt. Soc. Am.* **54,** 132.
Piseri, L., Powell, B. M., and Dolling, G. (1973). *J. Chem. Phys.* **58,** 158.
Swan, P. R. (1962). *J. Polym. Sci.* **56,** 403.
Tasumi, M., and Krimm, S. (1967). *J. Chem. Phys.* **46,** 755.
Tasumi, M., Shimanouchi, T., and Miazawa, T. (1962). *J. Mol. Spectr OSC* **9,** 261.
Tasumi, M., Shimanouchi, T., and Miazawa, T. (1963). *J. Mol. Spectr OSC* **11,** 422.
Tobin, M. C. (1955). *J. Chem. Phys.* **23,** 891.
Wunderlich, B. (1973–1980). "Macromolecular Physics," Vol. 1–3. Academic Press, New York.

CHAPTER 3

Submillimeter Solid-State Physics

S. Perkowitz

Department of Physics
Emory University
Atlanta, Georgia

I. Introduction

The far-infrared–submillimeter (submm) portion of the electromagnetic spectrum, from 10 to 200/cm, has proven most useful for probing basic properties of matter. Although this wavelength range is traditionally a difficult one in which to work, it serves so well for the study of condensed matter that an immense amount of solid-state spectroscopy has occured in the last eighty years. From the pioneering work of Rubens around the turn of the century to current investigations of elaborate solid-state systems, submm spectroscopy has thrown important light on almost every aspect of solid behavior. In addition there are benefits for applications. Submm spectroscopy can characterize materials in accurate and nondestructive ways, and solid-state devices play an important role in the development of a complete submm technology. In this chapter we review both basic and

71

applied aspects of submm solid-state physics. So vast is the subject, however, that we shall treat only two classes of solids, semiconductors and supercon- ductors. These classes represent most of the important features of solid behavior including crystalline and lattice effects, free-carrier phenomena, impurity behavior, and quantum-mechanical effects.

Our aim in this review is to provide workers in the submm field with some insight into the physical processes in solids that are significant at submm energies. Therefore this chapter will presuppose little background in solid- state theory. We shall develop this theory as needed for each solid-state process with emphasis on two aspects: an indication of typical process energies and a description of the mechanism that couples each process to photons. This coupling will almost always arise from a classical charged particle – electric field interaction. Each of our theoretical discussions will be followed by examples of actual measurements in as many materials as is practical. To make the review tractable and up to date, many of our examples will be drawn from work done in the last ten years.

Before turning to submm solid-state physics, a proper definition of submm spectroscopy and a survey of experimental methods should be made, for the difficulties of submm research make an appreciation of the experimental techniques important. A review of submm solid-state physics is also a review of improvements in submm spectroscopic methods.

II. Survey of Experimental Methods

The usual meaning of *optical spectroscopy* is the measurement of light intensity as a function of photon energy when the radiation is reflected, absorbed, or transmitted by a sample. Much submm work in solids has been such optical spectroscopy. In other cases submm researchers have used techniques borrowed from microwave science, such as cavity and wave- guidelike methods. Our review will deal primarily with these optical and microwave approaches. We shall not deal with other types of spectroscopy, which although important in submm work are of a less general nature. These include submm magnetospectroscopy, in which a magnetic field is applied to the sample to gain additional information; and inelastic processes exemplified by Raman scattering, which is difficult to carry out at submm energies.

A recurring theme in submm spectroscopy is the difficulty of the tech- niques. The outstanding problem is that there exists no submm source that has both high power and continuous tunability. The sources that work well in other parts of the spectrum fail badly in the submm range. For instance, blackbody sources at temperatures near 2000 K have their peak output near 10 μm and can provide adequate power over the near and middle infrared.

At the other end of the spectrum, it is possible to build microwave tubes of excellent quality that, although not greatly tunable can provide large amounts of stable power at an exact frequency. Blackbody radiation is miserably weak in the submm range, however, and the construction of a waveguide device to produce radiation of, say, 0.5-mm wavelength is a difficult task.

These limitations have sometimes made submm spectroscopy appear more of an art-form than a science. Excellent work has been done with blackbody sources in submm grating spectrometers, because long wavelength gratings are cheap and easy to make, but the greatest care had to be taken to filter out higher orders of radiation and to maximize the signal-to-noise ratio as far as possible. Even so, some anomalous effects have shown up in the most careful work. This picture changed some twenty years ago with the realization that the best spectrometer for submm use was not the grating instrument but rather an interferometric spectrometer based on Michelson's familiar design. Excellent reviews of interferometric spectroscopy appear elsewhere (Bell, 1972; Chantry, 1971) and we shall give only the briefest account.

In an interferometer a beam of light from a source is divided by a beam splitter, traverses two different optical paths, and then is recombined. The resulting interferogram signal, which depends on the variable path length difference, is just the Fourier transform of the intensity spectrum, intensity versus light frequency. This awkwardness in the form of the output data, which requires that a computer be used to extract the desired information, is more than offset by the fact that the signal-to-noise ratio in the interferometer is much higher than can be achieved in a grating device for the same measurement time. The improvement arises because (1) the detector sees many frequencies at a time rather than only one as in the grating instrument (the Jacquinot advantage) and (2) the interferometer's resolution depends only on the total path difference and not on slit sizes, so the slits can be made as large as desired for no loss of power with high resolution (the Fellgett advantage). These commanding features of the Fourier system have made it the preferred system for submm spectroscopy, and much of the solids work we shall describe was carried out with a Fourier spectrometer.

Even the best spectrometer needs a good detector, however, and the subject of submm detectors also has a full history. Many of these detectors are solid-state devices and themselves represent an important application of submm solid-state physics. Later on we shall say more about the relevant solid-state processes, but in this section we concentrate on the technical details that determine detector performance in research applications.

In characterizing a detector, a useful measure of its sensitivity is the noise equivalent power (NEP), defined as the amount of radiation power that, when incident on the detector, produces an rms voltage output equal to the rms noise voltage output. (The NEP depends on the bandwidth of the detector electronics. All NEP's quoted here are for an assumed 1 Hz bandwidth.) Other important parameters are the detector response time and the wavelength range to which the detector responds.

The bulk of submm solid-state spectroscopy has been carried out with two types of thermal detectors, Golay cells and cooled bolometers, and with several types of photoconductive devices. The Golay detector (Golay, 1947) contains a small chamber filled with an inert gas. Modulated radiation entering the chamber heats the gas at the modulating frequency. The resulting periodic pressure change flexes a small membrane, and these distortions are detected by an auxiliary optical system. This device generates a noise level only three or four times worse than the theoretical prediction at 300 K, yielding an NEP of 10^{-10} W. Because the detecting principle is thermal, the Golay cell can be used for any wavelength range by proper choice of window material. Its main drawbacks are its sensitivity to vibration and the slowness of its response to radiation.

Cooled bolometers are solid-state devices with large temperature coefficients of resistance at cryogenic temperatures. One type has been made from commercial carbon resistors (Boyle and Rodgers, 1959) and yielded an NEP of 6×10^{-12} W when operating at 2 K. Other cooled bolometers have been made from germanium containing various impurities. A Low-type bolometer (Low, 1961), of germanium doped with gallium, has produced an NEP of 5×10^{-13} W and a response time of 400 μsec. These thermal devices work over a broad wavelength range.

Photoconductive detectors are also cooled. Their operation depends on a change in resistance of a semiconductor caused directly by submm photons. Such changes can arise from the intrinsic gap behavior of semiconductors, from the impurity behavior, or from free-carrier properties, all of which will be described later. Depending on the process, the wavelength range for detection may be limited. Germanium doped with various impurities has functioned as an impurity photoconducting detector for frequencies above 100/cm (Robinson, 1973). Typical NEP's are 10^{-11} W.

Very recently, spectroscopic possibilities have been broadened by the introduction of various submm lasers. It has long been known that certain gas lasers—HCN at 337 μm, for example—produce one or a few submm lines, often at high power. However, spectroscopy as we have defined it requires continual frequency coverage over the range of interest. This need has been partially met with the introduction of the optically pumped submm laser invented by Chang and Bridges (1970).

In the pumped laser, an exciting line (usually provided by a CO_2 laser) pumps a collection of gas molecules into an excited vibrational state. As the excited system relaxes, transitions occur between rotational levels, producing output at submm energies. Because there are nearly a hundred CO_2 pumping lines and dozens of pumpable gases, there is a vast array of powerful submm lines, over 800 at last count (Gallagher *et al.*, 1977; Rosenbluh *et al.*, 1976; Yamanaka, 1976). This set of lines provides enough coverage to carry out solid-state spectroscopy, although there would be difficulties in the gas phase where absorption structure is very narrow. The laser lines are 10 to 1000 times more powerful than the blackbody radiation source in a Fourier spectrometer and often can yield excellent data even with less sensitive detectors. A detailed study of submm laser methods is given by Bean and Perkowitz (1979).

As we describe different submm solid-state applications, we shall indicate the experimental techniques that were used. In the case of superconductor work especially, it will be possible to see the steady improvement in data quality as new instrumental methods have become available.

III. Submillimeter Semiconductor Physics

A. INTRODUCTION AND SEMICONDUCTOR THEORY

Submillimeter radiation has been used to study almost every imaginable aspect of semiconductor behavior. In this section we shall present examples of such submm probing, following a brief review of semiconductor physics. Each effect will be illustrated by examples chosen from the semiconductor literature. Much of the physics given here serves as the basis for the submm characterization of semiconductors, a topic to be covered in a later section.

Semiconductors comprise a very large class of materials. The earliest examples were the elemental semiconductors germanium and silicon, which lie in group IV of the periodic table. Later much interest and effort were put into binary compound materials constructed from elements in groups III and V, II and VI, and IV and VI (examples are, respectively, GaAs, CdTe, and PbTe). Most recently very complex ternary and quaternary semiconducting alloys have been of interest, such as $Pb_xSn_{1-x}Te$ and $In_xGa_{1-x}As_yP_{1-y}$. In addition some more exotic combinations, such as the I–III–VI chalcopyrites, show semiconducting behavior.

Semiconductors in some sense lie midway between metals and insulators when viewed as conductors of electricity. Metals conduct well at all temperatures because they contain free charge carriers — electrons — with densities approaching $10^{23}/cm^3$. Insulators contain negligible free carriers. Semiconductors can have free-carrier densities between 10^9 and $10^{21}/cm^3$, and the density can vary strongly as a function of temperature. Thus a semiconductor can approach insulating or metallic behavior depending on circumstances.

The reason that the carrier density varies so greatly and depends heavily on temperature lies in the most important single feature of semiconductor physics, the existence of the band gap. This gap is a quantum-mechanical effect, but the cause of its appearance can be sketched out in simple terms.

Like any other crystalline solid, a semiconductor contains a regular spatial array of its constituent atoms. The electrons associated with this array have energy levels that differ from the electron levels of a single atom. Consider an isolated hydrogen atom. Its single electron can exist in any of the discrete Bohr levels. However, if some 10^{23} of these atoms are placed in an array with a spacing of angstroms, each electron can interact with the other atoms. Then, to satisfy the Pauli exclusion principle, the discrete levels must broaden to contain all the available electrons, thus producing a continuous band of energies rather than the sharp allowed level of a single atom. Each successive electron enters the band at a higher energy until the last electron is accommodated at the energy called the Fermi energy.

A second quantum-mechanical feature of a periodic array of atoms is the band gap. The electrons, which can interact with many ionic cores, act as nearly free carriers and can be described quantum mechanically as modified plane waves. These waves interact with a periodic potential due to the ion array. It can be shown as a general result of wave theory that for some energies the electron waves propagate through the array with no difficulty, but for other ranges of energies they are severely attenuated. These latter energies constitute a forbidden energy range or gap of size E_g between two allowed bands.

Figure 1a illustrates the concept of bands and band gap for a typical semiconductor. The electrons below the gap are the inner valence electrons, which are not free to move. Electrons that exist above the gap are mobile and can contribute to electrical conduction. Figure 1b shows a different perspective, illustrating that electrons above the gap have an energy relationship given by

$$E = p^2/2m^* = \hbar^2 k^2/2m^*, \tag{1}$$

where p is the electron momentum, k the electron wave vector (the reciprocal of the de Broglie wavelength), and m^* the effective mass. Although the conduction-band electrons are mobile, they undergo interactions with the ion sites. These interactions can be included, within a certain approximation, if the free electron mass m_0 is replaced by the effective mass m^* in the electronic equations of motion. The value of m^* is a consequence of the particular band structure of each semiconductor and can be found from submm data, as we shall see later.

Although the conduction band is available for conduction electrons, the number of such electrons is large only under special conditions. One way to increase their concentration is to excite valence electrons thermally across

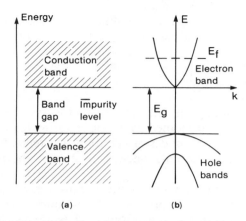

FIG. 1 (a) Schematic representation of semiconductor conduction and valence bands, the band gap, and an impurity level. (b) Energy versus wave-vector diagram showing the quadratic relationship for electrons and holes, multiple-hole bands, and the Fermi energy E_F.

the band gap. Because typical semiconductor gap energies are about 1 eV = 12,000 K, thermal excitation will produce only a few conduction electrons at ordinary temperatures. A noteworthy feature of such excitations is that an electron departing the valence band leaves behind a void, a *hole,* that acts like a positively charged current carrier. These holes obey the same effective mass model as electrons and have the interesting feature that multiple-hole bands can exist, as shown in Fig. 1b.

A more effective means than thermal excitation for increasing the conductivity of a semiconductor is to dope the material by adding impurities. Electrons associated with such impurities are less tightly bound than are the valence electrons, because the impurity is less tightly locked into the semiconductor structure. Effectively, the impurity electrons lie at energy levels close to the conduction band, as is shown schematically in Fig. 1a, and can be excited into that band in great numbers at room temperatures or even below.

Another mechanism for exciting electrons across the gap is optical. A photon of energy E_g can be absorbed by a valence electron which is then raised to the conduction band. For most elemental and binary semiconductors E_g lies in the visible or UV region, but for some of the more complex systems E_g is a submm energy, as we will see later. Thus submm radiation can be used to probe the very basic band parameter E_g as well as m^*.

This simple picture of bands, gaps, doping, and free carriers defines many of the basic properties of a semiconductor and can be used to predict the electrical and optical behavior. In addition, semiconductors share some of the basic properties of all solid crystalline matter such as the existence of lattice vibrations. We shall now describe these and other interactions in detail.

B. LATTICE VIBRATIONS

Like other crystalline materials, semiconductor crystals consist of a three-dimensional array of ions joined by strong restoring forces. Such an array has characteristic normal modes of vibration that represent movement of the entire lattice. These normal modes can be shown to be quantized and are called *phonons*.

For many important phonon processes, however, a classical picture is adequate. In this picture, phonons in a three-dimensional array can function like longitudinal or transverse waves. They are also characterized by whether they are optical or acoustic modes. In the former, a dipole moment exists that can couple to electromagnetic radiation; but in the latter, coupling dipole moments do not exist. *Transverse optical* (TO) phonons can thus couple to electromagnetic radiation. In a simple picture the resonant frequency of such a mode (known as the *reststrahlen frequency*) is given by

$$\omega_T = 2\pi(k/M)^{1/2}, \tag{2}$$

where k is the interionic coupling constant and M an appropriate reduced ionic mass. For many typical semiconductor systems, the frequency ω_T occurs in the submm range. As in any harmonic oscillator, resonant absorption of energy occurs at ω_T, as can be seen by examining the typical dielectric function

$$\varepsilon = \varepsilon_\infty + \frac{(\varepsilon_0 - \varepsilon_\infty)\omega_T^2}{\omega_T^2 - \omega^2 + i\omega\Gamma}, \tag{3}$$

where ε_∞ and ε_0 are the high-frequency and static dielectric constants, Γ a damping constant, and ω the optical frequency.

Not only the TO mode may be seen at submm frequencies. When several phonons of different frequencies are excited, new modes are created at the sum and difference frequencies. Even if each contributing phonon has an energy above the submm range, the multiphonon difference modes can interact with submm radiation.

It is instructive to examine the response of a phonon system described by Eq. (3). For simplicity, we set $\Gamma = 0$; and in fact, phonon damping is usually quite low. Then we can simply plot ε as shown in Fig. 2. The resonant behavior at ω_T is clear. Another significant frequency is the *longitudinal optical* (LO) frequency $\omega_L = (\varepsilon_0/\varepsilon_\infty)^{1/2}\omega_T$, where $\varepsilon = 0$.

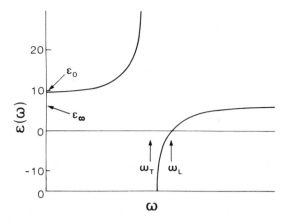

FIG. 2 The phonon dielectric function versus frequency for the case of zero damping.

To find the optical properties of a sample described by Eq. (3) we need to calculate the complex index of refraction, given by

$$n - ik = \varepsilon^{1/2}. \tag{4}$$

Then the reflection coefficient R and absorption coefficient α are given by

$$R = [(n - 1)^2 + k^2]/[(n + 1)^2 + k^2] \tag{5}$$

and

$$\alpha = 2\omega k/c, \tag{6}$$

where Eq. (5) holds for a semiinfinite sample with radiation normally incident.

Figure 3 is a plot of the calculated reflection curve for our hypothetical phonon system. The TO and LO frequencies appear clearly in the reflection plot. The characteristic frequencies also are visible in absorption work.

Lattice work was carried out in the III–V semiconductors by Hass and Henvis (1962) who observed the TO mode at 184.7/cm in InSb using a conventional grating instrument. This early submm result was accurate to no more than 5 to 10%. Their reflectivity data, shown in Fig. 4, shows the general shape predicted by the theory. In other early work, a member of the IV–VI lead-chalcogenide family, PbS, was examined by Geick (1964) between 5 and 250/cm. He obtained a very low value of ω_T, 70/cm, and a very large static dielectric constant. Similar behavior was seen later in the related compound PbTe. Kinch and Buss (1972) used a Fourier spectrometer and a cooled bolometer to measure the reflectivity of bulk PbTe,

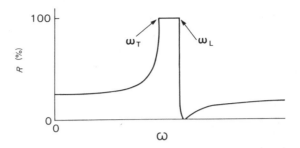

FIG. 3 The reflectivity versus frequency curve calculated from the phonon dielectric function of Fig. 2.

obtaining $\omega_T = 18$/cm and $\varepsilon_0 = 1300$. Burkhard *et al.* (1976) were able to observe a very clear TO mode in the reflectivity of a thin PbTe film and determined $\omega_T = 18.2$/cm and $\varepsilon_0 = 1380$.

Prior to 1967 many workers examined submm absorption in CdTe and observed, in addition to a primary TO resonance at 140/cm, many subsidiary bands. Bottger and Geddes (1967) identified these secondary bands as arising from multiphonon processes. Stolen (1969) observed phonon differ-

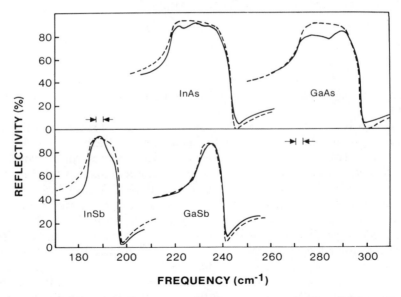

FIG. 4 Reflectivity results for InSb and other III–V compounds. The solid curves are the experimental results and the dashed curves are the theoretical fits from Eqs. (3)–(5). The TO and LO frequencies are clearly observable. (After Hass and Henvis, 1962.)

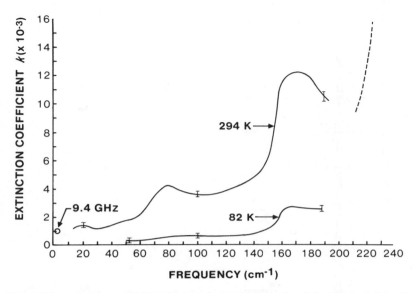

FIG. 5 The extinction coefficient k [where $k = (\alpha/\text{cm})/(2\,\omega/\text{cm})$] for high-resistivity GaAs at two different temperatures. The structure below 200/cm is due to phonon difference processes. The dashed line is an extrapolation of the TO absorption at 268/cm. (After Stolen, 1969.)

ence bands in GaAs, with absorption structure as shown in Fig. 5. His later analysis (Stolen, 1975) showed that both two- and three-phonon difference processes contributed. Slack and Roberts (1971) obtained submm data in a variety of compounds including CdTe, GaAs, and ZnTe and identified two-phonon processes also.

CdTe has remained an interesting material for submm investigation. Vodop'yanov et al. (1974) studied this material and concluded that the optical phonon frequencies depend somewhat on the damping constants. They obtained TO and LO frequencies of 140 and 167/cm, respectively. Batalla and co-workers (1977) used a Fourier spectrometer to measure a very complete set of data between 20 and 440/cm. They were able to correlate their results in detail with neutron scattering data, and were also able to assign every observed peak to some phonon combination mode. Their precision was generally greater than any that was previously obtained for CdTe.

The detailed knowledge of CdTe was useful in dealing with a more complex system, the mixed semiconductor $Hg_{1-x}Cd_xTe$, which combines characteristics of the two constituent compounds HgTe and CdTe (Baars and Sorger, 1972). For all values of x the mixed system showed peaks at the

S. PERKOWITZ

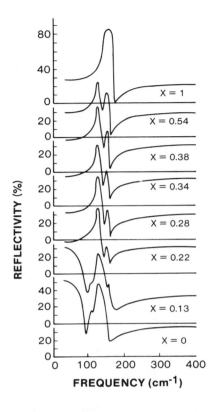

FIG. 6 Reflectivity of $Hg_{1-x}Cd_xTe$ for different values of x at 77 K, showing the transition from the reststrahlen pattern of pure CdTe to that of pure HgTe. (After Baars and Sorger, 1972.)

reststrahlen frequencies for the binary compounds, with the relative magnitudes of these contributions changing with x, as shown in Fig. 6. An additional band near 100/cm was observed in HgTe and for the alloy up to $x = 0.54$. The two TO frequencies and the associated LO frequencies were shown to vary linearly with x, and the 100/cm band was shown by its temperature dependence to arise probably from a subtractive phonon process. Another interesting result was that the high-frequency dielectric constant varied smoothly with x between the values for HgTe and CdTe.

A quaternary semiconducting system reaches new heights in lattice complexity. Recently Perkowitz and co-workers (Amirtharaj *et al.*, 1980) determined the reflectivity of thin film $In_{1-x}Ga_xAs_yP_{1-y}$ between 20 and 410/cm. This film was grown on a substrate made of the semiconductor InP. The presence of the two alloying parameters x and y makes it possible to tune the band gap of the film to a desired value while maintaining a lattice spacing that matches that of the InP. The lattice match is achieved when $y/x = 2.2$.

We used a Fourier spectrometer and Golay detector to obtain the reflectivity of five samples with y between 0.22 and 0.66. The observed TO and

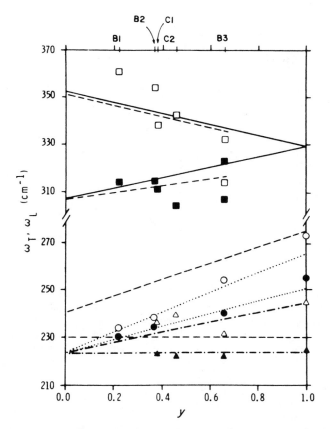

FIG. 7 The TO and LO frequencies for $In_{1-x}Ga_xAs_yP_{1-y}$ plotted versus y for the fixed ratio $y/x = 2.2$. The mixed modes are identified in terms of the end-point modes from which they originate, as follows:■□—, InP-like; ●○··· , GaAs-like; and ▲△—·—·—·, InAs-like. Each label at the top of the figure (B1, B2, etc.) represents a sample with a particular value of y. Samples B1, B2, and B3 were grown in a different laboratory than were C1 and C2. Samples B2 and C1 are almost identical in composition but show large spectral differences. The lines denote the behavior predicted by an impurity-mode model which is seen to work reasonably well. The similar Raman results of Pinczuk *et al.* (1978) are shown by the dashed line. (Reproduced from Amirtharaj *et al.*, 1980.)

LO frequencies are shown in Fig. 7 as a function of y. The complex behavior can best be understood in terms of the end-point compounds, and the modes are labelled accordingly.

Complexities in lattice behavior occur not only in mixed crystal systems but also in crystals with complex structures. An example is $CdAs_2$, a semiconducting compound of the II–V type. The unusual distorted tetrahedral structure of this material leads to a more complex spectrum than we

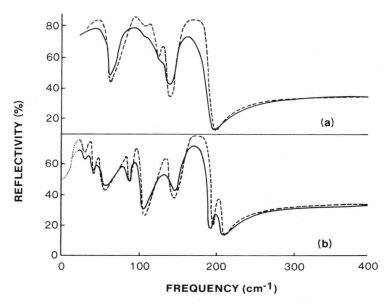

FIG. 8   The complex reflectivity behavior of Sb_2Se_3 showing the large number of optical modes. The results differ depending on the orientation of the electric field **E** of the incident radiation: (a) **E** parallel to **c** and (b) **E** parallel to **a** where **c** and **a** are primary crystallographic directions. There also is a substantial temperature dependence, as can be seen by comparing the dashed curve (90 K) with the solid curve (300 K). (After Petzelt and Grigas, 1973.)

have so far encountered (Gregora and Petzelt, 1972). The reflectivity depends on the polarization of the incident light and shows several peaks. These results could be fitted with the sum of several resonant terms like those in Eq. (3).

Even more complex lattice behavior is seen in the isomorphous group of compounds Sb_2S_3, Bi_2S_3, and Sb_2Se_3 which exhibit a complicated ribbon-like structure (Petzelt and Grigas, 1973). The internal structure is such that there are two weakly bonded directions and one strongly bonded direction, leading to strong anisotropy in the properties. The structural complexity leads to large numbers of optical modes as shown in Fig. 8. A detailed analysis of the data gives the degree of polarity for the different compounds and suggests that observed temperature anomalies are due to a phase transition that does not change the symmetry.

C.   FREE CARRIERS

As we have seen, a semiconductor can contain so-called "free" carriers in addition to the valence electrons bound into its basic crystallographic structure. Because such carriers can be accelerated by an electric field, they can absorb energy from an incoming radiation field and hence produce an important optical effect. As we shall show, the energy absorption generally increases with the wavelength of the radiation, so free-carrier absorption is an important submm effect even for a lightly doped semiconductor.

The effective-mass approximation gives a simple way to discuss free-carrier optical absorption. In that approximation, the equation of motion for a free carrier—to be definite, an electron of charge e—is

$$m^*\left(\frac{d\mathbf{v}}{dt} + \frac{\mathbf{v}}{\tau}\right) = e\mathbf{E}e^{i\omega t}, \tag{7}$$

where $\mathbf{v}$ is the electron velocity, $\mathbf{E}e^{i\omega t}$ the applied electric field, and τ a scattering time, a simple way of including the interaction of the free electron with any other feature in the semiconductor that may drain energy from it. In Eq. (7) we shall treat τ as a constant, but in fact, it is a much more complex quantity. Its value depends on how the electron interacts with phonons, with impurity sites, or with other electrons; and in general it is energy- and temperature-dependent. For some interactions it cannot even be well defined.

Equation (7) can be used to derive the optical properties of an assembly of free electrons, because the electron current $\mathbf{J}$ can be obtained from the definition $\mathbf{J} = ne\mathbf{v}$, where n is the density of electrons. Then Eq. (7) and the definition $\mathbf{J} = \sigma\mathbf{E}$ yield the ac conductivity

$$\sigma = ne^2\tau/m^*(1 + i\omega\tau). \tag{8}$$

From Maxwell's equations, the conductivity gives rise to the dielectric function

$$\varepsilon = 1 - \omega_p^2/\omega(\omega - i/\tau), \tag{9}$$

which can be used to obtain any desired optical property. Here ω_p, the electron plasma frequency, is given by $\omega_p^2 = 4\pi ne^2/m^*$. It is the characteristic frequency of a collective mode of electron oscillation called a *plasmon*. From Eqs. (4), (6), and (9) we can write the electron absorption coefficient

$$\alpha \propto \omega_p^2\tau/(1 + \omega^2\tau^2). \tag{10}$$

For many semiconductors, $\omega\tau \gg 1$ and Eq. (10) takes the limiting form $\alpha \propto \omega^{-2}$. This semiclassical result does not agree with observations of electron behavior in the middle infrared, where it is often found that $\alpha \propto \omega^{-3}$.

The answer to this discrepancy lies in the quantum-mechanical nature of the problem. The simplistic equation-of-motion approach for electron behavior in a radiation field is true only when the quantum nature of the light can be ignored, i.e., when the photon energy $\hbar\omega$ is less than a typical electron energy E. This, however, is often not the case, and the theory of electron absorption must include this feature. In such a calculation it is also important to specify in detail the microscopic nature of the electron-scattering processes; it is not sufficient to lump all scattering behavior under the catch-all quantity τ. Rather, each process must be identified and an appropriate interaction Hamiltonian written for each case.

These quantum-mechanical derivations were carried out by several workers including Dumke (1961), Haga and Kimura (1963, 1964), and others. The most complete and coherent exposition has been given in a series of papers by Jensen (1971, 1973a, b, 1977, and 1978). These developments were successful in explaining why a nonquadratic dependence was observed at wavelengths in the middle infrared, and why a quadratic dependence held in the submm region. The reason is that submm photons do not generally meet the condition $\hbar\omega > E$, which is satisfied in the middle infrared. Thus it may be that submm optical behavior of electrons can be described by the relatively simple classical theory represented by Eqs. (7) to (10). Perkowitz (1969) showed this conjecture to be true in detail for GaAs by examining the quantum-mechanical representation for the three major processes of impurity scattering, acoustic-phonon scattering, and optical phonon scattering. It was found that the classical form would hold for the electronic optical behavior for frequencies below 200/cm — that is, in the submm range — for carrier concentrations above 10^{17} cm^{-3}.

The classical dependence on ω^{-2} was illustrated experimentally for GaAs by Sobotta (1970a, 1970b), Perkowitz (1971), and most elegantly and completely by Zitter and As'saadi (1974). Their submm absorption results came from a Fourier spectrometer used to measure transmission through thin bulk GaAs samples. Their results, shown in Fig. 9, clearly confirmed the wavelength dependence.

The conclusion that the classical picture holds true in the submm range has made it easy to analyze free-carrier behavior. In the limit where $\omega\tau \gg 1$, for instance, Eq. (9) becomes

$$\varepsilon = 1 - \omega_p^2/\omega^2. \tag{11}$$

From this expression and Eqs. (4) and (5) it can be shown that the reflectivity has a minimum at the frequency $\omega = \omega_p$. This *plasma edge* gives a simple way to measure carrier concentration in a semiconductor. The classical model also makes it possible to find the free-carrier mobility, defined as the carrier velocity per unit electric field and given by $\mu = e\tau/m^*$.

So far we have identified two characteristic frequencies, ω_T for the phonons and ω_p for the free carriers. In a semiconductor that contains both, one might expect spectral features at ω_T and ω_p; but in fact, features appear

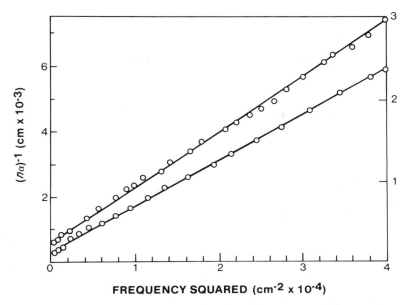

FIG. 9 The quantity $(n\alpha)^{-1}$, where n is the index of refraction and α is the absorption coefficient, versus frequency-squared for two samples of GaAs with different carrier concentrations, at $T = 300$ K. The index of refraction contributes no significant frequency dependence to the result and the plot shows the straight-line behavior predicted by the classical free-electron model. Upper line, $n = 1.0 \times 10^{16}$ cm^{-3} (left scale). Lower line, $n = 4.9 \times 10^{16}$ cm^{-3} (right scale). (After Zitter and As'Saadi, 1974).

at two new frequencies $\omega_\pm$ which result from coupling between the longitudinal phonons and the carriers, as predicted in theory by Varga (1965) and by Singwi and Tosi (1966). These theorists showed that the dielectric function for the combined electron–phonon system is

$$\varepsilon = \varepsilon_\infty + \frac{(\varepsilon_0 - \varepsilon_\infty)\omega_T^2}{\omega_T^2 - \omega^2 + i\omega\Gamma} - \frac{\varepsilon_\infty \omega_p^2}{\omega(\omega - i/\tau)}. \tag{12}$$

The reflectivity has a minimum approximately at $\varepsilon = 0$ which occurs at the frequencies

$$2\omega_\pm^2 = (\omega_L^2 + \omega_p^2) \pm \{(\omega_L^2 - \omega_p^2)^2 + 4\omega_L^2\omega_p^2[1 - \varepsilon_\infty/\varepsilon_0)]\}^{1/2} \tag{13}$$

in the limit where $\tau \to \infty$ and $\Gamma \to 0$, that is, with no damping. Thus the coupled-mode frequencies are identified with two minima in the reflectivity.

Although the first experimental observation of these modes by Olson and Lynch (1969) was just beyond the submm range, for many semiconductors

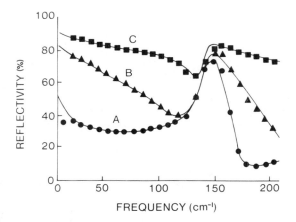

FIG. 10   The reflectivity of CdTe for three different carrier concentrations. The individual symbols represent experimental results, and the solid lines are theoretical fits from Eq. (12): curve A, $n = 5.1 \times 10^{16}\,\mathrm{cm^{-3}}$; curve B, $n = 4.1 \times 10^{17}\,\mathrm{cm^{-3}}$; curve C, $n = 1.3 \times 10^{18}\,\mathrm{cm^{-3}}$. The minima in curve A at 60 and 180/cm are the plasmon–phonon minima predicted by Eq. (13). As the carrier concentration and hence ω_p increases in curves B and C, the minima move to higher frequencies. The progress of the low-frequency minimum can be seen whereas the high-frequency minimum quickly moves beyond the frequency limit of the experiment. (After Perkowitz and Thorland, 1974.)

ω_- and ω_+ fall in the submm region. An example of this behavior is shown in Fig. 10, taken from our work in CdTe (Perkowitz and Thorland, 1974). The question of damping in the coupled-mode region is also of much interest. J. and A. Mycielski (1978) have calculated that the interaction of radiation with the collective plasmon modes produces energy dissipation over and above the electron scattering we have considered. This process is especially significant for materials with large values of the static dielectric constant ε_0, such as the lead chalcogenides. Burkhard *et al.* (1978) measured the submm reflectivity of thin films of PbTe on BaF$_2$ substrates in the coupled-mode region to study this dissipative process. They obtained a strong frequency dependence in the scattering time, as shown in Fig. 11, which could be described by the Mycielski model. Kim and Spitzer (1979) have considered the very complex question of coupled plasmon–phonon modes in the ternary system Ga$_{1-x}$Al$_x$As, for frequencies between 200 and 600/cm, but no such study has yet been made in the submm region.

D.   IMPURITIES

Whether it is intentionally doped or not, no semiconductor is entirely free of impurities or lattice defects. An impurity atom may be substitutional, replacing one of the lattice atoms at its site, or it may be interstitial, lying

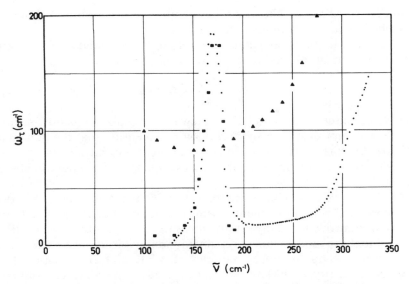

FIG. 11   The electron damping parameter $\omega_\tau = 1/\tau$ calculated from reflectivity data for thin-film PbTe on a BaF$_2$ substrate at three temperatures: ●$T = 5.0$ K, ■$T = 77$ K, and ▲$T = 300$ K. The large low-temperature peak near 150/cm occurs in the region of the coupled plasmon–phonon frequency. (Reproduced from Burkhard et al., 1978.)

between the ideal lattice sites. A defect arises when an atom is missing from the ideal lattice, removed by radiation or other means. Submm photons can probe these departures from lattice perfection.

A simple model serves to show why a semiconductor impurity has energy levels that lie in the submm range. Suppose one of the lattice atoms in the semiconductor silicon is replaced by an atom of arsenic. Each silicon atom in the pure crystal is connected to its four closest neighbors by covalent bonds equal in number to the valence number of silicon. The arsenic atom, however, with its valence of five, enters the basic four-bond structure with one valence electron left over. This extra electron is comparatively loosely bound and can easily enter the conduction band, so the arsenic acts as a *donor* impurity. (An impurity with a valence of three would readily accept an electron from the conduction band and so is an *acceptor* impurity.)

To find the binding energy of the donor electron, we note that it moves in the Coulomb field of an impurity atom with a net nuclear charge of one. Thus the donor-electron–impurity-atom system is like a hydrogen atom, with two modifications. First, because the system exists within a crystal, the Coulomb interaction is reduced by the amount $1/\varepsilon$ where ε is the dielectric constant of the crystal. Second, the donor electron interacts with the remainder of the lattice like any other conduction electron and should be described by the effective mass m^*.

These modifications can readily be inserted into the standard Bohr theory for a hydrogen atom, leading to a donor ionization energy

$$E_d = \frac{13.6(m^*/m_0)}{\varepsilon^2} \quad \text{eV}. \tag{14}$$

For typical values of m^* and ε, E_d is a few milli-electron-volts, so ionization can be accomplished by a submm photon. To ensure that the donors are deionized before radiation is applied, the experiment must be done at low temperatures, where the ionization process will appear as a peak in absorption at E_d.

The hydrogen model has some important inadequacies, and in fact does not work well for the elemental semiconductors, but it has functioned semiquantitatively for many submm measurements. Summers et al. (1970) showed that the model worked reasonably well in GaAs. They determined absorption spectra for doped samples at 4.2 K and also measured the photoconductivity. In the latter work the cooled sample was incorporated in a circuit that detected the increase in electrical current when the impurities were ionized. (This process is the basis of the whole class of impurity photoconductive submm detectors mentioned earlier.) The samples were subjected to mass-spectrographic analysis, and the dopants were known to be Ge, Si, Se, or S in concentrations near 10^{15} cm^{-3}. All measurements used a commercial Fourier-transform spectrometer at a resolution of 1/cm.

Figure 12 shows the absorption data, in which the main features are identified as arising from the $1s \rightarrow 2p$ hydrogen transition. The photoconductivity results correlate with the absorption data. Summers et al. used the values $m^*/m_0 = 0.0665$ and $\varepsilon = 12.5$ in Eq. (14) to predict $E_d = 5.79$ meV. Their measured values ranged between 5.8 and 6.1 meV. These values are only slightly higher than the hydrogen prediction; but they include a small (1 to 4%) central-cell correction that accounts for the fact that the impurity potential is not quite Coulombic near the impurity. The authors also felt that overlap between the impurity wave functions may have led to a discrepancy, because the hydrogen model assumes that each atom is perfectly isolated.

Other basic impurity research has been done by various workers. For instance, Chamberlain et al. (1971) used submm photoconductivity to observe the $1s \rightarrow 2p$ transition in InP. The exciting radiation was provided by a Fourier spectrometer. They deduced not only the energy E_d but the electron effective mass, obtaining $m^*/m_0 = 0.085 \pm 0.005$ in good agreement with other results.

The hydrogen model can be shown to work well for shallow donors — those with ionization energies near the bottom of the conduction band —

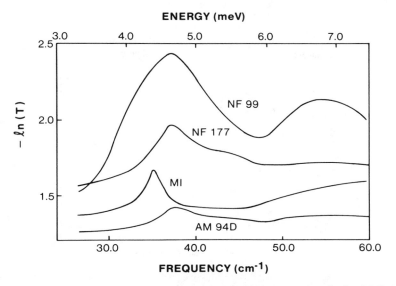

FIG. 12 Absorption expressed as $-\ln$ (transmission) for four samples of epitaxial GaAs at 4.2 K. The peaks near 35 to 37/cm arise from the $1s \rightarrow 2p$ hydrogen transition. The structure starting at about 48/cm may be associated with the photoionization threshhold. (After Summers et al., 1970.)

but deep donors, with large ionization energies, are not so tractable. These impurities are more tightly bound in the crystal structure and have strongly localized wave functions that are not amenable to the simple hydrogen effective-mass analysis. Kazanskii et al. (1979) have investigated these questions of localization in germanium. Their radiation source was a backward wave oscillator working between 3.6 and 17/cm, and their sample temperature ranged between 4.2 and 20 K.

Kazanskii et al. measured the submm absorption α for Ge samples doped with the deep impurity Be and the shallow impurity Ga. For each impurity type, α showed the same smooth monotonic increase with frequency. This behavior is indicative of a situation in which the impurities are sufficiently dense to interact. However, a given concentration of Be produced less than one-tenth the absorption as the same concentration of Ga. The authors explain this difference as arising from the greater localization of the deep donors, which reduces the probability of electron transitions among them.

In other work in Ge, Kazanskii et al. (1977) showed that the peak in photoconductivity for gallium-doped germanium moved to lower energies as uniaxial stress was applied, as illustrated in Fig. 13. The behavior was explained by noting that stress changes the valence band structure and hence the impurity binding energy. A theory by Price (1961) predicts that energy

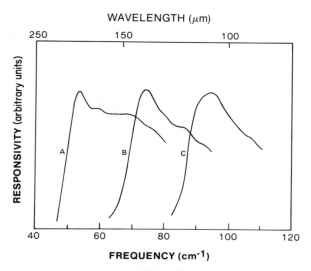

WAVELENGTH (μm)

FREQUENCY (cm^{-1})

Fig. 13 The behavior of the photoconductive peak in Ga-doped germanium for different applied uniaxial stresses. A, 6.6×10^3 kg/cm^2; B, 1.7×10^3 kg/cm^2; C, unstressed. (After Kazanski and Richards, 1977.)

decreases with uniaxial stress, although the authors made no quantitative comparison.

Submm impurity work is also of importance for materials characterization; we shall discuss this applied aspect later.

E. Band Structure: Effective Masses and Band Gaps

We have briefly described the origin of the band structure of a semiconductor and how this structure determines the semiconducting properties. We mentioned also that two of the most important band parameters are the effective mass m^* and the gap energy E_g, both of which can be found from submm measurements.

The effective mass enters into the submm properties of a semiconductor through the combination n/m^* that appears in the definition of the plasma frequency. Although n and m^* cannot be separately determined from a measurement of ω_p, n is often known from another technique, the dc Hall measurement. This situation has been exploited in recent submm work in GaAs (Chandrasekhar and Ramdas, 1980) where reflectivity data between 10 and 450/cm were taken with a Fourier spectrometer. The carrier concentration was determined from Hall measurements in each of several samples, with concentrations between 1.8×10^{16} and 7×10^{18}/cm^3. The measurements were taken on different parts of each sample to give an average over inhomogeneities. From the analysis of the data it was possible to deduce accurate values of m^*/m_0 (average error ± 0.001). The data illustrated nonparabolic effects in GaAs — that is, it showed that the quadratic dependence in Eq. (1) is not followed — and it was also possible to observe the temperature dependence of m^*.

The second and more interesting band application of submm radiation has been in examining the band gap in the so-called *zero–narrow-gap* semiconductors. These are materials with gaps of the order of a few milli-electron-volts or less that may, in some cases, even take on the value zero. An example of such a material is HgTe, which has the so-called *inverted band structure*. Here the lowest conduction band and the highest valence band touch each other at $k = 0$ (see Fig. 1), whereas there is an energy space between the highest valence band and the next lowest one, which is called an *inverted band gap*. In such a system, electrons may undergo interband transitions at energies approaching zero. The determining energy is given only by the electron occupancy, because the Fermi energy determines how high the electrons must jump. For typical Fermi energies, such transitions fall in the submm range. Thus we must add interband absorption to our picture of lattice and free-carrier absorption.

The situation becomes even more interesting when the zero-gap semiconductor is combined with a conventional semiconductor with the usual gap to form a solid solution, for instance, $Hg_{1-x}Cd_xTe$. Then, as the concentration is changed from one end of the range to the other, the system alters from one with a negative gap to one with a zero gap (in which the conduction band and two valence bands all touch), and then to one with a conventional positive gap. Analysis of such a system is a complex problem in free-carrier, lattice, and band behavior in the submm range. Inverted band behavior was first observed in α-Sn and PbSnTe, but considerable submm work has been done in HgTe and its alloys $Hg_{1-x}Cd_xTe$ and $Hg_{1-x}Mn_xTe$.

Grynberg and co-workers (1972, 1974) used Fourier spectroscopy to examine the submm reflectivity of HgTe between 80 and 700/cm. They found that the general shape of the reflectivity curve could not be explained without adding an interband absorption term ε_{int}. The calculations for ε_{int} are fairly formidable and will not be reproduced here, but the general shape of the imaginary part of ε_{int}, which determines absorption, is given in Fig. 14. The peak absorption occurs in the submm range, and there is a strong temperature dependence — generally stronger than in either the lattice or free-carrier parts — because both the Fermi energy and the transition probability are heavily influenced by temperature. Inclusion of ε_{int} made it possible to fit data that could not be fitted with the free-carrier – lattice model only. The analysis was greatly complicated because both of the dominant phonon frequencies also lay in the submm range. However, it was possible

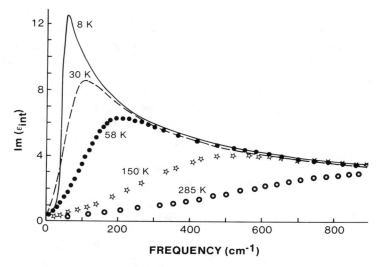

FIG. 14 The calculated imaginary part of the interband dielectric function versus frequency for HgTe at several temperatures. The absorption peak becomes large and more clearly defined and moves to submm energies as the temperature is decreased. (After Grynberg *et al.*, 1974.)

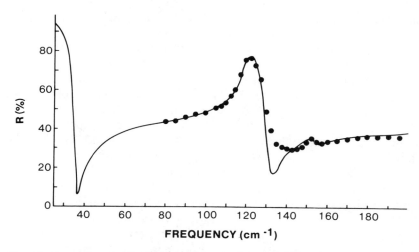

FIG. 15 Experimental reflectivity and theoretical fit (solid line) for $Hg_{0.87}Cd_{0.13}Te$ at 8 K. The fit was made using a constant interband term ($\Delta\varepsilon_{int} = 8$) and clearly does not describe the observed behavior around 140/cm. An interband term like that in Fig. 14 is required. (After Polian *et al.*, 1976.)

TABLE I

SUBMILLIMETER ANALYSIS OF $Hg_{1-x}Mn_xTe^a$

	Free-carrier parameters		Lattice parameters				Band parameters		
T (K)	n (10^{16} cm^{-3})	μ (10^3 cm^2/v-s)	ε_0	ε_∞	ω_T (cm^{-1})	Γ (cm^{-1})	m^*/m_0	E_g (meV)	E_F (meV)
5.5	0.086	103	15.2	10.6	118	8.9	0.0059	−39	6.4
30	0.20	98.0	15.2	10.5	118	9.5	0.0047	−25	15
77	1.7	92.6	19.6	13.8	118	12	0.0059	—	—
192	1.4	118	18.6	13.1	118	16	0.0032	—	—
295	3.0	49	13.7	9.9	120	17	0.0064	—	—

[a] Parameters were obtained from a fit to submm reflectivity data for a sample of $Hg_{0.91}Mn_{0.09}Te$. Parameters describing two additional lattice oscillators have been omitted. Statistical uncertainties in the free-carrier and lattice quantities and in m^*/m_0 are less than 5%. Uncertainties in the gap energy E_g and the Fermi energy E_F are about 25%. (Reproduced from McKnight et al. 1978a.)

to make a complete and mainly successful analysis of all these contributing effects.

Later Polian and co-workers (1976) extended the submm analysis of HgTe to the system $Hg_{1-x}Cd_xTe$ which has a zero gap at $x = 0.15$ and becomes a conventional positive-gap semiconductor above this value. Their analysis included an electron interband term with appropriate transitions for the negative and positive gap values. Their data could not be fitted without the inclusion of the interband term, as is clearly shown in Fig. 15.

McKnight, Amirtharaj, and Perkowitz (1978 a, b) have carried out a similar analysis for $Hg_{1-x}Mn_xTe$ which also spans the range between a negative and positive gap. Reflectivity data with a Fourier spectrometer were fitted to yield phonon and free-carrier parameters. In addition, because the interband term includes m^* but not n, we were able to calculate the first effective-mass values for this ternary. The wealth of data available from this submm analysis is shown in Table I. In fitting the interband term we used the very complete theory of Broerman (1972) which avoided some approximations made in the work on $Hg_{1-x}Cd_xTe$. We were able to obtain the imaginary part of the dielectric constant by a Kramers–Kronig transform of the reflectivity data. The plot of Im(ε) in Fig. 16 shows the onset of interband absorption, although the agreement between theory and experiment is not exact.

In recent unpublished work we have used a submm laser directly to

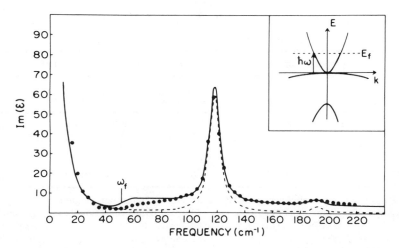

Fig. 16   Imaginary part of the dielectric function for a sample of $Hg_{0.91}Mn_{0.09}Te$ at 5.5 K.
The solid circles are derived from a Kramers–Kronig inversion of reflectivity data. The solid
and dashed lines come from a theoretical fit with and without an interband term, respectively.
The large peak at 118/cm arises from the TO mode absorption. As illustrated in the inset,
interband transitions do not occur until the photon energy equals or exceeds the Fermi energy
E_F. The onset of interband absorption in the data at about 52/cm fixes $E_F = \hbar\omega_F = 6.4$ meV.
(Reproduced from McKnight *et al.*, 1978a.)

measure transmission through a sample of $Hg_{1-x}Mn_xTe$, thus avoiding the
uncertainties of the Kramers–Kronig approach. Our preliminary results
agree well with theory.

IV.   Submillimeter Superconductor Physics

A.   INTRODUCTION AND SUPERCONDUCTOR THEORY

Although semiconductors form a rich field for the application of submm
spectroscopy, and can themselves contribute to submm technology, they are
rivalled in importance by superconductors. The basic phenomena of these
materials occur at submm energies, because of the innate nature of the
superconducting process. The most celebrated result of the Bardeen–Cooper–Schrieffer (BCS) theory, the appearance of a coupled pair of electrons
with a disassociation energy called the energy gap, is directly observable in
the submm range and was early confirmed by submm measurements.
Beyond the BCS theory in the so-called *strong coupling region,* the enhanced
connection between the electrons and phonons reflects the details of the
electron and phonon energy spectra. Some of these spectral details can
appear in the submm range, so that submm spectroscopy can shed light on

both the BCS theory and its extensions. In this introduction we shall review the basic theory of superconductivity in a qualitative fashion, to provide background for the experimental results to be described.

A superconductor has the property that below a critical temperature T_c its electrical resistance becomes zero. Thus current can be carried with no measurable losses due to heating. This is not the only important feature of superconductors — for instance, superconductors also have unusual magnetic properties — but lack of resistance is spectacular and its implications are easy to appreciate.

Many materials that qualify as superconductors are known. Some common elemental metals, including lead and aluminum, become superconducting at a T_c below 10 K. In the development of superconductors for practical applications it has been important to find materials with higher T_c's so that sophisticated cryogenics are not necessary. This search has led to the discovery of hundreds of superconducting compounds with T_c's above 10 K. In particular, one group known as the *A-15 compounds* (the name comes from the designation of their x-ray structure) has been found to be very rich in high-T_c materials. One A-15 compound, Nb_3Ge, has the highest T_c recorded to date, 23 K. This temperature can be achieved by refrigeration with liquid hydrogen rather than with expensive liquid helium.

Superconductivity was discovered in 1911, but its theoretical explanation was not given until 1957 by J. Bardeen, L. Cooper, and R. Schrieffer (1957), who shared the 1972 Nobel prize for their work. The BCS theory is heavily grounded in quantum mechanics, but we shall discuss the theory in simplified qualitative terms, giving few mathematical details.

It had long been observed in the study of superconductivity that some of the electrical results could be explained by assuming the existence of charge carriers with charge $2e$ rather than just the electron charge e. Because electrons strongly repel each other, however, the idea of a coupled pair of electrons was difficult to accept. A further clue to the understanding of superconductivity came with the observation that the transition temperature T_c varies as (ionic mass)$^{-1/2}$ (the isotope effect). This fact suggested that the lattice must be involved in what was heretofore thought to be a purely electron phenomenon; this in turn led to the understanding that a single electron, which distorts the lattice near it, can pair up with a second electron by means of this distortion. Bardeen, Cooper, and Schrieffer put this interaction on a firm theoretical basis using as the basic superconducting entity the concept of the Cooper pair — a team of two electrons that are spatially paired through the mediation of the lattice and that have opposite spins to form a boson-like state. In the BCS theory the pair is bound together by an amount of energy 2Δ, and can be broken apart by a photon of energy $\hbar\omega \geq 2\Delta$. The BCS theory further predicts that at temperatures far below T_c

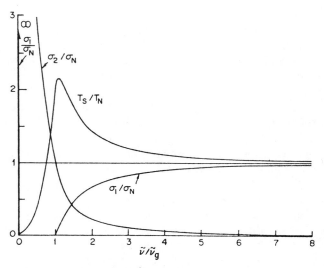

FIG. 17 Frequency dependence of the normalized conductivities σ_1/σ_N and σ_2/σ_N as calculated by Mattis and Bardeen (1958) and of the ratio of superconducting to normal transmission T_S/T_N. The frequency is plotted in units of the gap frequency $\tilde{\nu}_g = 2\Delta/h$. (Reproduced from Palmer and Tinkham, 1968.)

there is a simple relation $2\Delta = 3.53\ kT_c$. Because superconductors have $T_c \leq 23$ K, it is clear that a submm photon is just the right projectile to split up a Cooper pair.

Although the mathematical background for the BCS calculation is difficult, the mathematics describing the optical result are not so involved and are useful here. To find the optical properties of a superconductor it is only necessary to know the complex conductivity $\sigma(\mathbf{q}, \omega) = \sigma_1(\mathbf{q}, \omega) - i\sigma_2(\mathbf{q}, \omega)$ as a function of temperature, frequency, and radiation wave vector $\mathbf{q}$. The value of $\sigma(\mathbf{q}, \omega)$ has been calculated from the BCS theory by Mattis and Bardeen (1958) (see also Tinkham, 1970). The calculation is at its simplest and is done only in two limiting cases: the extreme anomalous limit $1/|\mathbf{q}| \ll \ell$, where ℓ is the electron spatial extent, and the extreme dirty limit $1/|\mathbf{q}| \gg \ell$. Fortunately, in both limits the ratio $\sigma(\mathbf{q}, \omega)/\sigma_N(\mathbf{q}, \omega)$, where σ_N is the normal-state conductivity, is the same and is independentof $\mathbf{q}$.

Figure 17 shows the Mattis–Bardeen result for σ_1/σ_N and σ_2/σ_N at a temperature far below T_c. The influence of the BCS gap energy 2Δ is seen clearly in σ_1/σ_N, which is zero for $\hbar\omega < 2\Delta$ (except at $\hbar\omega = 0$) and rises smoothly toward unity for $\hbar\omega \geq 2\Delta$. Because the power absorbed by a superconductor is proportional to σ_1, this behavior shows how a submm measurement can give a clear indication of the presence and size of the gap. The figure also shows the ratio of superconducting to normal state transmission T_S/T_N for a thin film. This ratio is often the experimentally measured quantity and has a highly visible peak near 2Δ.

The BCS theory is of major importance in explaining superconductivity, but it does not completely describe the class known as strong-coupled superconductors. In these materials the link between the electron and the lattice is exceptionally tight, and all the details of the electron–phonon interaction must be known, whereas in the BCS theory only an average interaction is used. The mathematical details of this interaction are contained in the electron–phonon spectral function $\alpha^2 F(\omega)$. Here $F(\omega)$ is the phonon density of states, giving the number of phonons available at frequency ω; and α^2 is a quantum-mechanical matrix element giving the actual electron–phonon coupling strength. This function enters into theories like the Eliashberg (1960) approach which avoids the BCS average-interaction approximation. It also enters into optical properties, because according to a formulation by Allen (1971) the absorption $A = 1 - R - T$ (where R is the reflection and T is the transmission) in either the superconducting or the normal state is given by an integral over the spectral function. Thus, if either $F(\omega)$ or $\alpha^2 F(\omega)$ has structure at submm energies — and this is often the case — then the submm behavior of the material will reflect this basic electron–phonon information. Strong-coupling materials often have values $2\Delta/kT_c > 3.5$, and this ratio can also be found by submm methods.

Submm spectroscopy is not the only technique used to probe the gap and the electron–phonon behavior, although it was one of the first. The method known as tunneling, in which dc current is applied to a junction of the form superconductor–insulator–superconductor or superconductor–insulator–normal metal, can give detailed electron–phonon and gap information and has been used effectively for elemental superconductors (McMillan and Rowell, 1969). However, it is difficult to construct reliable junctions for any of the A-15 materials, and here submm methods are uniquely useful. In addition, for any kind of superconductor, they provide important frequency-dependent information.

Submm superconductor measurements have their own difficulties. A superconductor, even in its normal state, is a highly conducting metal with nearly perfect reflection. Although the high conductivity effectively becomes infinite on cooling through T_c, the change in the reflection is only a fraction of a percent. The absorption changes more significantly, but this is an extremely small quantity. The transmission also changes sharply but is too small to measure except in the thinnest films.

These characteristics mean that the best available submm methods must be used to explore superconductivity. Even with excellent spectrometers and detectors, it has often been necessary to enhance the difference between the normal and superconducting states by recourse to cavity techniques. In the following sections we shall describe the submm experiments in some detail while discussing their implications for the understanding of superconductivity.

B. ELEMENTAL SUPERCONDUCTORS

Tinkham and his students and co-workers carried out much of the basic submm work that showed the presence of the energy gap. The earliest such work was in lead and tin in 1956 and just preceded the BCS theory; it gave clear evidence of the existence of an energy gap. Later work in elemental materials largely agreed well with BCS theory, although it also showed some deviations from it. We shall discuss some of the Tinkham papers to show the progress in detailed understanding of superconductivity. These early measurements preceded Fourier spectroscopy, and all were carried out with grating instruments optimized for submm use. Tinkham's results show that careful use of such spectroscopic systems can give excellent data, but they also highlight the great advantages of Fourier spectroscopy and submm lasers.

The earliest detailed report on lead and tin films is given by Glover and Tinkham (1957). Their data were taken with a submm monochromator designed for use between 0.1 and 0.75 mm with a mercury arc source. The instrument used 12-in. diameter optics that gave an f/1.5 aperture into the Golay detector. Lock-in detection at a chopping frequency of 10 Hz was used with a 60-sec time constant, and signals were typically averaged for one hour. Transmission measurements were made on films of Pb and Sn about 20-Å thick.

The results for T_S/T_N are shown in Fig. 18. Despite the extraordinary design of the submm monochromater, the uncertainties in the transmission are large. There are also significant uncertainties in the measured frequencies due to the frequency breadth of the spectral source. Nevertheless, the data are of sufficient quality to show clearly the gap behavior, although when this work was done the BCS theory was not available to fit the data.

A second set of data for Pb, Sn, In, and Hg was obtained by Ginsberg and Tinkham (1960). These results were much improved over the 1957 data. Although a grating spectrometer was still used, the optical coupling to the sample was improved by the use of light pipes to bring the submm radiation into the sample dewar. A very significant improvement was the replacement of the Golay cell by a cooled detector, a carbon bolometer. Ginsberg and Tinkham estimated the signal-to-noise ratio of the bolometer to be 50 times better than that of the Golay cell. The improvement in the data is apparent in a comparison between the results in Fig. 19 and those in Fig. 18. The results agreed qualitatively with the BCS theory.

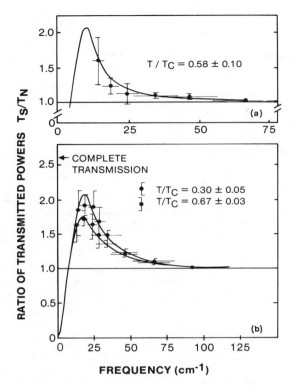

FIG. 18 Experimental results for the transmission ratio T_S/T_N for films of (a) Sn, for which $R_N = 176\ \Omega/\square$, and (b) Pb, for which $R_N = 192\ \Omega/\square$, where R_N is the normal state surface resistance. Despite the large uncertainties in the data, the peak that arises from the energy gap is visible. (After Glover and Tinkham, 1957.)

Palmer and Tinkham (1968) published a very refined analysis for Pb that shows in lucid form the presence of the gap, and that probably represents the high point in the use of gratings for submm spectroscopy. A new submm grating monochromator was coupled to a Cassegrain configuration that allowed simultaneous measurement of the sample reflection and transmission. The Cassegrain arrangement was large, with a 19-cm diameter secondary mirror. The submm radiation was detected by Low-type gallium-doped Ge bolometers that greatly improved the signal-to-noise ratio.

Because both reflection and transmission were measured it was possible to determine σ_1 and σ_2 separately. The result for σ_1 is shown in Fig. 20, together with a fit from the Mattis–Bardeen theory. The high quality of the data is evident, as is the good agreement with the BCS picture.

In the same work, however, Palmer and Tinkham noted that although σ_1 behaved as predicted by the BCS theory, the ratio T_S/T_N deviated appreci-

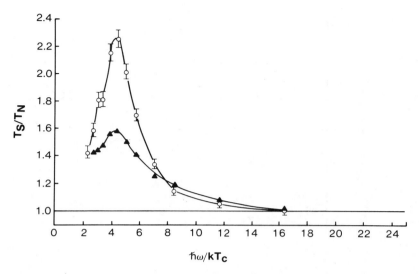

FIG. 19   Experimental results for T_S/T_N for two Pb samples: ▲, Pb 2, $R_N = 305\ \Omega/\square$; ○, Pb 3, $R_N = 197\ \Omega/\square$. The data are of much better quality than the earlier results shown in Fig. 18. (After Ginsberg and Tinkham, 1960.)

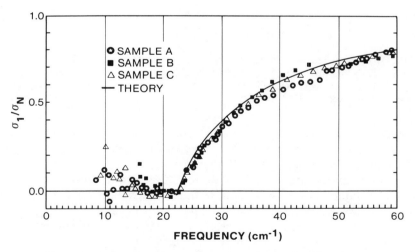

FIG. 20   Experimental results for σ_1/σ_N in three Pb films at 2 K. The solid line is a fit from the Mattis–Bardeen theory with $2\Delta = 22.5/\text{cm}$. (After Palmer and Tinkham, 1968.)

ably from the BCS result for photon energies below 2Δ. Because σ_1 is zero in this range, this discrepancy must come from σ_2. The difficulty was resolved by appealing to the strong-coupling theory of superconductivity, in which the electron–phonon interaction is treated more thoroughly than in the BCS approach. Palmer and Tinkham were able to explain the experimental anomalies by strong-coupling calculations, as carried out by Nam (1967a, b) and later by Shaw and Swihart (1968).

Further submm work in elemental superconductors gave more evidence of strong-coupling behavior. Joyce and Richards (1970) measured the absorptivity of bulk Pb with a Fourier-transform spectrometer. The detector was a doped-Ge thermometer mounted directly on the back of the 0.5-mm thick sample. With this system, it was possible to measure absorptivities of the order of 10^{-3} with errors of the order of 10%, a precision good enough to clearly show phonon-related structure in lead near 65 and 95/cm. These features were qualitatively analyzed in terms of the theory of Holstein (1964) that describes the low-temperature behavior of phonons in a metal.

Other more quantitative submm work followed in Pb. Brändli and Sievers (1972) were able to obtain quantitative agreement with the calculation of Shaw and Swihart, as shown in Fig. 21. They used samples arranged in a

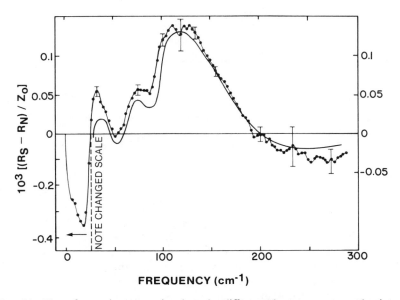

FIG. 21 Plot of quantity proportional to the difference between superconducting and normal absorption A_S-A_N in Pb. The structure at 55 and 90/cm arises from Holstein processes. The solid curve is derived from the calculation of Shaw and Swihart (1968). (After Brändli and Sievers, 1972.)

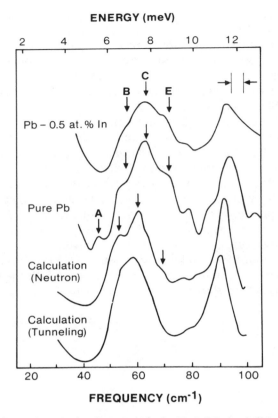

FIG. 22 Experimental results for $d(A_S - A_N)/d\omega$ for Pb + 0.5 atomic % In and for pure Pb compared with the phonon density of states computed from neutron scattering and tunneling data. The main phonon peaks occur near 60 and 90/cm. The arrows refer to smaller phonon structure. (After Farnworth and Timusk, 1974.)

parallel-waveguide geometry. Gavini and Timusk (1971) and Farnworth and Timusk (1974, 1976) used a nonresonant cavity whose inner walls were coated with lead foils or evaporated lead films. Radiation entering the cavity reflected from the lead an average of 100 times before exiting to a doped-Ge bolometer. These multiple reflections effectively amplified the small difference between superconducting and normal absorption $A_S - A_N$. The spectrometer was a Fourier system with light-pipe optics.

The data were of such good quality (the signal-to-noise ratio for a single spectrum was 2000 : 1) that they gave the derivative $d(A_S - A_N)/d\omega$ with an error of only 2%. Figure 22 shows this derivative compared with other experimental results for the phonon spectrum. There is good agreement in the phonon peaks near 60 and 90/cm, and apparently even the fine structure is meaningful.

These are only some of the highlights of submm work in elemental superconductors. Many other investigations have been made, for instance, on the role of different types of impurities in Pb and Nb (Wyder, 1976). This survey, however, has covered the two most basic areas—determination of the gap and of electron–phonon effects—accessible to submm spectroscopy. These phenomena are of equal importance in the next section, on superconducting compounds.

C. Compound Superconductors

Superconductivity was first observed in elemental materials, and the BCS theory was based on them. In recent years interest has turned to superconducting compounds, which can be made with higher T_c's and with other desirable properties that make them suitable for practical applications. Submm methods have served well in studying these compounds, especially the A-15 family. The history of these measurements has paralleled that for the elementals, moving from the simple determination of 2Δ to the determination of electron–phonon details.

An early measurement of an A-15 gap was made in Nb_3Sn by Bosomworth and Cullen (1967). They noted that several inconsistent values had been reported, with tunneling values on films almost always far below the BCS prediction whereas thermodynamic measurements on bulk samples gave $2\Delta/kT_c \geq 3.5$. These anomalies suggested that surface effects could be playing a role. Bosomworth and Cullen undertook a submm measurement to explore this possibility, using a Fourier spectrometer with light-pipe optics. The sample, a 100-μm thick film deposited on the inside surface of an insulating cylinder, formed part of the light pipe and acted as a nonresonant cavity with ten internal reflections from the sample surface. The radiation leaving the sample cylinder was detected by a carbon resistance thermometer at 1.7 K.

Figure 23 shows the percent difference $100(P_S - P_N)/P_N$, where P_S and P_N are the reflected powers in the superconducting and normal states, respectively. This quantity is proportional to the difference in absorption between the superconducting and normal states. The peak in curve A corresponds to the onset of absorption at $\hbar\omega = 2\Delta$, giving $2\Delta = 48 \pm 2$/cm and $2\Delta/kT_c = 3.77 \pm 0.16$. After the sample was mechanically polished, 2Δ decreased as shown in curve B, giving $2\Delta/kT_c = 1.9$. Finally the sample was etched in an acid solution to remove some 10 μm of its surface. This procedure restored 2Δ to 48/cm as shown in curve C. Thus submm spectroscopy coupled with the use of a nonresonant cavity gave an accurate value of 2Δ for this A-15 material and helped clarify the role of surface damage. Although the nonresonant cavity enhances the signal enormously, it has the

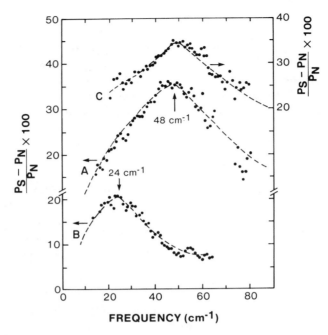

FIG. 23 Plot of quantity proportional to the absorption difference $A_S - A_N$ in Nb_3Sn. Curves A, B, and C show the variation of 2Δ with the surface state of the sample, as explained in the text. Note that curve C uses the right-hand vertical scale. (After Bosomworth and Cullen, 1967.)

disadvantage of being difficult to analyze optically. In all the experiments using nonresonant cavities the measured curves have been analyzed in terms of structure, but no quantitative fits have been made.

A similar nonresonant-cavity approach has been used by Tanner and Sievers (1973) to measure the submm absorption of V_3Si between 2 and 70/cm. They used single-crystal pieces with a total area of 15 cm² to cover part of the inner surface of a cylindrical cavity and estimated that there were several hundred reflections from the sample. The spectrometer was a Fourier instrument of the relatively unusual lamellar type (Bell, 1972) which gives superior performance at low frequencies. The detecting bolometer was also unusually sensitive, because it was operated at the very low temperature of 0.4 K in a ³He cryostat (Drew and Sievers, 1969).

This combination of nonresonant cavity, lamellar-grating interferometer, and ³He-cooled detector gave excellent data down to 2/cm with small errors and very high resolution. Figure 24 shows the experimental results for a quantity that is nearly proportional to the absorption difference $A_S - A_N$.

The features in the curve can be related to 2Δ, although Tanner and

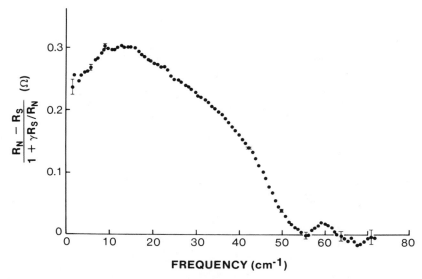

FIG. 24 Plot of a quantity nearly proportional to $A_S - A_N$ for bulk V_3Si. The data are of excellent quality down to 2/cm. An unequivocal value of 2Δ could not be derived from the curve. (After Tanner and Sievers, 1973.)

Sievers made no attempt to fit the data in detail. From the BCS theory and other considerations, they concluded that 2Δ lay between 14/cm (the value at the curve peak) and 50/cm (the value at maximum slope in the curve). Thus despite the excellence of the data, it was not possible to fix $2\Delta/kT_c$ any more closely than in the range 1.0 to 3.8. At least part of this uncertainty in interpretation arose because the optical quantity defined by the nonresonant cavity experiment is not well defined.

An experiment that did not require cavity enhancement but instead used a single pass through the superconducting sample would produce more tractable data but would require a signal-to-noise ratio superior even to that achieved with the 0.4-K detector. Such a single-pass experiment was recently carried out by Perkowitz and co-workers (McKnight *et al.*, 1979a). The necessary improvement in signal-to-noise ratio was achieved by use of a submm optically pumped laser as the spectroscopic source. The laser is described in detail elsewhere (Bean and Perkowitz, 1979). It provided so much power relative to a blackbody source that it was possible to obtain excellent data with ordinary Golay detectors; no cryogenic detection was required.

Two samples of V_3Si with thicknesses of 20 and 78 nm and T_c's of 15.0 and 15.7 K were measured. The experimental results for the transmission ratio T_S/T_N are shown in Fig. 25. The average percent error in these

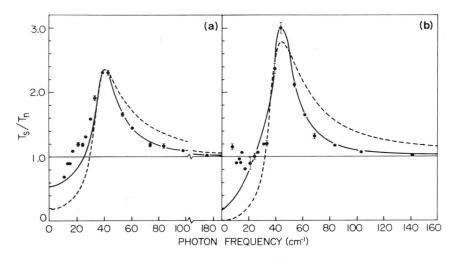

FIG. 25   Measured values of T_S/T_N and theoretical fits for two V₃Si films of thickness: (a)
20 nm and (b) 78 nm. Error bars not shown are smaller than the symbols. The peak connected
with the energy gap appears clearly. The dashed lines are the fits made in the extreme dirty and
extreme anomalous limits, and the solid lines are the improved fits made with the Leplae
formulation (see text). (Reproduced from McKnight *et al.*, 1979a.)

measurements is less than 2% in the determination of the transmission, an
excellent result for transmissions of the order of 0.01 to 1%. The spacing of
the available laser lines is sufficient to follow the general shape of the curve
and even to identify subsidiary structure, e.g., at 25 and 35/cm.

The results for T_S/T_N clearly have the shape predicted by the BCS theory;
but as the plots show, curves generated from the BCS result do not fit the
data well and give values $2\Delta/kT_c = 3.0$ and 2.9 for the two samples. Such
low values are unreasonable.

The difficulty lies in the fact that the V₃Si samples are in a regime
intermediate between the extreme anomalous and extreme dirty limits and
hence could not be adequately described by the Mattis–Bardeen results. To
obtain a good fit to the data it was necessary to apply the more complete
Leplae (1983) formulation of the BCS theory to give the correct intermediate values of $\sigma_1 - i\sigma_2$.

As Fig. 25 shows, the fit using the Leplae formulation is much improved,
although some of the low-frequency line shape remains unexplained. With
this new fit the values found for $2\Delta/kT_c$ are 3.50 ± 0.15 and 3.76 ± 0.1 for
the samples with $T_c = 15.0$ and 15.7 K, respectively. The result for the
higher T_c is the definitive one, so the submm measurement has led to the
conclusion that V₃Si is a strong-coupling material. This conclusion agrees

with that of Hauser *et al.* (1966) who obtained $2\Delta/kT_c = 3.8$ from tunneling data for V_3Si; it also agrees within experimental error with the tunneling results of Moore *et al.* (1979) who found $2\Delta/kT_c = 3.5 \pm 0.2$.

The accuracy in the submm laser measurement of 2Δ is the same as that obtained by Bosomworth and Cullen in Nb_3Sn. Because these authors used a cavity and a cooled detector, this comparison shows clearly the good performance that can be obtained in a single-pass geometry with a laser source and a room-temperature detector. In addition, the detailed analysis made possible by the single-pass mode has yielded values for the electron scattering time as well as for 2Δ.

Submm methods have also started to provide electron–phonon and other information for the A-15 materials. One such measurement in V_3Si, carried out by Perkowitz *et al.* (1976) was motivated by the theoretical work of Pickett and Allen (1973). They predicted that submm data would reflect details of the electron density of states and might also show the presence of the martensitic transformation. In this transformation the lattice structure alters from cubic to tetragonal at a characteristic temperature T_M; it occurs in some of the A-15 materials including V_3Si and may be related to the

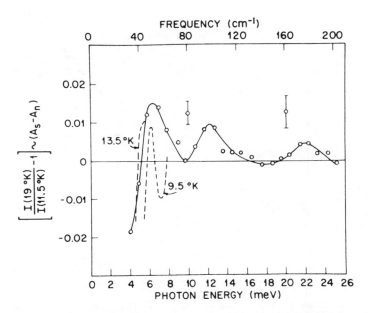

FIG. 26 Plot of quantity proportional to A_S-A_N for bulk V_3Si: O, present results; ---Tanner and Sievers (1973). The solid curve is drawn only to guide the eye. The structure is related to peaks in the electron–phonon spectral function $\alpha^2F(\omega)$. (Reproduced from Perkowitz *et al.*, 1976.)

superconducting behavior. Tanner and Sievers had searched for evidence of this effect in their work on V_3Si and saw none. However, not every sample of V_3Si undergoes the transformation and theirs had not been independently checked.

We used an excellent bulk sample of V_3Si with $T_c \cong 17$ K and a definite martensitic transformation at $T_M \simeq 21$ K. The sample, of about 2 cm² area, was examined in a single-pass reflection mode with a commercial Fourier spectrometer and a Golay detector. A variable-temperature system was used to set the sample temperature as desired between 11 and 300 K. Because we measured a reflected intensity I proportional to the reflection coefficient, the quantity $[I(T)/I(T') - 1]$ at two temperatures T and T' is very nearly proportional to the absorption difference $A(T') - A(T)$.

To search for the martensitic transformation, we set T' and T above and below T_M; but we saw no evidence of any effect to within 0.5% over the

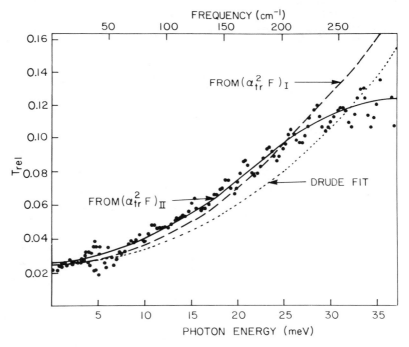

FIG. 27 Transmission through a film of V_3Si at 19 K. The apparent structure near 5 meV is an artifact, and the oscillations above 20 meV are incompletely cancelled interference fringes. The solid curve, which fits the data well, comes from an electron–phonon spectral function with peaks at 5, 18, and 42 meV. The less satisfactory dashed and dotted curves are derived from other spectral functions. (Reproduced from McKnight et al., 1979b.)

range 40 to 200/cm. When we set T' and T to straddle T_c, however, the difference $A_S - A_N$ displayed interesting behavior. The result is shown in Fig. 26. The errors are small enough to give confidence in the two peaks at 45 and 100/cm, although perhaps not in the third peak at 180/cm.

These observed structures are like those seen in Pb and ascribed to Holstein processes, which were mentioned earlier. A qualitative analysis of the data suggested that the structure arose from peaks in the electron–phonon spectral function at 6 ± 2 and 15 ± 2 meV. Later detailed analysis (De and Perkowitz, 1980) has confirmed this result. Using the Holstein theory, we have been able to fit the data using a spectral electron–phonon function with peaks at 5, 15, and 41 meV.

Further confirmation has come from work in thin-film V_3Si carried out by Perkowitz and co-workers (McKnight *et al.*, 1979b). In this work we examined two thin-film samples of V_3Si (the same two alluded to earlier) by Fourier spectroscopy between 5 and 280/cm. One set of transmission data, taken by Tanner at 19 K with two different spectrometers and cooled detectors, was especially suited for electron–phonon analysis. The very high resolution data are shown in Fig. 27 along with several fits. The best fit is

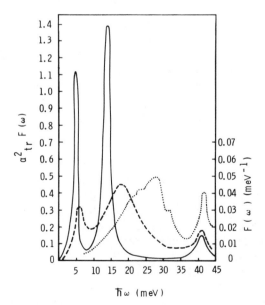

FIG. 28 Phonon density of states $F(\omega)$ (dotted line) obtained from neutron scattering; electron–phonon spectral function $\alpha^2 F(\omega)$ (dashed line) obtained from the thin-film normal-state V_3Si data shown in Fig. 27; and electron–phonon spectral function (solid line) obtained from the bulk, superconducting V_3Si data shown in Fig. 26. See text for a comparison of the curves.

based on the Holstein process with peaks in the spectral function at 5, 18, and 42 meV.

The spectral functions obtained from the bulk work and the film work are shown in Fig. 28, where the phonon function $F(\omega)$ is also displayed. Both electron–phonon spectral functions show a downward shift in the 28 meV peak in $F(\omega)$, and both show the emergence of a new peak near 5 to 6 meV. Recent work (Mitrović and Carbotte, 1982; Mitrović, 1982) suggests that this apparent low-frequency enhancement arises from fine structure in the electron density of states. This interpretation is now under investigation. The deviations in the details of the two curves for $\alpha^2 F(\omega)$ may represent the difference between normal and superconducting behavior or between bulk and film properties.

This variety of results shows the power of submm methods for the most fundamental studies of compound superconductors. The next section describes some useful applications of submm spectroscopy.

V. Submillimeter Semiconductor Characterization

A. INTRODUCTION

Semiconductor characterization means the measurement of parameters that vary from sample to sample in semiconductors. Such measurements are needed as indicators of sample purity and to provide specific values for device fabrication. The important parameters for semiconductors include carrier concentration and mobility, the number and type of impurity atoms, and the impurity ionization energies. Information about impurities can be helpful in tracking down contaminants during growth, and impurity and free-carrier data are used for device design.

Standard methods are available to determine the carrier concentration and mobility of bulk materials. However, there is increasing interest in semiconductor films that provide new capabilities such as the ability to tailor optical properties. Such films are usually grown on a substrate of another semiconductor, and characterization of the complex film–interface–substrate system is not yet a routine task.

Some applications require semiconductors of very high purity. Examples include nuclear detectors, which use ultrapure Ge; high-resistance substrates for semiconductor films, often made from high-purity GaAs; and numerous solid state electronic devices, which also use high-purity GaAs. It is difficult to detect and identify the shallow donors and acceptors that may contaminate these pure materials. Standard electrical techniques, such as Hall measurements as a function of temperature, and optical techniques, such as luminescence experiments, do not serve well.

Submm methods can contribute to all these characterization needs. They offer advantages over some of the conventional approaches or provide completely new approaches. In this section we shall illustrate some submm characterization methods, using the background in semiconductor theory developed earlier.

B. FREE-CARRIER CHARACTERIZATION: BULK AND FILM SEMICONDUCTORS

To help understand the usefulness of submm characterization of free-carrier concentration and mobility, we first review the conventional electrical measurements that are commonly used. These depend on the relations

$$\sigma = ne\mu \tag{15}$$

and

$$R_H = r_H/nec \tag{16}$$

to determine n and μ. Here, σ is the dc conductivity, R_H is the Hall constant, and r_H is a complex quantity describing the scattering processes in the material and their magnetic-field dependence. The value of r_H is often close to unity, although it can depart appreciably from this value.

The conductivity is measured from a straightforward voltage–current measurement, giving the product $n\mu$. The Hall constant arises from a measurement in which a magnetic field H is applied perpendicular to an impressed current I in the sample, producing a transverse Hall voltage V_H. The Hall constant R_H is given by the ratio V_H/HI; then n can be found if r_H is known. The calculation of r_H requires a detailed knowledge of the scattering processes in the sample. For simplicity r_H is often set equal to one with a resulting uncertainty in the Hall mobility.

The dc measurements also have practical drawbacks. They are poorly suited for dealing with very small samples or for determining sample homogeneity, i.e., point-to-point variation in concentration or mobility, because good electrical contacts must be applied. It is not always easy to make ohmic contacts for a semiconductor, especially one with a low carrier concentration, or to use contacts to probe small areas. For a sample consisting of a thin film on a substrate, the interpretation of dc data can become complicated because the substrate, or any conducting interface layer, can "short out" the dc measurements.

Optical measurements in the submm range can provide some answers to these problems. They give more flexibility in determining homogeneity both in the plane of the sample and perpendicular to it. They also make it possible to measure a carrier concentration and mobility that are less influenced by any assumptions about the scattering behavior and hence are more reliable, and they do away with the difficulties of making ohmic contacts.

Earlier we showed that the quantities n/m^* and $\tau = \mu m^*/e$ appear in the expression for the submm dielectric constant, Eq. (12). Thus provided m^* is known, n and μ can be found from various types of submm experiments. In transmission experiments, for example, the absorption coefficient α can be obtained if the reflectivity is known. A plot of $1/\alpha$ versus ω^2 should then give a straight line, from which ω_p and τ can be deduced according to Eq. (10) and Fig. 9. We have made such an analysis for GaAs samples with n between 3×10^{15} and 2×10^{16} cm^{-3} (Perkowitz, 1971). The transmission was measured between 20 and 210/cm with a Fourier spectrometer and a Golay detector. The reflectivity was not measured but was calculated in a simplified manner assuming only a lattice contribution. On this basis the optical values for n and μ agreed with the dc values to within 25% on the average, a discrepancy that is comparable to the total experimental uncertainty. One drawback in the transmission method is that it is difficult to use for very high values of n, where the sample transmission is negligible.

This difficulty is avoided in a second approach, which is to measure the submm reflectivity. A second advantage of the reflectivity method is that the sample thickness and other auxiliary information are not needed. A third advantage is that for reasons having to do with how τ enters into the optical expressions, reflection data are less sensitive to the type of scattering than are transmission data.

These features were examined by Perkowitz and Breecher (1974) in bulk GaAs with concentrations between 5×10^{15} and 4×10^{18} cm^{-3}. The reflectivity was measured between 20 and 200/cm with a commercial Fourier system and a Golay detector. A least-squares fit was made to each curve to yield values of ω_p and τ. Two different approaches were used in extracting n and μ. In the first, the presence of any type of energy-dependent scattering was ignored, τ was treated as a constant, and n and μ were calculated and compared on this basis. The overall agreement between electrical and optical values was within 25%. In the second, a realistic scattering model was assumed for GaAs that used a weighted average of ionized impurity and optical phonon scattering. On this basis, too, the overall agreement was within 25%. However, on moving between the two models, the calculated Hall mobility values changed by some 25% whereas the calculated optical mobilities altered by only 10%. Thus under both models the optical values agreed with dc values to within experimental uncertainty, but the optical results are less sensitive to scattering details and thus represent a more reliable quantity.

An even simpler method for characterizing n and μ from reflectivity data

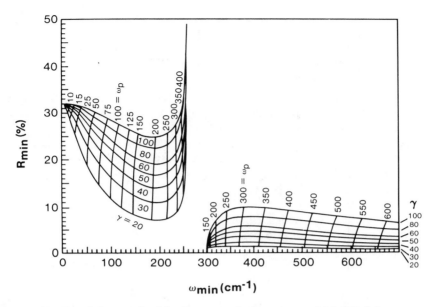

FIG. 29 The minimum reflectivity R_{min} versus the frequency at which the minimum occurs ω_{min} for GaAs. The plot has two separate branches because there are two distinct minima that, in the zero-damping limit, occur at the coupled plasmon–phonon frequencies $\omega_{\pm}$ given by Eq. (13). The plot shows a family of curves generated for specific values of ω_p and of $\gamma = 1/\tau$, as indicated. For any measured R_{min} and ω_{min}, the corresponding ω_p and γ can be read directly from the plot; n and μ can then be easily calculated. (After Holm et al., 1977.)

was presented by Perkowitz and Breecher (1973). In this approach, instead of making a reflectivity fit to the entire experimental curve, only the positions and heights of the low-frequency plasmon–phonon minimum were used to determine n and μ, requiring a measurement in only a limited region of the spectrum. The same samples investigated previously (Perkowitz and Breecher, 1974) were examined. It was found that the reflection-minimum method gave n and μ values within 4% of those found by the much more complex least-squares fit and was usable for concentrations between 5×10^{15} and 6×10^{17} cm^{-3}. This approach has led to a simple graphical method for evaluating n and μ, first presented by Perkowitz and Breecher and later refined by Holm et al. (1977), as shown in Fig. 29.

The reflectivity approach has been applied to other bulk materials of interest. In CdTe (Perkowitz and Thorland, 1974), agreement between electrical and optical parameters was excellent for samples with n between 1.4×10^{17} and 1.3×10^{18} cm^{-3}, although the concentration and mobility differed by a factor of two for $n = 5.1 \times 10^{16}$ cm^{-3}. However, in this case the resistivity values agreed. A similar good agreement was found for PbTe (Perkowitz, 1975) with n between 1.7×10^{17} and 1.5×10^{18} cm^{-3}, but mobilities were found to diverge by a factor of three for $n = 1.7 \times 10^{19}$ cm^{-3}. In general the reflection methods have proven useful in characterizing different semiconductors to an accuracy of some 20%, at least in the range 10^{16} to 10^{18}, with some incompletely explained discrepancies at the high and low ends of these ranges.

The same basic ideas apply to the determination of n and μ in an epitaxial layer, although the problem can be greatly complicated by the presence of the substrate and possible interface layers between the epilayer and the substrate. Tennant and Cape (1976) made a detailed analysis of the reflectivity of a structure composed of a low carrier density ($\sim 10^{16}$ cm^{-3}) p-type Pb$_{0.88}$Sn$_{0.12}$Te film on a high carrier density ($\sim 10^{18}$ cm^{-3}) Pb$_{0.78}$Sn$_{0.22}$Te substrate. Reflectivity spectra were made between 90 and 320/cm with a Fourier spectrometer.

The data were far more complex than would be obtained from a bulk sample. Above 120/cm, where the film was transparent, interference fringes appeared, showing directly the presence of the thin film. The reflection data were fitted first with a two-layer model that assumed an abrupt change in n at the interface. The dielectric constant for each layer was given by Eq. (12), and all the parameters were chosen to give the best fit.

As is seen in Fig. 30 the two-layer model fit the data well, below 230/cm. This fit also gave a value of film thickness (8 μm) in excellent agreement with that determined from a microscopic examination. (As would be expected, the spacing between the interference fringes relates directly to the film thickness). The fit above 230/cm is not acceptable, however.

To improve the high-frequency fit Tennant and Cape replaced the abrupt-interface model with one that allowed the carrier concentration to vary continuously over the interface region. Such graded regions may frequently occur because of the diffusion of carriers during the film-growth process. Three different grading models were used with linear, exponential, and hyperbolic tangent variation of n with distance. As Fig. 30 shows, the fit was improved with the graded models. All models yielded evidence of a graded region a few μm thick and gave a mean film thickness of 8 to 9 μm. In addition the fits gave a complete set of lattice and free-carrier parameters for both layers, with generally small statistical errors.

The use of submm radiation to gain such detailed information was pursued further by Amirtharaj, Bean, and Perkowitz (1977). We examined three different thin-film systems between 8 and 350/cm. Reflectivity data were obtained with a commercial Fourier system, and an optically pumped laser gave transmission data.

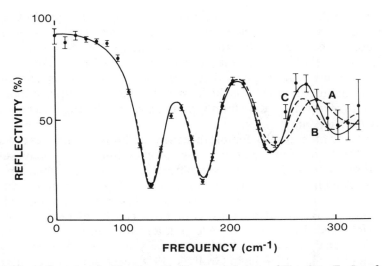

FIG. 30   Reflectivity of a $Pb_{0.88}Sn_{0.12}Te$ film on a substrate of $Pb_{0.78}Sn_{0.22}Te$. Interference fringes are apparent above 120/cm. The fitting curves arise from the following models: curve A. two-layer, abrupt interface; curve B. graded interface, linear variation; and curve C. graded interface, exponential variation. (After Tennant and Cape, 1976.)

The first sample consisted of a 2-μm thick film of PbSe on an NaCl substrate. Because there should be no significant grading between a semi-conductor and an insulator, this system should be well described by the abrupt two-layer model. We were able to fit the entire range of reflection data with this model. The values of the fit parameters, including the film thickness, agreed very well with independent determinations. The other two systems, $Pb_{0.79}Sn_{0.21}Te$ on a substrate of the same material and a 12-μm film of InAs on GaAs, were more complex. Neither could be entirely described by an abrupt model. The results were especially interesting for the InAs–GaAs system. The abrupt model returned a value of the substrate carrier concentration of 10^{16} cm^{-3} when it was applied to reflection data. This result was orders of magnitude higher than was expected for the high-purity GaAs. Further, laser transmission data gave an average transmission of 10% at low frequencies, but a carrier concentration of 10^{16} cm^{-3} through the whole substrate would give a transmission too small to measure. We conjectured that only a narrow region of high carrier concentration lay between the thin film and the substrate.

This conjecture was borne out by a more detailed analysis of the same data (Amirtharaj and Perkowitz, 1979). The data could be well fitted on the assumption that an interface layer about 0.7-μm thick existed. From a careful examination of the lattice resonance portion of the data we concluded also that this interface region consisted of the ternary $In_xGa_{1-x}As$. These results are remarkably consistent with the results of Wagner (1976), who used Auger electron spectroscopy and energy-dispersive x-ray analysis to determine the nature of the interface region in InAs–GaAs. He concluded that a ternary region exists, and extrapolation of his data yields a rough estimate of 2 μm for the thickness of this region. Of equal importance, the values of n and μ that we found from our submm data agreed reasonably well with Hall data.

These and other results, for instance the very thorough analyses of Palik *et al.* (1977), have shown that even for complex thin-film systems submm spectroscopy can yield information difficult or impossible to obtain by other means. The film thickness can be measured with great accuracy, the free-carrier parameters of the film and the substrate can be found individually, and information about interface conditions can be extracted by careful analysis, especially when both reflection and transmission data are available. The analysis of spatial inhomogeneity may benefit from new approaches like that of Hild and Grofcsik (1978), who solve Maxwell's equations for inhomogeneously doped layers to obtain the corresponding optical properties.

C. IMPURITY CHARACTERIZATION

We have already discussed how and why shallow impurities can be observed at submm energies. For many applications extremely small impurity concentrations must be detected. For example, ultrapure Ge can be made with as few as 10^9 impurities per cm^3. Here we describe the methods used to characterize such small numbers of contaminant atoms.

The simplest approach would be to carry out conventional submm absorption spectroscopy, but the absorption due to impurity ionization is too small to detect 10^9 cm^{-3} impurities. Far more sensitivity can be obtained from photoconductive methods. I have already noted how the excited electron (hole) from a donor (acceptor) contributes to the conductivity when it reaches the conduction (valence) band. The eventual excitation to the continuum always involves a phonon, and such processes are favored at higher temperatures. In opposition to this requirement, low temperatures are needed to produce many deionized impurities. As a result, photoconductive measurements of impurities are best made at some intermediate temperature, typically around 10 K.

This sensitive technique of photothermal excitation has been used by Haller and Hansen (1974) and Skolnick *et al.* (1974) in very pure Ge and Si. The former examined a variety of acceptors in over 50 samples of Ge with acceptor concentrations between 10^9 and 10^{12} cm^{-3}. To ensure meaningful

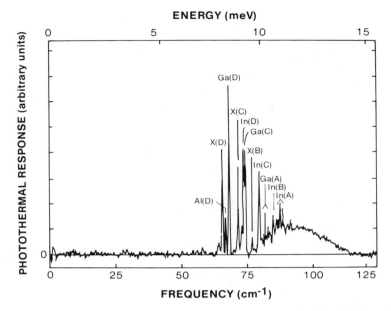

ENERGY (meV)

FIG. 31   Photothermal spectrum of ultrapure Ge with a net impurity concentration of 10^{11} cm^{-3}, at a temperature of 8 K. The peaks are labelled with the appropriate impurities Al, Ga, and In as well as the unknown "X." Each impurity produces more than one line; for example, there are four In lines, labelled A through D. (After Haller and Hansen, 1974.)

photoconductive responses care was taken to make good electrical contacts that remained operative and ohmic down to a temperature of 1.2 K. The contacting method was chosen to ensure that it added no additional impurities to the samples. A Fourier spectrometer, operating at the very high resolution of 0.25/cm, was used to produce photothermal spectra like that shown in Fig. 31.

It is apparent that high spectral resolution is essential because of the sharpness of the absorption structure. Comparison of these peaks to energy-level schemes for known impurities made it possible to determine the impurity types. The figure shows structure arising from Ga, In, and Al impurities, and from an unknown contaminant. Skolnick *et al.* used an extension of these same methods to identify simultaneously the donor and acceptor impurities in Si and Ge. They found P donors and B acceptors in their high-purity Si and observed unidentified peaks for Ge that probably arose from lattice defects rather than from chemical impurities.

Submm photoconductivity has also been used to study donors in GaAs by Cooke *et al.* (1978). Here the approach was to measure the central-cell corrections to the hydrogen theory for known dopants and then to use these central-cell corrections, or chemical shifts, as signatures for unknown impurities. Many of the data were taken with an applied magnetic field, which sharpened the observed structure so that the chemical shifts could be accurately measured. The combination of this work with the earlier results of Stillman *et al.* (1972) led to the tentative assignment of chemical shifts for eight dopants. As an application of these data, Cooke *et al.* examined unintentionally doped high-purity samples prepared by three different growth methods in several laboratories. All samples grown by the same technique exhibited nearly identical central-cell structures. For instance, all material grown by vapor-phase methods contained Si, C, and a third contaminant that was either Pb or a Ga vacancy.

Recent work in identifying donors in high-purity epitaxial GaAs has been carried out by Afsar *et al.* (1980). Here the photoconductive response was determined as a function of magnetic field, leading to a "signature curve" for each dopant. The measurements used a Fourier spectrometer with a resolution of 0.14/cm, and fields up to 16 T were employed. It was possible to produce signature curves for dopants including S, Si, Se, and Ge.

VI.   Summary: Recent Work and Future Projections

Theoretical discussions and examples drawn from the literature have shown how much fundamental condensed-matter physics occurs at submm energies. Even with the limitation of discussing only semiconductors and superconductors, we have had to omit many important areas. We list some of them here, together with recent references, to give the flavor of the work: amorphous semiconductors, like a-Ge, that show a characteristic submm absorption (Pidgeon *et al.*, 1978; Shen *et al.*, 1980; Trodahl *et al.*, 1980); quasi-two-dimensional behavior in layered semiconductor-like compounds, such as GaTe and 1T-TaSe$_2$, that have a very strong asymmetry in their bonding properties (Allakhverdiev *et al.*, 1979; Karecki and Clayman, 1979; Mead, 1980); a different kind of two-dimensional behavior in surface layers of Si (Allen *et al.*, 1977; Theis *et al.*, 1978); lattice instabilities and phase transitions in such materials as Pb$_{1-x}$Ge$_x$Te (Jantsch *et al.*, 1978; Nishi *et al.*, 1980); and inhomogeneous or granular superconductors, like Sn in a KCl matrix or NbN, a new class of materials that is poorly understood and needs much further study (Carr *et al.*, 1980; Perkowitz, 1982; Karecki *et al.*, 1983).

Although we have not reviewed these or other specialized areas, or work in metals and insulators, much of the solid-state theory and submm technique discussed here applies also to them. Additional reviews of various areas in submm solid-state physics (although not the most recent ones) can be found in several articles and monographs including Tinkham (1964), Perry *et al.* (1966), Hadni (1967), Kaplan (1967), Chantry (1971), Hughes (1971), Möller and Rothschild (1971), Robinson (1973), and Wyder (1976). Some recent solid-state work is described in Perkowitz (1976, 1979).

Our treatment of the applied aspects also has been limited. We have shown the potential of submm-characterization methods but have said little about submm solid-state devices. It is clear that solid-state components and materials will play an increasing role in submm technology. They will appear in a passive way — for instance, as windows, filters (Holah 1980a, b), couplers (Julien and Lourtioz, 1980), and absorbing coatings (Grammer *et al.*, 1979). More importantly, they function as active devices including detectors, modulators, and perhaps even sources.

Development of submm detectors continues at a heavy pace, with many devices based on both semiconducting and superconducting principles now under study. The prospect of heterodyne detectors is especially appealing. Tannenwald (1980) has reviewed developments in solid-state detectors and their suitability for heterodyne use. NEP's as low as 10^{-19} W (into a 1 Hz bandwidth) have been achieved in heterodyne configurations. Fan *et al.* (1977) have described a submm modulator and polarizer based on thin film VO_2. Several solid-state mechanisms are known that produce tunable submm power, for instance, emission from plasma oscillations in surface layers of Si (Tsui *et al.*, 1980) and cyclotron emission from InSb and other semiconductors (Gornik, 1979). Although none of these mechanisms produces more than a μW, the ability to tune continuously is important. Such solid-state sources may prove effective when coupled to improved detectors. Certain conference digests (Perkowitz 1976, 1979) contain reports on work in detectors and other components.

Because so many basic solid-state processes occur at submm energies, it is unlikely that submm methods will become obsolete even for the newest and most exotic condensed-matter systems. These techniques will continue to contribute to basic understanding and characterization of materials. Given the current trends in semiconductor research, the main contributions will probably come for materials with increasingly complex combinations of constituents, such as quaternary semiconductors and unusual film — substrate combinations like GaAs–Ge, only recently grown by molecular-beam epitaxy, and for systems of complex geometry, such as the superlattices made from stacked angstrom-thick layers of semiconductors. In superconductivity, much remains to be done even with the more mature materials such as the A-15 family, and the new granular systems offer enormous challenges. Submm research in these new areas will be supported by an improved submm technology. Much of this technology is now arising from solid-state research. The interaction between improved understanding of the solid state and its translation into submm-research devices, which in turn lead to even better understanding, will be fascinating to watch.

ACKNOWLEDGMENTS

The work reported from Emory University was supported by the Department of Energy, the Emory University Research Committee, the National Science Foundation, the Office of Naval Research, and the Research Corporation. I thank G. S. De for reading and commenting on a portion of this manuscript.

REFERENCES

Afsar, M. N., Button, K. J., and McCoy, G. L. (1980). *Int. J. IR/MM Waves* **1**, 145.
Allakhverdiev, K. R., Abbasov, A. N., Mekhtiev, T. R., and Nani, R. Kh. (1979). *Phys. Status Solidi B* **94**, K31.
Allen, P. B. (1971). *Phys. Rev. B* **3**, 305.
Allen, S. J., Tsui, D. C., and Logan, R. A. (1977). *Phys. Rev. Lett.* **38** 980.
Amirtharaj, P. M., Bean, B. L., and Perkowitz, S. (1977). *J. Opt. Soc. Am.* **67**, 939.
Amirtharaj, P. M., and Perkowitz, S. (1979). *Thin Solid Films* **62**, 357.
Amirtharaj, P. M., Holah, G. D., and Perkowitz, S. (1980). *Phys. Rev. B* **21**, 5656.
Baars, J., and Sorger, F. (1972). *Solid State Commun.* **10**, 875.
Bardeen, J., Cooper, L. N., and Schrieffer, J. R. (1957). *Phys. Rev.* **108**, 1175.
Batalla, E., Koteles, E. S., and Datars, W. R. (1977). *Can. J. Phys.* **55**, 2013.
Bean, B. L., and Perkowitz, S. (1979). *In* "Infrared and Millimeter Waves" (K. J. Button, ed.), Vol. 2, pp. 273–298. Academic Press, New York.
Bell, R. J. (1972). "Introductory Fourier Spectroscopy." Academic Press, New York.
Bosomworth, D. R., and Cullen, G. W. (1967). *Phys. Rev.* **160**, 346.
Bottger, G. L., and Geddes, A. L. (1967). *J. Chem. Phys.* **47**, 4858.
Boyle, W. S., and Rodgers, K. F. (1959). *J. Opt. Soc. Am.* **49**, 66.
Brändli, G., and Sievers, A. J. (1972). *Phys. Rev. B* **5**, 3550.
Broerman, J. G. (1972). *In* "Proceedings of the Eleventh International Conference on the Physics of Semiconductors, Warsaw, 1972," Vol. 2, p. 917. PWN, Warsaw.
Burkhard, H., Bauer, G., and Lopez-Otero, A. (1976). *Solid State Commun.* **18**, 773.
Burkhard, H., Bauer, G., and Lopez-Otero, A. (1978). *Phys. Rev. B* **18**, 2935.
Carr, G. L., Garland, J. C., and Tanner, D. B. (1980). *In* "Inhomogeneous Superconductors 1979 (Berkeley Springs, WV)" (D. U. Gubser, T. L. Francavilla, J. R. Leibowitz and S. A. Wolf, eds.), p. 288. AIP, New York.
Chamberlain, J. M., Ergun, H. B., Gehring, K. A., and Stradling, R. A. (1971). *Solid State Commun.* **9**, 1563.
Chandrasekhar, H. R., and Ramdas, A. K. (1980). *Phys. Rev. B* **21**, 1511.
Chang, T. Y., and Bridges, T. J. (1970). *Opt. Commun.* **1**, 423.
Chantry, G. (1971). "Submillimetre Spectroscopy." Academic Press, London.
Cooke, R. A., Hoult, R. A., Kirkman, R. F., and Stradling, R. A. (1978). *J. Phys. D* **11**, 945.
De, S. G., and Perkowitz, S. (1980). Unpublished.
Drew, H. D., and Sievers, A. J. (1969). *Appl. Opt.* **8**, 2067.
Dumke, W. P. (1961). *Phys. Rev.* **124**, 1813.
Eliashberg, G. M. (1960). *Sov. Phys.–JETP* **11**, 696.
Fan, J. C. C. *et al.* (1977). *Appl. Phys. Lett.* **31**, 11.
Farnworth, B., and Timusk, T. (1974). *Phys. Rev. B* **10**, 2799.
Farnworth, B., and Timusk, T. (1976). *Phys. Rev. B* **14**, 5119.

Gallagher, J. J., Blue, M. D., Bean, B. and Perkowitz, S. (1977). *Infrared Phys.* **17**, 43.
Gavini, A., and Timusk, T. (1971). *Phys. Rev. B* **3**, 1049.
Geick, R. (1964). *Phys. Lett.* **10**, 51.
Ginsberg, D. M., and Tinkham, M. (1960). *Phys. Rev.* **118**, 990.
Glover, R. E., and Tinkham, M. (1957). *Phys. Rev.* **108**, 243.
Golay, M. J. E. (1947). *Rev. Sci. Instr.* **18**, 357.
Gornik, E., (1979). *J. Magn. Magn. Mat.* **11**, 39.
Grammer, J. R., Alff, W. H., Blue, M. D. and Perkowitz, S. (1979). *In* "Thermophysics and Thermal Control" (R. Voskunta, ed.), p. 39. AIAA, New York.
Gregora, I., and Petzelt, J. (1972). *Phys. Status. Solidi B* **49**, 271.
Grynberg, M., and Le Toullec, R. (1972). *In* "Proceedings of the Eleventh International Conference on the Physics of Semiconductors, Warsaw, 1972," Vol. 2, p. 931. PWN, Warsaw.
Grynberg, M., Le Toullec, R., and Balkanski, M. (1974). *Phys. Rev. B* **9**, 517.
Hadni, A. (1967). "Essentials of Modern Physics Applied to the Study of the Infrared." Pergamon, Oxford.
Haga, E., and Kimura, H. (1963). *J. Phys. Soc. Japan* **18**, 777.
Haga, E., and Kimura, H. (1964). *J. Phys. Soc. Japan* **19**, 658.
Haller, E. E., and Hansen, W. L. (1974). *Solid State Commun.* **15**, 687.
Hass, M., and Henvis, B. W. (1962). *J. Phys. Chem. Solids* **23**, 1099.
Hauser, J. J., Bacon, D. D., and Haemmerle, W. H. (1966). *Phys. Rev.* **151**, 296.
Hild, E., and Grofcsik, A. (1978). *Infrared Phys.* **18**, 23.
Holah, G. D. (1980a). *Int. J. IR/MM Waves* **1**, 225.
Holah, G. D. (1980b). *Int. J. IR/MM Waves* **1**, 235.
Holm, R. T., Gibson, J. W., and Palik, E. D. (1977). *J. Appl. Phys.* **48**, 212.
Holstein, T. (1964). *Ann. Phys. (N.Y.)* **29**, 410.
Hughes, A. E. (1971). *Contemp. Phys.* **12**, 257.
Jantsch, W., Lopez-Otero, A., and Bauer, G. (1978). *Infrared Phys.* **18**, 877.
Jensen, B. (1971). *Solid State Commun.* **9**, 1587.
Jensen, B. (1973a). *Ann. Phys. (N.Y.)* **80**, 284.
Jensen, B. (1973b). *J. Phys. Chem. Solids* **34**, 2235.
Jensen, B. (1977). *Solid State Commun.* **24**, 853.
Jensen, B. (1978). *Phys. Status Solidi B* **86**, 291.
Joyce, R. R., and Richards, P. L. (1970). *Phys. Rev. Lett.* **24**, 1007.
Julien, F., and Lourtioz, J-M. (1980). *Int. J. IR/MM Waves* **1**, 175.
Kaplan, R. (1967). *Appl. Opt.* **6**, 685.
Karecki, D. R., and Clayman, B. P. (1979). *Phys. Rev. B* **19**, 6367.
Karecki, D. R., Carr, G. L., Perkowitz, S., Gubser, D. U. and Wolf, S. A. (1983). *Phys. Rev. B* (to be published).
Kazanskii, A. G., Richards, P. L., and Haller, E.E (1977). *Appl. Phys. Lett.* **31**, 496.
Kazanskii *et al.* (1979). *Sov. Phys.–Semicond.* **13**, 1198.
Kim, O. K., and Spitzer, W. G. (1979). *Phys. Rev. B* **20**, 3258.
Kinch, M. A., and Buss, D. D. (1972). *Solid State Commun.* **11**, 319.
Leplae, L. (1983). *Phys. Rev. B* **27**, 1911.
Low, F. J. (1961). *J. Opt. Soc. Am.* **51**, 1300.
Mattis, D. C., and Bardeen, J. (1958). *Phys. Rev.* **111**, 412.
McKnight, S. W., Amirtharaj, P. M., and Perkowitz, S. (1978a). *Solid State Commun.* **25**, 357.
McKnight, S. W., Amirtharaj, P. M., and Perkowitz, S. (1978b). *Infrared Phys.* **18**, 919.
McKnight, S. W., Bean, B. L., and Perkowitz, S. (1979a). *Phys. Rev. B* **19**, 1437.
McKnight, S. W., Perkowitz, S., Tanner, D. B., and Testardi, L. R. (1979b). *Phys. Rev. B* **19**, 5689.

McMillan, W. L., and Rowell, J. M. (1969). *In* "Superconductivity" (R. D. Parks, ed.), Vol. I, p. 561. Dekker, New York.

Mitrović, B. (1982). Private communication.

Mitrović, B., and Carbotte, J. P., (1982). *Phys. Rev. B* **26**, 1244.

Mead, D. G. (1980). *Int. J. IR/MM Waves* **1**, 309.

Möller, K. D., and Rothschild, W. G. (1971). "Far-Infrared Spectroscopy." Wiley, New York.

Moore, D. F., Zubeck, R. B., Rowell, J. M., and Beasley, M. R. (1979). *Phys. Rev. B* **20**, 2721.

Mycielski, J., and Mycielski, A. (1978). *Phys. Rev. B* **18**, 1859.

Nam, S. B. (1967a). *Phys. Rev.* **156**, 470.

Nam, S. B. (1967b). *Phys. Rev.* **156**, 487.

Nishi, S., Kawamura, H., and Murase, K. (1980). *Phys. Status Solidi B* **97**, 581.

Olson, C. G., and Lynch, D. W. (1969). *Phys. Rev.* **177**, 1231.

Palik, E. D., Holm, R. T., and Gibson, J. W. (1977). *Thin Solid Films* **47**, 167.

Palmer, L. H., and Tinkham. M. (1968). *Phys. Rev.* **165**, 588.

Perkowitz, S. (1969). *J. Appl. Phys.* **40**, 3751.

Perkowitz, S. (1971). *J. Phys. Chem. Solids* **32**, 2267.

Perkowitz, S. (1975). *Phys. Rev. B* **12**, 3210.

Perkowitz, S. (ed.) (1976). "Digest of the Second International Conference on Submillimeter Waves and Their Applications." IEEE, New York.

Perkowitz, S. (ed.) (1979). "Digest of the Fourth International Conference on Infrared and Millimeter Waves and Their Applications." IEEE, New York.

Perkowitz, S. (1982). *Phys. Rev. B* **25**, 3420.

Perkowitz, S., and Breecher, J. (1973). *Infrared Phys.* **13**, 321.

Perkowitz, S., and Breecher, J. (1974). *In* "Digest of the First International Conference on Submillimeter Waves, Atlanta, 1974," p. 193. IEEE, New York.

Perkowitz, S., and Thorland, R. H. (1974). *Phys. Rev. B* **9**, 545.

Perkowitz, S., Merlin, R. and Testardi, L. R. (1976). *Solid State Commun.* **18**, 1059.

Perry, C. H., Geick, R., and Young, E. F. (1966). *Appl. Opt.* **5**, 1171.

Petzelt, J., and Grigas, J. (1973). *Ferroelectrics* **5**, 59.

Pickett, W. E., and Allen, P. B. (1973). *Solid State Commun.* **12**, 677.

Pidgeon, C. R. *et al.* (1978). *Infrared Phys.* **18**, 923.

Pinczuk, A., Worlock, J. M., Nahory, R. E., and Pollack, M. A. (1978). *Appl. Phys. Lett.* **33**, 461.

Polian, A., Le Toullec, R., and Balkanski, M. (1976). *Phys. Rev. B* **13**, 3558.

Price, P. J. (1961). *Phys. Rev.* **128**, 713.

Robinson, L. C. (1973). "Physical Principles of Far-Infrared Radiation." Academic Press, New York.

Rosenbluh, M., Temkin, R. J., and Button, K. J. (1976). *Appl. Opt.* **15**, 2635.

Shaw, W., and Swihart, J. C. (1968). *Phys. Rev. Lett.* **20**, 1000.

Shen, S. C., Fang, C. J., Cardona, M., and Genzel, L. (1980). *Phys. Rev. B* **22**, 2913.

Singwi, K. S., and Tosi, M. P. (1966). *Phys. Rev.* **147**, 658.

Skolnick, M. S. *et al.* (1974). *Solid State Commun.* **15**, 1403.

Slack, G. A., and Roberts, S. (1971). *Phys. Rev. B* **3**, 2613.

Sobotta, H. (1970a). *Phys. Lett.* **32A**, 4.

Sobotta, H. (1970b). *Ann. Physik* **25**, (7) 411.

Stillman, G. E., Wolfe, C. M., and Korn, D. M. (1972). *In* "Proceedings of the Eleventh International Conference on the Physics of Semiconductors, Warsaw, 1972," Vol. 2, p. 863. PWN, Warsaw.

Stolen, R. H. (1969). *Appl. Phys. Lett.* **15**, 74.

Stolen, R. H. (1975). *Phys. Rev. B* **11**, 767.

Summers, C. J., Dingle, R., and Hill, D. E. (1970). *Phys. Rev. B* **1**, 1603.

Tannenwald, P. E. (1980). *Int. J. IR/MM Waves* **1**, 159.

Tanner, D. B., and Sievers, A. J. (1973). *Phys. Rev. B* **8**, 1978.

Tennant, W. E., and Cape, J. A. (1976). *Phys. Rev. B* **13**, 2540.

Theis, T. N., Kotthaus, J. P., and Stiles, P. J. (1978). *Solid State Commun.* **26**, 603.

Tinkham, M. (1964). *Science* **145**, 240.

Tinkham, M. (1970). *In* "Far-Infrared Properties of Solids" (S. S. Mitra and S. Nudelman, eds.), p. 223. Plenum, New York.

Trodahl, H. J., Fee, M., Livick, N., and Buckley, R. G. (1980). *Solid State Commun.* **35**, 551.

Tsui, D. C., Gornik, E., and Logan, R. A. (1980). *Solid State Commun.* **35**, 875.

Varga, B. B. (1965). *Phys. Rev.* **137**, A1896.

Vodop'yanov, L. K., Vinogradov, E. A., Kolotkov, V. V., and Mityagin, Yu. A. (1974). *Sov. Phys.-Solid State* **16**, 912.

Wagner, N. K. (1976). *Thin Solid Films* **38**, 353.

Wyder, P. (1976). *Infrared Phys.* **16**, 243.

Yamanaka, M. (1976). *Rev. Laser Eng.* **3**, 57.

Zitter, R. N., and As'Saadi, K. (1974). *J. Phys. Chem. Solids* **35**, 1593.

CHAPTER 4

Review of the Theory of Infrared and Far-Infrared Free-Carrier Behavior in Semiconductors*

B. Jensen

Department of Physics
Boston University
Boston, Massacusetts

I. Introduction

The classical Drude theory can be used to extract the mobility and the free-carrier density of electrons in a semiconductor from an analysis of the reflectance and transmission data in the far infrared (Holm *et al.,* 1977; Fan, 1967; Palik and Holm, 1979; Perkowitz and Thorland, 1974). It is known, however, that the Drude expression for the optical conductivity, or imaginary part of the complex dielectric constant, becomes invalid at frequencies just below the fundamental absorption edge in many polar semiconducting compounds (Dumke *et al.,* 1970; Dutt *et al.,* 1976, 1977; Kung *et al.,* 1974; Spitzer and Whelan, 1959). This departure from the λ^2 classical regime can

* This research was supported by the U.S. Department of Energy under contract DE–AC02–79ER10444,A000

be described in terms of a frequency-dependent electron-scattering rate that reduces to a constant in the far infrared and gives the result of a quantum-mechanical calculation at high frequencies when used in the Drude formula for the optical conductivity.

The departure from the classical theory arises when the energy of the incident photon becomes large compared with the average electron energy, which is of the order of $k_0 T$ or E_f for nondegenerate and degenerate materials, respectively. The absorption coefficient must then be calculated quantum mechanically, using the methods of time-dependent perturbation theory and an appropriate model of the band structure (Dumke, 1961; Haga and Kimura, 1964; Jensen, 1973; Kane, 1957). The quantum-mechanical treatment predicts the approximate λ^3 dependence of the absorption coefficient that is observed in GaAs, InP, InAs, CdTe, and ZnSe at high frequencies (Dumke et al., 1970; Dutt et al., 1976, 1977; Kung et al., 1974; Spitzer and Whelan, 1959). This wavelength dependence is characteristic of electron scattering via interaction with longitudinal polar optical mode phonons (Jensen, 1973), which is the dominant mechanism in lightly doped materials. A significant contribution from charged impurity scattering increases the power of the wavelength dependence of the observed absorption coefficient.

The quantum-mechanical result can also be derived from the equation of motion of the quantum density matrix. It is known that the quasi-classical Boltzmann transport equation, which gives the Drude result, becomes invalid when photon energies become large compared with electron energies. The quantum-mechanical derivation of the Boltzmann transport equation from the density matrix equation of motion has been treated by several authors (Argyres, 1961; Jensen, 1975; Kohn and Luttinger, 1957; Price, 1966); a summary of this derivation as given by Jensen (1975) can be found in Appendix A. These treatments give the usual Boltzmann equation in the limit of low frequencies and elastic scattering mechanisms.

The result found using second-order time-dependent perturbation theory can also be obtained using a quantum-mechanical extension of the Boltzmann transport equation to calculate the optical conductivity or the absorption coefficient (Jensen, 1975, 1978, 1979). An advantage of the latter method is that the quasi-classical result is recovered at low frequencies for elastic scattering mechanisms. One can then identify the electron scattering rate and obtain an expression for it, as a function of frequency for the various scattering mechanisms, that reduces to the usual expression in the appropriate limit. When this is done for polar scattering, one finds an expression that becomes frequency dependent at high frequencies and reduces to a constant that is a function of experimentally available parameters and gives the dc mobility at low frequencies in lightly doped materials.

In Section II of this chapter the theoretical results are given, and the transition to the quasi-classical limit at low frequencies is demonstrated. Section III gives results for polar scattering, and Section IV discusses these in terms of an absorption coefficient proportional to a scattering cross section. This is the formulation that arises naturally from the use of time-dependent perturbation theory. Section V develops the formulation in terms of a frequency-dependent electron scattering rate, which follows from the use of the generalized Boltzmann equation. The relation of the scattering rate to the electron cross section is demonstrated.

In Section VI, the analysis is repeated for impurity scattering. Section VII examines the high-concentration low-temperature limit and gives an approximate analytical expression for the low-frequency impurity scattering rate, which can be used to determine the dc mobility in this limit. Theoretical results are summarized in Appendix B for the total conductivity, which includes all scattering mechanisms. Section VIII compares theoretical results with experimental data for a number of compounds.

II.   Theoretical Results

The derivation of the high-frequency optical conductivity from the equation of motion of the quantum density matrix is discussed in detail by Jensen (1975). One obtains, for polar optical mode scattering, the following result:

$$\sigma = (e^2/3w^2V) \sum_{\substack{kk' \\ Q=|k'-k|}} (S_{kk'}/\hbar w)|\mathbf{v}_k - \mathbf{v}_{k'}|^2$$
$$\times f_{k'}^0 (1 - f_k^0)(e^{\beta \hbar w} - 1)[(n_Q^0 + 1)\delta(e_{k'k} - \hbar w^+)$$
$$+ n_Q^0 \delta(e_{k'k} - \hbar w^-)]$$
$$= (c/4\pi)n\alpha. \tag{1}$$

In Eq. (1), n_Q^0 is the equilibrium occupation number at temperature $T = 1/k_0\beta$ for phonons of wave vector Q, $f_{k'}^0$ is the equilibrium electron Fermi distribution function, v_k is the velocity of an electron in the Bloch state characterized by wave vector k, and k and k' denote the wave vectors in the initial and final electron states, respectively. Also, $e_{k'}$ and e_k denote the final and initial electron energies measured relative to the conduction band edge, and $\hbar w$ and $\hbar w_Q$ are the energies of the photon and phonon. We define

$$e_{k'k} = e_{k'} - e_k, \tag{2}$$

$$\hbar w^\pm = \hbar w \pm \hbar w_Q \tag{3}$$

$$S_{kk'} = (2\pi/\hbar)|V_{k'k}|^2 \rightarrow \tfrac{1}{2} \sum_{s_i s_f} (2\pi/\hbar)|V_{k'k}|^2, \tag{4}$$

and $V_{k'k}$ in Eq. (4) is essentially the Fourier transform of the scattering interaction $H'(r)$ between electron states characterized by wave vectors k' and k. The average over initial and sum over final electron spin states in Eq. (4) is made in the case of spin degeneracy. The crystal volume is denoted by V.

On solving Eq. (1) for $n\alpha$, where n is the real part of the refractive index and α is the absorption coefficient, one obtains the exact quantum-mechanical expression for $n\alpha$ that has also been calculated using the methods of time-dependent perturbation theory (Jensen, 1973).

Using the principle of detailed balance and the properties of the delta function, one can write Eq. (1) in the following form (Jensen, 1973):

$$n\alpha = 4\pi\sigma/c = (4\pi e^2/w^2 c) \sum_{kk'}^{Q=|k'-k|} \tfrac{1}{3}(S_{kk'}/V\hbar w)|\mathbf{v}_k - \mathbf{v}_{k'}|^2$$
$$\times (f_k^0 - f_{k'}^0)(F_+(w, w_Q)\delta(e_{k'k} - \hbar w^+) + F_-(w, w_Q)\delta(e_{k'k} - \hbar w^-)), \quad (5)$$

where

$$F_\pm(w, w_Q) = \frac{\sinh(\hbar w/2k_0 T)}{2 \sinh(\hbar w_Q/2k_0 T) \sinh(\hbar w^\pm/2k_0 T)}. \quad (6)$$

The quasi-classical limit, which occurs when $\hbar w^\pm \ll k_0 T, E_f$, where E_f is the Fermi energy measured relative to the conduction band edge, is obtained from Eq. (1) on making the following replacements:

$$\delta(e_{k'k} - \hbar w^\pm) \cong \delta(e_{k'k}),$$
$$n_Q^0 = (e^{\beta\hbar w} - 1)^{-1} \cong k_0 T/\hbar w_Q \cong n_q^0 + 1,$$
$$(e^{\beta\hbar w} - 1)/\hbar w \cong 1/k_0 T,$$
$$(1/k_0 T)f_{k'}^0(1 - f_k^0)\delta(e_{k'k}) \cong (-\partial f_k^0/\partial e_k)\,\delta(e_{k'k}),$$
$$-\partial f_k^0/\partial e_k \cong \delta(e_k - E_f), \quad (7)$$
$$|\mathbf{v}_k - \mathbf{v}_{k'}|^2\,\delta(e_{k'k}) \cong 2v_k^2(1 - y)\,\delta(e_{k'k}),$$
$$y = \mathbf{v}_k \cdot \mathbf{v}_{k'}/v_k v_{k'}.$$

Using Eq. (7) in Eq. (1) one obtains

$$\sigma \cong (4e^2/3w^2 V) \sum_{k'k} [S_{kk'}(k_0 T/\hbar w_Q)](1 - y)$$
$$\times \delta(e_{k'k})\,v_k^2\,(-\partial f_k^0/\partial e_k). \quad (8)$$

From Eq. (4), one can write

$$W_{kk'} = S_{kk'}(k_0 T/\hbar w_Q), \quad (9)$$

where $W_{kk'}$ is the transition rate for an electron to be scattered from state k to state k' by absorption or emission of a phonon in the limit $\hbar w_Q \ll k_0 T$. The relaxation time τ is defined as

$$1/\tau = \int [2V/(2\pi)^3]\, d^3k'\delta(e_{k'k})W_{kk'}(1 - y). \tag{10}$$

On converting from summation to integration in the usual manner, one finds the following expression for σ:

$$\sigma = \tfrac{2}{3}(e^2/w^2)\int d^3k/(2\pi)^3(-\partial f_k^0/\partial e_k)(v_k^2/\tau) \tag{11}$$

for $\hbar w\pm \ll k_0 T, E_f$.

For the case of spherical, but not necessarily parabolic, bands one has for the electron concentration

$$n_e = [2/(2\pi)^3](4\pi/3)k_f^3, \qquad E_f \gg k_0 T, \tag{12}$$

where k_f is the wave vector at the Fermi surface. Also, the electron velocity and the wave vector are related as

$$\mathbf{v_k} = (1/\hbar)(de_k/dk)\,\hat{\mathbf{k}} = \hbar\mathbf{k}/m^*, \tag{13}$$

$$m^* = \begin{cases} m_n & \text{parabolic bands,} \\ m_n(1 + 2e_k/G) & \text{hyperbolic bands,} \end{cases} \tag{14}$$

where $1 + 2e_k/G = \gamma \cong 1$, $e_k \ll G$, and m_n and G are the effective mass and band gap energy at $k = 0$. The case of hyperbolic bands applies to the III–V and II–VI semiconducting compounds which can be described by the band structure of the Kane theory (Kane, 1957).

Using Eqs. (12)–(14) in Eq. (11), one recovers the quasi-classical high-frequency result

$$\sigma = \sigma_0/w^2\tau^2, \\ \sigma_0 = n_e e\mu = n_e e^2\tau/m^*, \tag{15}$$

where σ_0 is the dc conductivity and μ the mobility.

The quasi-classical limit is thus obtained for the case of deformation potential scattering at room temperature whenever $\hbar w \ll k_0 T$, because the energy of the acoustic phonon is negligible and Eq. (11) holds. Using the appropriate expression for H' (Jensen, 1975) and neglecting the overlap wave-function factor of the Kane theory, one finds

$$1/\tau = m^* k_f D^2 k_0 T/\hbar^3 \rho s^2\pi, \tag{16}$$

which is the usual result (Bardeen and Shockley, 1950), where D is the deformation potential constant, ρ the crystal density, and s the longitudinal velocity of sound.

For polar semiconducting compounds, deformation potential scattering is not expected to be the dominant scattering mechanism at room temperature, and one must examine the cases of polar optical mode scattering and impurity scattering.

III. Polar Scattering

Polar optical mode scattering is the most important scattering mechanism in lightly doped III–V and II–VI compounds. The energy of the optical phonon is not small compared with $k_0 T$ at room temperature, however, and the optical conductivity does not reduce to the quasi-classical limit of Eq. (10), although the λ^2 dependence is obtained for $\hbar w \ll k_0 T$, $\hbar w_Q$.

For polar scattering, one has (Ehrenreich, 1959; Matz, 1967)

$$S_{kk'} = (2\pi/\hbar)(g/|\mathbf{k}' - \mathbf{k}|^2 V)G(k, k', y), \tag{17}$$

where V is the crystal volume, taken as unity, and

$$G(k, k', y) = \tfrac{1}{2}\sum_{ss'} |(U_{k's'}|U_{ks})|^2. \tag{18}$$

The quantity $G(k, k', y)$ is the Kane overlap wave-function factor involving the cell periodic part of the Bloch function U_{ks}, which reduces to unity at the band edge. The constant g, which is known from experimentally determined parameters, is defined by

$$g = 2\pi e^2 \hbar w_Q[(1/e_\infty) - (1/e_0)], \tag{19}$$

where e_0 and e_∞ are the static dielectric constant and the dielectric constant at high frequencies, respectively.

Use of Eqs. (17) through (19) in Eq. (1) gives the optical conductivity for polar scattering. This becomes, in the limit $\hbar w \ll k_0 T$, $\hbar w_Q$,

$$\sigma = (e^2/3w^2 V) \sum_{kk'} S_{kk'}|\mathbf{v}_k - \mathbf{v}_{k'}|^2 (1/k_0 T)f_k^0(1 - f_k^0)$$
$$\times [(n_Q^0 + 1)\delta(e_{k'k} - \hbar w_Q) + n_Q^0\delta(e_{k'k} + \hbar w_Q)]. \tag{20}$$

On taking the limit $\hbar w_Q \ll k_0 T$, E_f in Eq. (20), one obtains

$$\sigma \rightarrow (4e^2/3w^2 V) \sum_{kk'} [S_{kk'}(k_0 T/\hbar w_Q)](1 - y)\delta(e_{k'k})$$
$$\times v_k^2(-\partial f_k^0/\partial e_k), \qquad \hbar w, \hbar w_Q \ll k_0 T, E_f. \tag{21}$$

But this is the same as Eq. (8) and gives the quasi-classical Boltzmann result in Eq. (15), with τ representing the relaxation time for polar scattering.

Because the condition $\hbar w_Q \ll k_0 T$ does not hold at room temperature in most polar semiconducting compounds, one instead has Eq. (20) as the low-frequency result. This, however, is seen to be proportional to λ^2. Defining, for any function of $e_{k'}$,

$$f_{k'_\pm} = f_{k'}(e_{k'})|_{e_{k'} = e_k + \hbar w \pm \cong e_k \pm \hbar w_Q}, \qquad \hbar w \ll k_0 T, \qquad (22)$$

Eq. (20) can be written

$$\sigma = \sigma_+ + \sigma_-, \qquad (23)$$

where

$$\sigma_\pm = \frac{e^2}{3w^2} \int \frac{d^3k}{(2\pi)^3} v_{k'_\pm}^2 \left(\frac{1}{k_0 T}\right) f_{k'_\pm}^0 (1 - f_k^0)(1/\tau_\pm), \qquad (24)$$

$$\frac{1}{\tau_+} = \int \frac{2V}{(2\pi)^3} d^3k' \frac{|v_k - v_{k'}|^2}{2v_{k_+}^2} S_{kk'}(n_Q^0 + 1)\delta(e_{k'k} - \hbar w_Q),$$

$$\frac{1}{\tau_-} = \int \frac{2V}{(2\pi)^3} d^3k' \frac{|v_k - v_{k'}|^2}{2v_{k_-}^2} S_{kk'} n_Q^0 \delta(e_{k'k} + \hbar w_Q). \qquad (25)$$

In the limit of high temperatures, the quasi-classical result Eq. (15) is obtained with

$$\frac{1}{\tau_\pm} \to \int \frac{2V}{(2\pi)^3} d^3k'(1 - y) S_{kk'}(k_0 T/\hbar w_Q)\delta(e_{k'k}) = \frac{1}{\tau}$$

$$e_{k'} = e_k \pm \hbar w_Q \cong e_k. \qquad (26)$$

Assuming parabolic bands and setting the factor $G(k, k'_\pm, y)$ in Eq. (18) equal to unity, one then has

$$1/\tau = (g/\pi\hbar^2 v_k)(k_0 T/\hbar w_Q). \qquad (27)$$

The two relaxation times $\tau_\pm$ correspond to the fact that the absorption and emission of a photon can each be accompanied by the absorption or emission of a phonon. These two relaxation times are equal only when the phonon energy is small compared with $k_0 T$. If the optical phonon energy is not small compared with the electron energy, which is of the order of E_f or $k_0 T$ for degenerate and nondegenerate materials, respectively, the integration over electron energies in Eq. (25) must be performed. The λ^2 dependence of the low-frequency limit is still obtained, however.

In the limit of low temperatures and $\hbar w > \hbar w_Q$, one has from Eq. (5) using Eq. (6)

$$\sigma = \frac{e^2}{w^2} \sum_{\substack{kk' \\ Q=|k'-k|}} \frac{S_{kk'}}{V\hbar w} \left(\frac{|v_k - v_{k'}|^2}{3}\right) (f_k^0 - f_{k'}^0)\delta(e_{k'k} - \hbar w^-). \qquad (28)$$

Defining

$$\frac{1}{\tau(w, T)}\bigg|_{k_0 T/\hbar w - \ll 1} = \frac{1}{\tau(w, 0)}$$

$$= \int \frac{V d^3 k'}{(2\pi)^3} \left(1 + \frac{v_{k'}^2}{v_k^2} - \frac{2\mathbf{v}_k \cdot \mathbf{v}_{k'}}{v_k^2}\right) S_{kk'} \delta(e_{k'k} - \hbar w^-)$$

$$\rightarrow \int \frac{2V}{(2\pi)^3} d^3 k'(1 - y) S_{kk'} \delta(e_{k'k}), \qquad \hbar w - \ll E_f, \qquad (29)$$

one can write Eq. (28) as

$$\sigma(w, 0) = \frac{e^2}{3w^2} \int \frac{d^3 k}{(2\pi)^3} \frac{(f_k^0 - f_{k'}^0)}{\hbar w} \frac{v_k^2}{\tau(w, 0)}. \qquad (30)$$

On setting the overlap wave-function factor $G(k, k'_+, y)$ equal to unity in Eq. (29), one obtains, in the parabolic limit,

$$1/\tau(w, 0) = (g/\pi \hbar^2 v_{k'}) k'^2_-/k^2 \sim (g/\pi \hbar^2 v_{k_f}), \qquad \hbar w^- \ll E_f. \qquad (31)$$

Since $k'_-/k \simeq 1$ in the parabolic limit, the order of magnitude of the relaxation time for polar scattering is determined by the factor $(g/\pi \hbar^2 v_{k'})$.

IV.   Formulation in Terms of a Cross Section

On substituting Eqs. (17) through (19) into Eq. (5), one obtains the following expressions for $\sigma_\pm$:

$$\sigma_\pm = \frac{F_\pm(w, w_Q)}{\hbar w} \left(\frac{eh}{m_n w}\right)^2 \frac{2\pi g}{3\hbar} \int\int \frac{d^3 k d^3 k'}{(2\pi)^6} D(k, k', y)$$

$$\times (f_k^0 - f_{k'}^0) \delta(e_{k'k} - \hbar w^\pm). \qquad (32)$$

In Eq. (32), $D(k, k', y)$ is a dimensionless factor, appropriate to the band structure of the Kane theory, (Kane, 1957) that reduces to unity in the case of parabolic bands and neglect of the overlap wave-function factor in Eq. (18). It contains the angular and energy dependence of the second-order matrix element for the quantum transition (Jensen, 1973), neglecting interband terms, and is given by

$$D(k, k'_\pm, y) = \frac{1}{\gamma_\pm^2} \frac{|\hat{\mathbf{v}} k'_\pm - \mathbf{v}_k/v k'_\pm|^2}{|\hat{\mathbf{k}}'_\pm - \mathbf{k}/k'_\pm|^2} G(k, k'_\pm, y), \qquad (33)$$

where

$$\gamma_\pm = (1 + 2e_{k'}/G)|_{e_{k'} = e_k + \hbar w^\pm} \qquad (34)$$

and $\hat{\mathbf{k}}'$ and $\hat{\mathbf{v}}_{k'}$ are unit vectors in the direction of $\mathbf{k}'$. The evaluation of

$D(k, k'_\pm, y)$, using the band structure of the Kane theory, is given in Appendix C.

Defining the dimensionless variables

$$u = e_k/G, \qquad X = \hbar w/G, \qquad X_0 = \hbar w_Q/G,$$
$$X^\pm = X \pm X_0, \qquad u_\pm = u + X^\pm, \tag{35}$$

and the dimensionless angle averaged quantity

$$d(u, u_\pm) = (1/4\pi) \int d\Omega' \, D(u, u_\pm, y) \tag{36}$$

and also the density of states $g(u) \, du$ for spherical (but not necessarily parabolic) spin-degenerate bands,

$$2 d^3k/(2\pi)^3 = (d\Omega/4\pi)g(u) \, du, \tag{37}$$

$$g(u) \, du = (1/V_c)h(u) \, du, \tag{38}$$

where, in Eq. (38), the constant V_c has the dimensions of volume and $h(u) \, du$ is a dimensionless function, one can write Eq. (32) in the form

$$\sigma_\pm = (c/4\pi)n\Sigma_0(F_\pm(w, w_Q)/X^3)(I_\pm/V_c). \tag{39}$$

The quantity $I_\pm$ in Eq. (39) is a dimensionless integral, and $I_\pm/V_c$ becomes proportional to the carrier concentration in the high-energy quantum limit as follows:

$$I_\pm/V_c = (1/V_c) \int_0^\infty du \, h(u)h(u_\pm) \, d(u, u_\pm)[f^0(u) - f^0(u_\pm)]. \tag{40}$$

When $X^\pm \gg u$, one has $f^0(u_\pm) \cong 0$ and $u_\pm \cong X^\pm$, because the initial electron energy is small compared with the energy of the photon. Hence, evaluating $h(u_\pm) \, d(u, u_\pm)$ at $u = 0$, $u_\pm = X^\pm$, and removing it from the integrand, one can write

$$I_\pm/V_c \cong h(X^\pm) \, d(0, X^\pm) \left[\int_0^\infty du \, g(u)f^0(u) \right] = n_e f(X^\pm),$$
$$f(X^\pm) = h(X^\pm) \, d(0, X^\pm), \tag{41}$$

because the integral in square brackets in Eq. (41) is the electron density n_e.

The product of the index of refraction n and the absorption coefficient α is proportional to the conductivity and also to the imaginary part of the dielectric constant e_2. One has

$$n\alpha = n(\alpha_+ + \alpha_-) = (4\pi\sigma/c) = (w/c)e_2, \tag{42}$$

where

$$n\alpha_{\pm} = (n\Sigma_0/X^3)F_{\pm}(w, w_Q)(I_{\pm}/V_c) = n(n_e\Sigma_{\pm}). \qquad (43)$$

Thus $\alpha_{\pm}$ is proportional to a cross section per electron $\Sigma_{\pm}$ that becomes independent of the electron energy at high photon energies and is only a function of frequency, that is,

$$\begin{aligned}
n\Sigma_{\pm} &= (n\Sigma_0/X^3)F_{\pm}(w, w_Q)I_{\pm}/n_eV_c \\
&\rightarrow (n\Sigma_0/X^3)F_{\pm}(w, w_Q)f(X^{\pm}), \qquad X^{\pm} \gg u.
\end{aligned} \qquad (44)$$

It is shown in detail in Jensen (1973) that

$$f(X^{\pm}) = \beta(X^{\pm})(1 + X^{\pm})/(1 + 2X^{\pm}), \qquad (45)$$

$$1/(1 - \beta^2(X^{\pm})) = \gamma^2(X^{\pm}) = (1 + 2X^{\pm})^2 \qquad (46)$$

if one neglects the contribution of interband terms. A more general expression is obtained if they are included.

The constant $n\Sigma_0$ in Eq. (44) is given by

$$\begin{aligned}
n\Sigma_0 &= (e^2/4\pi\hbar c)[(8\pi/3)(\hbar c_0 g/G^3)] = (e^2/4\pi\hbar c)\phi_p, \\
\phi_p &= (8\pi/3)R_P^2, \qquad R_p^2 = \hbar c_0 g/G^3,
\end{aligned} \qquad (47)$$

where

$$c_0 = \alpha_0 c, \qquad \alpha_0 = (G/2m_nc^2)^{1/2}. \qquad (48)$$

Hence the constant factor in Eq. (47) is determined in terms of the band-gap energy G, the effective mass m_n and the polar scattering interaction parameter g, which are experimentally known quantities. The Kane theory gives the band structure, determining $I_{\pm}$ in Eq. (40) and hence $n\alpha$, as follows. The density of states $g(u)\, du$ is given in Eq. (38), where

$$\begin{aligned}
h(u)\, du &= \beta(u)\gamma^2(u)\, du, \\
\beta(u) &= 2u^{1/2}(1 + u)^{1/2}/(1 + 2u), \\
\gamma(u) &= 1/[1 - \beta^2(u)]^{1/2} = 1 + 2u, \\
h(u_{\pm}) &= \beta(u_{\pm})\gamma^2(u_{\pm}) = \beta_{\pm}\gamma_{\pm}^2,
\end{aligned} \qquad (49)$$

and

$$\begin{aligned}
1/V_c &= 2/\pi^2 \lambda_c^3, \\
\lambda_c &= \hbar/m_nc_0 = \hbar/m_n\alpha_0 c.
\end{aligned} \qquad (50)$$

With Eqs. (49) and (50) and the expression for $d(u, u_{\pm})$ given in Appendix C, the integral $I_{\pm}$ in Eq. (40) can be integrated numerically and inserted into Eq. (39) to obtain the optical conductivity for all frequencies from $w\tau > 1$ to the fundamental absorption edge. On using the fact that the scattering rate $1/\tau$ coverages to a constant value for $\hbar w \lesssim k_0 T$, which gives the dc conductivity in Eq. (15), one may write the conductivity for all frequencies as follows:

$$\sigma = \sigma_0/w^2\tau^2\eta = \sigma_0/(1 + w^2\tau^2) \longrightarrow \begin{cases} \sigma_0, & w\tau \ll 1, \\ \sigma_0/w^2\tau^2, & w\tau \gg 1, \end{cases} \quad (51)$$

where

$$\eta = 1 + (w\tau)^{-2}.$$

V.   Formulation in Terms of a Relaxation Time

The optical conductivity in Eq. (39) and the discussion in section IV are formulated in terms of an electron-scattering cross section. To relate the scattering cross section to the generalized relaxation time, we write, from Eq. (1),

$$\sigma_\pm = \frac{e^2}{3w^2} \int \frac{d^3k}{(2\pi)^3} \, v_{k_\pm}^2 \, f_k^0(1 - f_{k_\pm}^0) \frac{(1 - e^{-\beta\hbar w})}{\hbar w} \frac{1}{\tau_\pm(w, k)}, \quad (52)$$

$$1/\tau_\pm(w, k) = \frac{2}{(2\pi)^3} \int d^3k' \frac{|\mathbf{v}_{k'} - \mathbf{v}_k|^2}{2v_{k_\pm}^2} \, W_{kk'}^{p,(\pm)}\delta(e_{k'k} - \hbar w^\pm), \quad (53)$$

where

$$W_{kk'}^{p,(+)} = (2\pi/\hbar)(g/|\mathbf{k}' - \mathbf{k}|^2)G(k, k', y)(n_Q^0), \quad (54)$$
$$W_{kk'}^{p,(-)} = (2\pi/\hbar)(g/|\mathbf{k}' - \mathbf{k}|^2)G(k, k', y)(n_Q^0 + 1).$$

In Eq. (54), $W_{kk'}^{p,(\pm)}$ is the transition probability for electron scattering from state k to state k' via interaction with an optical phonon. Equation (52) reduces to Eq. (24) with $1/\tau_\pm$ given by Eq. (26) in the high-temperature, low-frequency limit as

$$f_k^0(1 - f_{k_\pm}^0)\,(1 - e^{-\beta\hbar w})/\hbar w \; \delta(e_{k'k} - \hbar w^\pm) \longrightarrow (-\partial f_k^0/\partial e_k)\delta(e_{k'k}),$$
$$\hbar w^\pm \ll k_0 t, E_f. \quad (55)$$

The generalized relaxation time in Eq. (53) can be arranged to give

$$1/\tau_\pm(w, k) = (g/\pi\hbar^2 v_{k_\pm})(n_Q^0 + {}^0_1)[\gamma^2(u_\pm)\, d(u, u_\pm)], \quad (56)$$

where the factor in square brackets reduces to unity for the case of parabolic bands. Equation (56) is the appropriate generalization of Eq. (27), and is the frequency- and energy-dependent relaxation time for polar optical mode scattering.

The Kane theory gives the following relations between the electron velocity v_k and wave vector k where $e_k/G = u$:

$$v_k = c_0\beta(u) = \hbar k/m_n\gamma(u), \qquad v_{k'_\pm} = c_0\beta(u_\pm) = \hbar k'_\pm/m_n\gamma(u_\pm), \quad (57)$$

$$k = (m_n c_0/\hbar)\beta\gamma = (1/\lambda_c)\beta\gamma. \tag{58}$$

Hence in Eq. (56) one has

$$1/\tau_\pm(w, k) = (g/\pi\hbar^2 c_0)(n_Q^0 + {0 \atop 1})\{[\gamma^2(u_\pm)/\beta(u_\pm)]d(u, u_\pm)\}. \tag{59}$$

The polar-scattering rate is proportional to the constant

$$1/\tau_p = g/\pi\hbar^2 c_0. \tag{60}$$

The quantity

$$\mu_p = e\tau_p/m_n \tag{61}$$

gives the order of magnitude of the dc mobility in many polar semiconductors as shown for various compounds in Jensen (1979).

Inserting Eq. (59) into Eq. (52), one can show that

$$\sigma_\pm = (n_e e^2/m_n w^2)1/\tau_\pm = (c/4\pi)n_e n\Sigma_\pm, \tag{62}$$

where $1/\tau_\pm$ is defined as follows:

$$1/\tau_\pm = \langle 1/\tau_\pm \rangle = (1/\tau_p)(F_\pm(w, w_Q)/X)I_\pm/12n_e V_c. \tag{63}$$

Hence from Eq. (42), in terms of the imaginary part of the dielectric constant, one has

$$4\pi\sigma_\pm = cn\alpha_\pm = we_{2\pm} = (w_p^2/w^2)1/\tau_\pm, \tag{64}$$

$$w_p^2 = 4\pi n_e e^2/m_n. \tag{65}$$

One can thus express the optical conductivity in the standard form in terms of an effective relaxation time that is frequency dependent when $\hbar w$ is not small compared with $k_0 T$ and plays the role of the relaxation time in the Drude theory. The effective relaxation time is proportional to g^{-1} and τ_p is related to $n\Sigma_0$, which determines the electron cross section as

$$n\Sigma_0 = (1/12n_e)(w_p^2/w_g^2)(1/c\tau_p) = e^2/4\pi\hbar c[(8\pi/3)R_p^2], \tag{66}$$

$$R_p^2 = \hbar c_0 g/G^3, \qquad w_g = G/\hbar.$$

VI. Impurity Scattering

Polar optical mode scattering is the dominant scattering mechanism in pure or lightly doped materials. As the carrier concentration is increased through an increase in the doping, the contribution of impurity scattering becomes important. The contribution to the optical conductivity arising from impurity scattering is obtained by replacing H' in Eq. (A3) in Appendix A with the interaction for ionic scattering and repeating the calculation outlined previously. One finds

$$\sigma^I = \frac{2e^2}{3w^2} \int \frac{d^3k}{(2\pi)^3} \, v_{k_+}^2 \frac{(f_k^0 - f_{k_+}^0)}{\hbar w} \, [1/\tau^I(w, k)], \tag{67}$$

where the relaxation time $\tau^I(w, k)$ appropriate to impurity scattering is given as

$$1/\tau^I(w, k) = \int d^3k' \frac{2V}{(2\pi)^3} \frac{|\mathbf{v}_{k_+} - \mathbf{v}_k|^2}{2v_{k_+}^2} \, W_{kk'}^I \delta(e_{k'k} - \hbar w), \tag{68}$$

$$W_{kk'}^I = \frac{2\pi}{\hbar} |V_{k'k}^I|^2, \tag{69}$$

$$|V_{k'k}^I|^2 = \frac{N_i}{V} \frac{(4\pi e^2/e_0)^2}{|Q^2 + a_s^2|^2} \, G(k, k', y). \tag{70}$$

The quantity $G(k, k', y)$ given in Eq. (18) is the overlap wave function previously discussed that reduces to unity at $k = 0$. N_i is the number of singly charged impurities per unit volume, $Q^2 = |\mathbf{k}' - \mathbf{k}|^2$, and $a_s = 1/r_s$, where r_s is the screening length. Screening can be neglected when $a_s^2/Q^2 \ll 1$, and its neglect gives an upper limit on the contribution of impurity scattering. Screening is discussed in Section VII.

In the limit that $\hbar w \ll E_f$, the relaxation time in Eq. (68) assumes the form

$$1/\tau^I(w, k_f) \cong 1/\tau^I = \int [2V/(2\pi)^3] d^3k' \delta(e_{k'k}) W_{kk'}^I (1 - y). \tag{71}$$

The optical conductivity in Eq. (67) becomes, in the same limit,

$$\sigma^I = \sigma_0^I/(w\tau^I)^2, \qquad \hbar w \ll E_f,$$
$$\sigma_0^I = n_e e^2 \tau^I/m^*, \tag{72}$$

which is the quasi-classical expression. The relation

$$f_k^0 - f_{k_+}^0 \cong -\hbar w \partial f_k^0/\partial e_k, \qquad v_{k_+}^2 \cong v_k^2, \tag{73}$$

which is valid for $\hbar w \ll E_f$, has been used in obtaining Eq. (72).

The optical conductivity for impurity scattering, which is elastic, thus reduces to the quasi-classical result in the low-frequency limit.

In the intermediate-frequency range, one has a frequency- and energy-dependent relaxation time in Eq. (68) that can be written

$$1/\tau^I(w, k) = g^I/\pi \hbar^2 v_{k_+} [\gamma^2(u_+) d^I(u, u_+)], \tag{74}$$

where

$$d^I(u, u_+) = 4 \left[\frac{1}{\beta^2(u_+) \, \gamma^2(u_+)} \right] \int \frac{d\Omega'}{4\pi} \frac{D(k, k'_+, y)}{|\hat{k}'_+ - \mathbf{k}/k'_+|^2}, \tag{75}$$

and $D(k, k'_+, y)$ is given in Eq. (33). We define, in analogy with Eq. (60),

$$1/\tau_0^I = g^I/\pi\hbar^2 c_0, \tag{76}$$

$$g^I = 8N_iV_c(e^2/e_0)^2/\hbar_c. \tag{77}$$

The expression in square brackets in Eq. (74) is a dimensionless integral that is related to the angle-averaged second-order matrix element for the quantum transition, as in the case for polar optical mode scattering discussed previously. Also, for impurity scattering

$$k'|_{e_k = e_k + \hbar w} = k'_+, \tag{78}$$

$$\beta_+^2 = \beta^2(u_+) = 1 - (1/\gamma_+^2) = 4u_+(1 + u_+)/(1 + 2u_+)^2, \tag{79}$$

$$u_+ = u + X, \tag{80}$$

and u and X are defined in Eq. (35). One can write the expression for the generalized relaxation time for impurity scattering as

$$1/\tau^I(w, k) = (1/\tau_0^I)(\gamma_+^2/\beta_+) \, d^I(u, u_+). \tag{81}$$

The optical conductivity for impurity scattering can then be written, using Eqs. (38) and (49), as

$$\sigma^I = (n_e e^2/m_n w^2 \tau_0^I) \, (1/6n_e V_c X)$$

$$\times \int_0^\infty du \, h(u)h(u_+) \, d^I(u, u_+)[f^0(u, a, b) - f^0(u_+, a, b)] \tag{82}$$

where n_e is the electron concentration and $n_e V_c$ a dimensionless quantity defined by

$$n_e V_c = \int_0^\infty du \, h(u)f^0(u, a, b), \tag{83}$$

$$f^0(u, a, b) = 1/(e^{b(u-a)} + 1), \tag{84}$$

$$b = G/k_0 T, \qquad a = E_f/G. \tag{85}$$

If the relaxation rate for impurity scattering is defined as

$$\langle 1/\tau^I \rangle = 1/\tau_0^I, \qquad (I_+^I/6n_e V_c X) = 1/\tau^I, \tag{86}$$

where

$$I_+^I = \int_0^\infty du \, h(u)h(u_+) \, d^I(u, u_+)[f^0(u, a, b) - f^0(u_+, a, b)]$$

$$= I_+^I(a, b, X), \tag{87}$$

one can write the generalized optical conductivity as

$$\sigma^I = (n_e e^2/m_n w^2)(1/\tau^I), \qquad w\tau^I \gg 1. \tag{88}$$

The integral over the electron energy I_+^I, for impurity scattering, is obtained from the analogous integral I_+ in Eq. (40) for polar optical mode scattering on replacing the angle-averaged quantity $d(u, u_+)$ in Eq. (40) by the expression $d^I(u, u_+)$ in Eq. (75), appropriate to impurity scattering.

The functions $d(u, u_+)$ and $d^I(u, u_+)$ are given in Appendixes C and D, respectively. These can be inserted into Eqs. (40) and (87) and integrated numerically to obtain the quantities I_+ and I_+^I and then the optical conductivities.

In the high-frequency limit $X \gg u, u_+ \cong X$, one has

$$I_+^I/n_e V_c \cong [h(X)\,d^I(0, X)]\,\frac{1}{n_e}\int_0^\infty g(u)f^0(u)\,du$$

$$= h(X)\,d^I(0, X) = f^I(x)/4, \tag{89}$$

where, as shown in Jensen (1973),

$$f^I(x) = \beta(X)/X(1 + 2X) \tag{90}$$

and hence in Eq. (86),

$$\langle 1/\tau^I \rangle \rightarrow 1/\tau_0^I\,(f^I(X)/12X), \qquad u \ll X. \tag{91}$$

A more general expression is obtained in Eq. (90) if intermediate states in other bands are included (Jensen, 1973).

The optical conductivity can be expressed in terms of a cross section per electron Σ^I as

$$4\pi\sigma^I/c = n\alpha^I = n_e n\Sigma^I = n_e(n\Sigma_0^I/X^3)(2I_+^I/n_e V_c), \tag{92}$$

$$n\Sigma_0^I = (e^2/4\pi\hbar c)(8\pi/3)R_I^2 = (1/12n_e)(w_p^2/w_g^2 c\tau_0^I), \tag{93}$$

$$R_I^2 = \hbar c_0 g^I/G^3, \tag{94}$$

$$\phi_I = (8\pi/3)R_I^2. \tag{95}$$

The optical conductivity due to impurity scattering thus assumes the form of the usual Drude expression if written in terms of an effective relaxation time $\langle 1/\tau^I \rangle = 1/\tau^I$ that is related to the cross section per electron as

$$n\Sigma^I = (1/n_e)(w_p^2/w^2 c)1/\tau^I. \tag{96}$$

The constant g^I in Eq. (77) determines an electron cross section $(e^2/\hbar c)\,\phi_I$

and relaxation time $1/\tau_0^I$ characteristic of the scattering interaction where

$$(e^2/\hbar c)\phi_I = (8\pi/3)(w_p^2/w_g^2)(1/n_e c\tau_0^I). \tag{97}$$

VII. Impurity Scattering at High Concentrations and Low Temperatures

The total conductivity σ, which is the sum of the conductivities due to polar optical mode and impurity scattering, is proportional to the total effective relaxation rate $1/\tau$ and the total absorption coefficient α where

$$4\pi\sigma = 4\pi(\sigma_+ + \sigma_- + \sigma^I) = (w_p^2/w^2)1/\tau = cn\alpha, \qquad w\tau \gg 1, \tag{98}$$

$$1/\tau = 1/\tau_+ + 1/\tau_- + 1/\tau^I. \tag{99}$$

Equation (98) gives the high-frequency limit in which $w\tau \gg 1$. The transition to the low-frequency regime, $w\tau \le 1$, is given by

$$\sigma = \sigma_0/w^2\tau^2\eta = \sigma_0/(1 + w^2\tau^2) \rightarrow \begin{cases} \sigma_0, & w\tau \ll 1, \\ \sigma_0/w^2\tau^2, & w\tau \gg 1, \end{cases} \tag{100}$$

where

$$\begin{aligned} \eta &= 1 + (w\tau)^{-2}, \\ \sigma_0 &= n_e e^2\tau/m_n \rightarrow n_e e^2\tau/m^*, \qquad k_0 T/E_f \ll 1. \end{aligned} \tag{101}$$

The generalized relaxation rate in Eq. (99) can now be calculated by numerical integration for a given sample as a function of carrier concentration n_e and E_f/G, where E_f is the Fermi energy measured relative to the conduction band edge. A negative value of E_f/G signifies a Fermi energy lying below the conduction band edge. The ratio E_f/k_0T is a measure of degeneracy with large positive values corresponding to degenerate samples. At 300 K, this quantity ranges from -5.0 to 27.9 in GaAs for carrier concentrations from 3.4×10^{15} to 8.9×10^{19} cm^{-3}, from -4.70 to 2.61 in InP for carrier concentrations from 4.69×10^{15} to 2.13×10^{18} cm^{-3}, and from -2.78 to 8.41 in InAs for carrier concentrations from 6.19×10^{15} cm^{-3} to 3.40×10^{18} cm^{-3}. Many samples considered cannot be classified as degenerate or nondegenerate but lie in a transition region, and the problem of screening for impurity scattering is somewhat ambiguous in these cases. The screening radius that is chosen affects the magnitude of the low-energy ($\hbar w \lesssim k_0 T$) rate for impurity scattering and hence affects the mobility found for samples with high carrier concentrations in which impurity scattering is important. It has been recognized that deviations between computational methods at high carrier concentrations are due to the way in which the ionized impurity scattering rate is corrected for degeneracy. Littlejohn et al., 1978; Rode, 1971.

It was desirable to avoid determining the low-frequency mobility at high carrier concentrations by the uncertain choice of screening for impurity scattering. Polar scattering presents no difficulty because screening is not required to prevent the divergence of an integral and is not important at the concentrations in which polar scattering is the dominant mechanism. Therefore screening was neglected in this case.

For the case of impurity scattering the divergence of the Coulomb integral can be prevented by either of two methods, both of which give similar results at high carrier concentrations where impurity scattering dominates. They differ from each other by a factor of order unity associated with the logarithmic term that contains the uncertainty regarding the correction for degeneracy. The first is the usual method of using a screening parameter as follows.

For $\hbar w \lesssim k_0 T$ one can assume that the Coulomb scattering matrix element is screened and determine the impurity scattering mobility in the limit that $\hbar w \to 0$ using the relaxation rate in Eqs. (86) and (87). The effect of screening is included by replacing $r_+ = 1 + (k/k'_+)^2$ by $r_+ = 1 + (k/k'_+)^2 + (a_s/k'_+)^2$ in the expression for the cross section and scattering rate in Appendixes D and C where a_s is the inverse screening radius in Eq. (70). If one takes the limit $k/k'_+ \to 1$ and notes

$$d_0/(r_+^2 - s_+^2) \to J_1^+/(r_+^2 - s_+^2) \to 1/(1 + k/k'_+)^2 \to 1/4 \qquad (102)$$

in Eq. (D3), one finds (where $a = E_f/G$)

$$\frac{1}{2} \beta_+^2 \gamma_+^4 \, d^I(a, a_+) = Q^I(a, a_+)$$

$$= \left(\frac{1}{2} \ln \frac{4k_+^2}{a_s^2} + \frac{1}{2} \right) \left[1 - \frac{\delta_c}{(\ln 4k_+^2/a_s^2 + 1)} \right], \qquad (103)$$

where $Q^I(a, a_+)$ is defined by the above equation and

$$1/\tau^I = \frac{1}{\tau_0^I} \frac{(\frac{1}{2} \ln 4k_+^2/a_s^2 + \frac{1}{2})}{12 n_e V_c} \left[1 - \frac{\delta_c}{\ln 4k_+^2/a_s^2 + 1} \right]. \qquad (104)$$

Alternatively, one can put $a_s = 0$ in Eq. (70) and cut off the photon energy at $k_0 T$. In this case one recovers for $k_0 T/E_f \ll 1$ from Appendix D the result

$$\frac{1}{2} \beta_+^2 \gamma_+^4 d^I(a, a_+) = Q^I(a, a_+)$$

$$= \left(\ln \frac{4a}{X_c} + \frac{1}{2} \right) \left[1 - \frac{\delta_c}{2 \ln 4a/X_c + 1} \right] \qquad (105)$$

and

$$\frac{1}{\tau^I} = \frac{1}{\tau_0^I} \frac{(\ln 4a/X_c + \frac{1}{2})}{12n_e V_c} \left[1 - \frac{\delta_c}{2 \ln 4a/X_c + 1} \right], \tag{106}$$

where $X_c = k_0 T/G$, $a = E_f/G$, and the term in square brackets is proportional to δ_c, where

$$\delta_c = 2\{1 - [1 - a/\gamma(a)]^2\} \tag{107}$$

is a small correction. The scattering rate is essentially determined by the coefficient of the term in square brackets in (104) and (106) in the limit $\hbar w \leq k_0 T$ and these are approximately equal if one assumes the scattering matrix element is screened by the electrons in the thermal layer of the Fermi surface. In this case one can write

$$a_s^2 = C(4\pi e^2/e_0)[n_e/(2E_f/3)](k_0 T/E_f), \tag{108}$$

$$a_s^2/4k_f'^2(a) = a_s^2 \lambda_c^2/4\beta^2(a)\gamma^2(a) = A_0(k_0 T/4E_f), \tag{109}$$

where C in Eq. (108) is a constant of order unity that depends on the details of the band structure and would involve the correction for degeneracy. Taking $C = 4$, one has

$$A_0 = \tfrac{1}{2} m_n \bar{w}_p \lambda_c^2/E_f a(1 + a). \tag{110}$$

For GaAs at T = 77 K, one has the following typical values: $a = 0.24$, $n_e = 2.25 \times 10^{19}$ cm^{-3}, $E_f = 0.362$ eV, $\bar{w}_p = 3.01 \times 10^{14}$ sec^{-1}, $G = 1.51$ eV, $\lambda_c = 1.21 \times 10^{-7}$ cm, and $\tfrac{1}{2} m_n \bar{w}_p^2 \lambda_c^2 = 0.0794$ eV. Using these values in Eq. (110), one finds $A_0 = 0.737 \sim 1$.

In summary, if one either uses a screening parameter where the effective electrons are those in the thermal layer of the Fermi surface or neglects screening and cuts off the Coulomb integral when the photon energy equals the average thermal energy $k_0 T$ in a mode of the radiation field, the impurity scattering rate in the limit $k_0 T \ll E_f$ is given by

$$\frac{1}{\tau^I} = \frac{1}{\tau_0^I} \frac{(A_1 \ln(4a/A_0 X_c) + \frac{1}{2})}{12n_e V_c} \left[1 - \frac{\delta_c}{2A_1 \ln 4a/A_0 X_c + 1} \right], \tag{111}$$

where

$$\delta_c = 2\{1 - [1 - a/\gamma(a)]^2\}, \qquad X_c = k_0 T/G, \qquad a = E_f/G,$$
$$6n_e V_c = \beta^3(a)\gamma^3(a) = 8[a(1 + a)]^{3/2}, \tag{112}$$

and

$$\begin{aligned} A_0 &= A_1 = 1, && \text{cut off at } k_0 T, \\ A_0 &\sim 1, A_1 = 1/2, && \text{screening.} \end{aligned} \tag{113}$$

Other choices may be estimated using Eq. (104). In the following numerical calculations, the cutoff will be used because it eliminates the uncertainty associated with A_0 and specifies $A_0 = 1$. This value gives good agreement with experimental results without adjustable parameters. The factor $12n_e V_c$ in Eq. (106) reduces to $2\beta^3(a) = 2(v_f/c_0)^3$ in the parabolic limit $a \ll 1$, and one recovers the usual v^{-3} dependence of the impurity scattering rate. The characteristic volume V_c is given for the various compounds in Table 1 for temperatures of 300 K and 77 K, and it is assumed that the values of V_c at 77 K and 4 K are equal. The value of V_c for a given compound is a function

TABLE I

VALUES OF EXPERIMENTAL QUANTITIES AND THEORETICAL QUANTITIES THAT ARE FUNCTIONS OF EXPERIMENTAL QUANTITIES[a]

Quantity[b]	InP	GaAs	InAs	InAs*
G (eV)	1.35	1.43	0.36	—
m_n/m	0.073	0.071	0.024	0.028
$\hbar w_Q$ (eV)	0.0450	0.0349	0.0299	—
e_∞	10.6	11.1	11.6	—
e_0	15.0	13.1	14.3	—
$(e_\infty^{-1} - e_0^{-1})$	0.02767	0.01350	0.01619	—
α_0	1/234	1/224	1/266	1/287
g (eV^{2}cm)	1.13×10^{-9}	4.25×10^{-10}	4.39×10^{-10}	—
$1/\tau_p$ (sec^{-1})	6.46×10^{12}	2.33×10^{12}	2.86×10^{12}	3.09×10^{12}
$e\tau_p/m_n$ (cm^2/Vs)	3.73×10^3	1.06×10^4	2.59×10^4	2.03×10^4
$1/V_c$ (cm^{-3})	1.09×10^{20}	1.14×10^{20}	2.62×10^{18}	3.30×10^{18}
λ_c (cm)	1.23×10^{-7}	1.21×10^{-7}	4.26×10^{-7}	3.94×10^{-7}
g^1/N_i (eV^{2}cm^4)	5.42×10^{-29}	6.95×10^{-29}	7.16×10^{-28}	4.50×10^{-28}
$1/\tau_0^i$ (sec^{-1})	$2.97 \times 10^{-7}N_i$	$3.81 \times 10^{-7}N_i$	$4.66 \times 10^{-6}N_i$	$3.16 \times 10^{-6}N_i$
μ (cm^2/Vs)[c]	4.53×10^3	8.30×10^3	4.69×10^{4e}	3.78×10^{4e}
μ (cm^2/Vs)[d]	4.5×10^3	8.6×10^3	3.00×10^{4f}	—

Quantity[g]	InP	GaAs	InAs	InSb
G (eV)	1.41	1.51	0.41	0.225
$1/V_c$ (cm^{-3})	1.27×10^{20}	1.05×10^{20}	4.16×10^{18}	6.40×10^{17}
g^1/N_i (eV^{2}cm^4)	4.87×10^{-29}	7.15×10^{-29}	5.26×10^{-28}	9.45×10^{-28}
$1/\tau_0^i$ (sec^{-1})	$2.67 \times 10^{-7}N_i$	$3.92 \times 10^{-7}N_i$	$3.42 \times 10^{-6}N_i$	$5.93 \times 10^{-6}N_i$

[a] From Pankove, 1971; Madelung, 1964; Long, 1968; Iwasa et al., 1964. $\alpha_0 = (G/2m_n c^2)^{1/2}$ is taken as temperature independent.
[b] $T = 300$ K.
[c] Theoretical value.
[d] Experimental value.
[e] Theoretical value for $n_e = N_i = 1.6 \times 16^{16}$ cm^{-3}.
[f] Experimental value for $n_e = 1.7 \times 10^{16}$ (Harman et al., 1956).
[g] $T = 77$ K.

only of G and m_n or, equivalently, of m_n and $\alpha_0 = c_0/c$. Equations (106) and (107) thus involve only experimentally known quantities.

Equation (106) gives only the lowest order approximation. To obtain a more accurate result for the optical conductivity and scattering rate for impurity scattering one can write from Eq. (67)

$$\sigma^I = (2e^2/3w^2) \int d^3k/(2\pi)^3 v_{k_+}^2 \delta(e_k - E_f) 1/\tau^I(T, k), \tag{114}$$

where Eq. (73) has been used for $\hbar w \sim k_0 T \ll E_f$ and

$$\frac{1}{\tau^I(T, k)} = \int d^3k' \left(\frac{2V}{(2\pi)^3}\right) \frac{|\mathbf{v}_{k_+} - \mathbf{v}_k|^2}{2v_{k_+}^2} W_{kk'}^I \delta(e_{k'k} - k_0 T), \tag{115}$$

$$W_{kk'}^I = (2\pi/\hbar)|V_{k'k}^I|^2. \tag{116}$$

Defining $u = e_k/G$ and performing the integration over k using Eq. (38), one obtains, where $a_+ = a + X_c$,

$$\sigma^I = (e^2/w^2 m_n)(1/6V_c)h(a) \beta^2(a_+) [1/\tau^I(X_c, a)]. \tag{117}$$

In the limit $k_0 T/E_f = x_c/a \ll 1$, one has

$$n_e = 2 \int_0^{k_f} \frac{d^3k}{(2\pi)^3} = \frac{1}{V_c} \int_0^a h(u) \, du = \frac{1}{6V_c} \beta^3(a)\gamma^3(a), \tag{118}$$

Hence

$$\sigma^I = (n_e e^2/m_n w^2)(1/\tau_D^I), \tag{119}$$

where the impurity scattering rate is given by

$$1/\tau_D^I = 1/\gamma(a)\tau^I(X_c, a) = (1/\tau_0^I)[\gamma(a)/\beta(a)] \, d^I(a, a_+). \tag{120}$$

The effective mass at the Fermi surface is

$$m^* = m_n \gamma(a) = m_n(1 + 2a), \tag{121}$$

and this is automatically accounted for if one uses $1/\tau_D^I$ in Eq. (120) as the Drude limit of the impurity scattering rate.

The function $d^I(a, a_+)$ is given in Appendix D. If one expands this as a series in powers of $k_0 T/E_f = X_c/a$ and neglects terms of order X_c/a and higher, one recovers Eq. (106) as the Drude limit for the scattering rate in Eq. (119).

The dc mobility due to impurity scattering is given by

$$\mu^I = e\tau_D^I/m^* = e\tau_D^I/m_n\gamma(a), \tag{122}$$

and the Joule heating is (Jensen, 1975)

$$\langle \mathbf{J} \cdot \mathbf{E} \rangle_w = \sigma^I \langle E^2 \rangle_w = n_e(\tfrac{1}{2}m_n v_w^2)/\tau_D^I, \tag{123}$$

where

$$\langle E^2 \rangle_w = E^2/2, \qquad v_w = eE/m_n w. \qquad (124)$$

VIII.   Comparison with Experimental Results

Tables I–IV and Figures 1–9 show results for samples of different carrier concentrations for the compounds considered and compare results obtained with various experimental data and with theoretical results obtained using a screening parameter (Rode, 1971). In Fig. 2, a comparison of results calculated as discussed above is made with various experimental data and with the theoretical results of Rode (1971), using numerical integration and the experimental parameters listed in Table I for InP. Because compensation lowers the mobility, and calculations have been done for pure uncompensated materials for which $N_i = n_e$, only the highest experimental values, for which compensation is assumed to be a minimum, are shown at low concentrations where compensation is expected to be most important. At high concentrations, a spread of the experimental values is shown because a

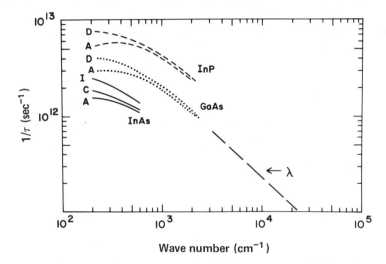

FIG. 1 Electron scattering rate versus wave number for various samples of InAs, GaAs, and InP. For each material, sample A includes the contribution of polar optical-mode scattering only. The remaining samples are for uncompensated materials for which $N_i = n_e$. The material parameters listed in Table 1 are used with $m_n = 0.028\, m$ for InAs. Polar optical-mode scattering dominates at high frequencies and is approximately independent of carrier concentration for light to moderately doped materials, whereas impurity scattering strongly affects the low-frequency ($\hbar w \simeq k_0 T$) limit. The dashed line shows the high-frequency limit of the wavelength dependence for polar optical-mode scattering.

comparison of theory with experiment and with other theoretical methods that use a screening parameter is of interest. Rode calculated the dc mobility from the Boltzmann equation with Fermi statistics using the Kane theory, neglecting screening of polar scattering, and including a screening parameter in the treatment of impurity scattering. The differences resulting from the treatment of impurity scattering shown in Fig. 2 are typical of differences with other methods of calculation that use a screening parameter. If the low-concentration polar-scattering mobilities coincide approximately, as is the case in Fig. 2, use of the screening parameter gives a lower mobility at intermediate concentrations and a higher mobility for heavily doped materials where impurity scattering becomes important. Higher mobilities at high carrier concentrations would be achieved by putting $A_0 > 1$ in Eq. (111). Theoretical curves calculated with the screening parameter are flat relative to the curves calculated using the $k_0 T$ cutoff. These curves exhibit a rather abrupt decrease when the transition from predominantly polar to predominantly impurity scattering occurs. The two methods give similar results for InP at low concentrations, and the experimental data shown span the region between and including both theoretical curves. At higher concentrations the bulk of the experimental data appears to lie below the curve for screened impurity scattering. The concentration dependence of the mobility for InP is similar to that of GaAs. The use of a screening parameter predicts a

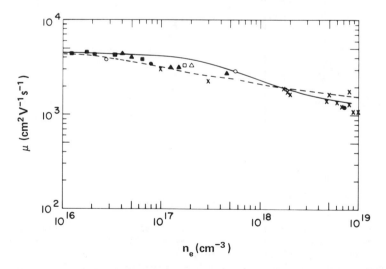

FIG. 2 Comparison of values of the mobility for InP at 300 K calculated as discussed in the text (solid line) with the theoretical values (dashed line) and the maximum experimental values (■, □, ○, ●, ▲, △, x) from Rode, 1971. The dashed line is the mobility calculated in Rode, 1971, Fig. 14, for a compensation ratio of 1.

higher mobility at high concentrations than does the $k_0 T$ cutoff ($A_0 = 1 = A_1$), which gives results quite reasonably in agreement with experiment. Numerical results for the scattering rate as a function of frequency are given in Tables II and III. The concentration dependence at high n_e is shown in Table V. The effect of compensation on mobility is shown for InP in Fig. 3.

In Figs. 4 and 5 a similar comparison is made for InAs. In this case, the theoretical results agree at high concentrations and, because of the charac-

TABLE II

THEORETICAL VALUES OF THE ELECTRON SCATTERING RATE IN THE LOW-FREQUENCY LIMIT AND THE CORRESPONDING MOBILITY FOR SAMPLES OF VARYING CARRIER CONCENTRATIONS

Sample	m^*/m_n	n_e (cm^{-3})	$1/\tau$ (sec^{-1})	$e\tau/m^*$ cm^2/V · sec
InAs[a]				
A	1	6.19×10^{15}	1.52×10^{12}[b]	4.07×10^{4}[b]
B	1	1.60×10^{16}	1.66×10^{12}	3.78×10^{4}
C	1	3.96×10^{16}	1.81×10^{12}	3.46×10^{4}
D	1	5.07×10^{16}	1.87×10^{12}	3.35×10^{4}
E	1	8.08×10^{16}	2.04×10^{12}	3.07×10^{4}
F	1.02	9.02×10^{16}	2.05×10^{12}	3.00×10^{4}
G	1.06	1.12×10^{17}	2.18×10^{12}	2.72×10^{4}
H	1.10	1.37×10^{17}	2.27×10^{12}	2.51×10^{4}
I	1.20	2.16×10^{17}	2.50×10^{12}	2.09×10^{4}
J	1.30	3.16×10^{17}	2.62×10^{12}	1.84×10^{4}
GaAs				
A	1	3.4×10^{15}	3.0×10^{12}[b]	8.3×10^{3}[b]
B	1	2.8×10^{16}	3.3×10^{12}	7.1×10^{3}
C	1	6.3×10^{16}	3.6×10^{12}	6.7×10^{3}
D	1	1.4×10^{17}	4.0×10^{12}	6.2×10^{3}
E	1	3.4×10^{17}	5.1×10^{12}	4.9×10^{3}
H	1.10	2.2×10^{18}	1.0×10^{13}	2.2×10^{3}
I	1.14	3.4×10^{18}	1.2×10^{13}	1.9×10^{3}
K	1.26	8.7×10^{18}	1.3×10^{13}	1.5×10^{3}
L	1.48	2.5×10^{19}	1.5×10^{13}	1.2×10^{3}
InP				
A	1	4.69×10^{15}	5.38×10^{12}[b]	4.53×10^{3}[b]
B	1	2.25×10^{16}	5.51×10^{12}	4.41×10^{3}
C	1	6.27×10^{16}	5.75×10^{12}	4.23×10^{3}
D	1	4.13×10^{17}	7.57×10^{12}	3.22×10^{3}
F	1.02	6.17×10^{17}	8.42×10^{12}	2.83×10^{3}
G	1.06	1.23×10^{18}	1.03×10^{13}	2.22×10^{3}
H	1.10	2.13×10^{18}	1.20×10^{13}	1.85×10^{3}

[a] $m_n/m = 0.028$.
[b] Polar.

TABLE III

THE FREQUENCY DEPENDENCE OF THE ELECTRON SCATTERING RATES FOR SAMPLES
CHARACTERIZED IN TABLE II[a]

							Sample			
Material	X	λ^{-1}/cm	A	B	C	D	E	G	I	J
InAs[b]	0.07	203	1.54	1.66	1.81	1.87	2.04	2.18	2.50	2.62
	0.08	232	1.54	1.65	1.77	1.82	1.95	2.07	2.35	2.47
	0.10	290	1.51	1.58	1.67	1.71	1.81	1.90	2.13	2.25
	0.20	581	1.12	1.15	1.19	1.21	1.26	1.30	1.43	1.52

Material	X	λ^{-1}/cm	A	B	C	D	E	F	G	H
GaAs	0.02	230	3.0	3.3	3.6	4.0	5.1	5.5	6.3	10.0
	0.04	461	2.8	2.9	3.0	3.3	3.9	4.1	4.4	7.3
	0.06	692	2.4	2.4	2.5	2.6	3.0	3.1	3.3	5.4
	0.09	1040	1.8	1.9	1.9	2.0	2.2	2.3	2.4	3.8
	0.20	2310	1.0	1.0	1.0	1.0	1.1	1.1	1.1	1.5

Material	X	λ^{-1}/cm	A	B	C	D	F	G	H
InP	0.02	218	5.38	5.51	5.75	7.57	8.42	10.3	12.0
	0.04	438	5.79	5.84	5.93	6.62	7.01	8.03	9.18
	0.06	654	5.08	5.12	5.18	5.67	5.94	6.65	7.52
	0.08	871	4.38	4.40	4.45	4.83	5.04	5.60	6.31
	0.10	1090	3.82	3.83	3.87	4.15	4.31	4.76	5.34
	0.20	2180	2.28	2.28	2.29	2.38	2.43	2.57	2.76

[a] Values of $(1/\tau) \times 10^{-12}$/sec for samples characterized in Table II. Samples A include the contribution of polar optical mode scattering only. All other samples are for uncompensated materials with $N_i = n_e$.
[b] $m_n = 0.028$ m.

teristic shapes of the curves, are quite far apart at lower concentrations where the highest experimental data again lie between the two curves. The effect of a choice of experimentally available parameters on the magnitude of the mobility is illustrated by plotting the mobility for two values of the effective mass at the band gap, 0.024 m and 0.028 m, respectively, in Fig. 4. The decreasing contribution of polar scattering to the total electron-scattering rate is indicated by plotting the mobility for polar scattering in Fig. 5. Results calculated using numerical integration for the two different values of the effective mass are given in Tables II, III, and IV. A comparison is made between the results calculated using the exact expression and numerical integration, and the approximation Eq. (106) for impurity scattering in Table VI.

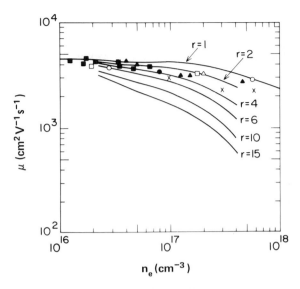

FIG. 3 Comparison of the same experimental data shown in Fig. 2 for the mobility of InP at 300 K, with the theoretical mobility curves for compensated materials with compensation ratios of $r = N_i/n_e = 1, 2, 4, 6, 10, 15$. The curve with $r = 1$ (uncompensated) is identical to the solid curve shown in Fig. 2. The highest values of the experimental data shown lie on the curve for uncompensated InP, and the remaining data correspond to theoretical compensation ratios of $N_i/n_e \lesssim 6$. The mobility for compensated materials is calculated according to the method described in Jensen (1979).

Table VII gives the results calculated for InSb from Eq. (106) at $T = 77$ K for the parameters listed. A mobility of 1.37×10^5 cm²/V sec for impurity scattering is found for $n_e = 1.37 \times 10^{16}$ cm⁻³, compared with an experimental mobility of $\sim 2 \times 10^5$ cm²/V sec for $n_e = 1 \times 10^{16}$ cm⁻³ and a mobility calculated using screening of approximately the same magnitude (Rode, 1971).

GaAs is the material that gives the most satisfactory agreement with the calculated theoretical results. As shown in Fig. 6, the alternative theoretical calculation (Brooks-Herring, dashed line) differs from that calculated (solid line) in a manner similar to that observed and discussed above in the case of InP. A maximum experimental mobility of 8500 to 8600 cm²/V sec has been reported (Long, 1968; Madelung, 1964; Pankove, 1971), compared with a mobility of 8300 cm²/V sec calculated here for polar scattering in the purest sample. An extended comparison of theory and experiment is given elsewhere for GaAs (Jensen, 1978), but Figs. 6–8 summarize the basic findings of interest. Agreement is obtained at low concentrations between the two theoretical methods for GaAs, as it is for InP. The difference at high

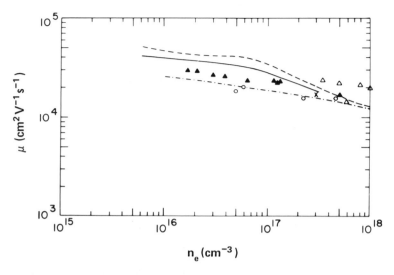

FIG. 4 Comparison of the mobility calculated here for uncompensated InAs at 300 K with the theoretical values ($— \cdot — \cdot — \cdot — \cdot —$) and with the highest experimental values for a given concentration ($\triangle$, $\blacktriangle$, $\bigcirc$) from Rode(1971). A recent experimental value x Chang $et\ al.$ (1977) is also shown. The theoretical curve from Rode(1971) is that for which the compensation ratio is unity. Higher compensation ratios give lower mobilities, as shown in Fig. 3. The two theoretical curves, calculated using different experimental values of the effective mass m_n are —, $m_n/m = 0.028$ (Pankove, 1971); and ----, $m_n/m = 0.024$ (Madelung, 1964). The mobility for polar scattering is more sensitive to the value of the effective mass than is the electron scattering time. The two quantities vary as $(m_n)^{-3/2}$ and $(m_n)^{-1/2}$, respectively. The values for the static and high-frequency lattice dielectric constants used by Rode are also slightly different. The effect of the variation in the value of the experimentally available parameters on the curves calculated here is to raise or lower the curve by an amount of the order of magnitude shown; but the basic shape is not altered. Because the results of Rode and the values calculated here coincide at high concentrations, they are necessarily farther apart at low concentrations due to the basic difference in shape.

concentrations is apparently due to the treatment of the impurity scattering screening parameter. The curve calculated here, using the lower limit of $n_q^0 \hbar w \rightarrow k_0 T$ on the photon energy in impurity scattering and omitting the screening parameter, falls below the mobility curves calculated using a screening parameter and gives a better fit to the experimental data at high concentrations (solid circles and square). This is shown for alternative sets of theoretical and experimental data in Fig. 7, where the effective mass is taken as m_n to show the concentration dependence of τ alone, and the effective mass at the Fermi level is used to calculate the mobility curve in Fig. 6. Use of m^* rather than m_n gives better agreement at high concentrations, as one would expect. The experimental data in Fig. 7 show the maximum experi-

TABLE IV

ELECTRON SCATTERING RATES AND MOBILITIES FOR InAs CALCULATED USING A
DIFFERENT REPORTED EXPERIMENTAL VALUE OF THE EFFECTIVE ELECTRON MASS

InAs[a]	$n_e = N_i$ (cm^{-3})	$1/\tau$ (sec^{-1})	μ (cm^2/V sec)
A	6.19×10^{15}	1.43×10^{12}	5.13×10^4
B	1.60×10^{16}	1.56×10^{12}	4.69×10^4
C	3.96×10^{16}	1.73×10^{12}	4.22×10^4
D	5.07×10^{16}	1.81×10^{12}	4.05×10^4
E	8.08×10^{16}	1.99×10^{12}	3.68×10^4
F	9.02×10^{16}	2.04×10^{12}	3.52×10^4
G	1.12×10^{17}	2.15×10^{12}	3.22×10^4
H	1.37×10^{17}	2.26×10^{12}	2.95×10^4
I	2.16×10^{17}	2.52×10^{12}	2.42×10^4
J	3.16×10^{17}	2.73×10^{12}	2.06×10^4

λ^{-1}/cm[b]	A	B	C	D	E	F	G	H	I	J
203	1.43	1.56	1.73	1.81	1.99	2.04	2.15	2.26	2.52	2.73
232	1.43	1.54	1.68	1.74	1.89	1.94	2.03	2.13	2.36	2.61
290	1.40	1.48	1.58	1.63	1.75	1.78	1.85	1.93	2.13	2.27
581	1.04	1.07	1.12	1.14	1.19	1.21	1.24	1.28	1.39	1.49

[a] At 300 K; $m_n/m = 0.024$; $\alpha_0 = 1/266$; $1/\tau_p = 2.86 \times 10^{12}$/sec; $e\tau_p/m_n = 2.59 \times 10^4$ cm^2/V·sec; $\mu_{polar} = 5.13 \times 10^4$ cm^2/V·sec; $1/\tau = 1.43 \times 10^{12}$/sec (polar scattering only, Sample A).
[b] $(1/\tau(w)) \times 10^{-12}$/sec

mental value for a given concentration. The two alternative theoretical calculations shown differ from the results obtained here in the manner previously discussed. Both give a flatter curve, and the higher mobilities at high carrier concentrations are an effect one would expect from inclusion of a screening parameter with $A_0 > 1$ and $A_1 = 1/2$ in Eq. (111).

Poth et al. (1978) have discussed the mobility in GaAs at 77 K and have noted that the experimental mobilities at high concentrations were below theoretically predicted values that use a screening parameter and were independent of a wide variety of conditions of material preparation. This cast doubt on the assumption that all such highly degenerate materials were electrically compensated, though it was noted that compensation might play a role at low concentrations where an increased scatter in mobility values is observed. In Fig. 8, the experimental data of Poth et al. are plotted along with the impurity scattering mobility as calculated from the analytical expression. As is shown, the analytical expression gives good agreement with the experimental data in the region where $k_0 T/E_f \ll 1$ and the approximation is valid, and also exhibits a fall below 10^3 cm^2/V sec at slightly higher

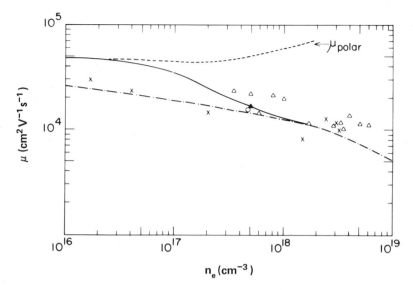

FIG. 5 Theoretical and experimental results for the mobility of uncompensated InAs at high concentrations, where the results of Rode and those obtained here coincide: — Theory $N_i = n_e$, — · — Theory (Rode, 1971), ▲, △, ○ Exp. (Rode, 1971), ✕✕✕ Exp. (Harman et al., 1956). The polar mobility μ_{polar}, which becomes increasingly unimportant at these concentrations, is also shown. The solid curve is calculated for $m_n = 0.024\ m$. Numerical results are given in Table VI.

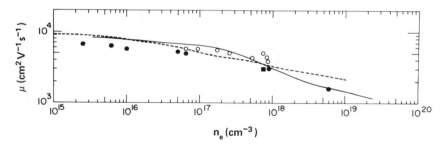

FIG. 6 Mobility versus carrier concentration for GaAs. Theoretical results calculated here (solid curve) are compared with theoretical and experimental results given in Dingle et al. (1978, Fig. 2). The solid square and circles are experimental results for bulk GaAs, and the open circles are the electron mobilities in the GaAs layers of modulation-doped GaAs–AlGaAs superlattices, in which the effect of impurity scattering is thought to be greatly reduced, Dingle et al. (1978). The dashed line is a theoretical result (Brooks-Herring) from Dingle et al. (1978). The experimental values at low concentrations lie below the highest reported experimental mobilities for $n_e < 10^{16}$ cm^{-3}, where polar scattering should determine the mobility in pure samples and may indicate the presence of compensation.

TABLE V

Comparison of Exact and Approximate Results for the Impurity Scattering Rate for Degenerate InP[a]

$a = E_\mathit{fl}/G$	n_e (cm^{-3})	$1/\tau_T$ (sec^{-1})	$1/\tau_p$ (sec^{-1})	$1/\tau^I$ (sec^{-1})	$1/\tau^I$† (sec^{-1})[b]	μ_T (cm^2/V · sec)
0.01	6.17×10^{17}	8.42×10^{12}	4.98×10^{12}	3.45×10^{12}	(3.30×10^{12})	2.83×10^3
0.03	1.23×10^{18}	1.03×10^{13}	4.64×10^{12}	5.69×10^{12}	(6.18×10^{12})	2.22×10^3
0.05	2.13×10^{18}	1.20×10^{13}	4.24×10^{12}	7.74×10^{12}	(7.46×10^{12})	1.85×10^3
0.09	(4.47×10^{18})	(1.29×10^{13})	(4.00×10^{12})		(8.89×10^{12})	(1.60×10^3)
0.13	(8.18×10^{18})	(1.38×10^{13})	(4.00×10^{12})		(9.75×10^{12})	(1.40×10^3)
0.24	(2.36×10^{19})	(1.51×10^{13})	(4.00×10^{12})		(1.11×10^{13})	(1.09×10^3)
0.45	(7.66×10^{19})	(1.55×10^{13})	(3.00×10^{12})		(1.25×10^{13})	(0.826×10^3)
0.50	(9.45×10^{19})	(1.57×10^{13})	(3.00×10^{12})		(1.27×10^{13})	(0.773×10^3)

[a] At 300 K. Results obtained by numerical integration compared with those calculated using the approximate analytical formula. Other quantities were obtained by numerical integration. Quantities in parentheses were calculated using the approximate analytical formulas.

[b] The values of $1/\tau^I$ and $1/\tau^I$† are the numerically integrated and analytical values of the impurity scattering rates, respectively. The polar scattering rate, calculated as discussed in Section III, is denoted by $1/\tau_p$, and the total scattering rate is $1/\tau_T$ where $1/\tau_T = 1/\tau^I + 1/\tau_p$. The total mobility, including the contributions of polar plus impurity scattering, is given by $\mu_T = e\tau_T/m^*$.

TABLE VI

Comparison of the Exact and Approximate Results for the Impurity
Scattering Rate for Degenerate InAs[a]

Sample	$a = E_f/G$	$1/\tau^I$ (sec^{-1})	$1/\tau^I\dagger$ (sec^{-1})	n_e (cm^{-3})
K	0.25	2.50×10^{12}	$2.39 \times 10^{12}\dagger$	0.7254×10^{18}
L	0.30	2.70×10^{12}	$3.04 \times 10^{12}\dagger$	0.9605×10^{18}
M	0.40	2.90×10^{12}	$3.27 \times 10^{12}\dagger$	1.5350×10^{18}
N	0.45	2.90×10^{12}	$3.37 \times 10^{12}\dagger$	1.8600×10^{18}
O	0.50	3.25×10^{12}	$3.46 \times 10^{12}\dagger$	2.3560×10^{18}
P	0.60	3.43×10^{12}	$3.60 \times 10^{12}\dagger$	3.4000×10^{18}

Sample	$1/\tau_p$ (sec^{-1})	$1/\tau_T$ (sec^{-1})	μ_T (cm^2/V sec)	$\mu^I\dagger$ (cm^2/V sec)
K	0.862×10^{12}	3.37×10^{12}	1.45×10^4	$1.69 \times 10^4\dagger$
L	0.776×10^{12}	3.47×10^{12}	1.32×10^4	$1.51 \times 10^4\dagger$
M	0.626×10^{12}	3.53×10^{12}	1.15×10^4	$1.24 \times 10^4\dagger$
N	0.551×10^{12}	3.45×10^{12}	1.12×10^4	$1.15 \times 10^4\dagger$
O	0.536×10^{12}	3.79×10^{12}	0.965×10^4	$1.06 \times 10^4\dagger$
P	0.471×10^{12}	3.90×10^{12}	0.855×10^4	$0.925 \times 10^4\dagger$

[a] At 300 K. Comparison of results obtained by numerical integration with those calculated using the approximate analytical formula. Quantities that are crossed (†) are obtained by use of the approximate analytical expression. Agreement between the two methods of calculation is expected to improve with decreasing temperatures and increasing carrier concentrations. Mobilities are calculated using $m^* = (1 + 2a)m_n$, the effective mass at the Fermi surface, and uncompensated samples with $N_i = n_e$ are considered. The impurity scattering mobility is $\mu^I = e\tau^I/m^*$, and the total mobility is $\mu_T = e\tau_T/m^*$ where $1/\tau_T = 1/\tau_p + 1/\tau^I$.

concentrations. The impurity scattering mobility is also plotted at 300 K and at 4 K to show the reasonable temperature dependence obtained from Eq. (106). For comparison, the total numerically integrated mobility due to polar plus impurity scattering is also shown at 300 K.

In Tables V, VI, and VIII the results for the exact numerically integrated scattering rate and the approximate analytical expression are further compared for InP, GaAs, and InAs. As is shown, the polar scattering rate decreases at high concentrations and the mobility is mainly determined by impurity scattering, which can be readily estimated with Eq. (106).

In Table IX, the mobility of the ternary compound $Ga_{0.47}In_{0.53}As$ is estimated by taking a linear combination of the scattering rates for InAs and GaAs for a given $N_i = n_e$ according to

$$1/\tau_{Ga_xIn_{1-x}As} = x(1/\tau_{GaAs}) + (1 - x)(1/\tau_{InAs}) \qquad (125)$$

and calculating the mobility

$$\mu_{Ga_xIn_{1-x}As} = e\tau_{Ga_xIn_{1-x}As}/m_n \qquad (126)$$

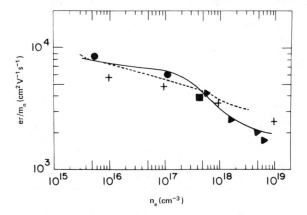

FIG. 7 Theoretical results (solid curve) for $e\tau/m_n$, $m_n = 0.071m$, for GaAs, are compared with the maximum reported experimental mobilities for a given carrier concentration, indicated by the solid symbols ($\bullet$, $\blacksquare$, $\blacktriangledown$), and with alternate theoretical calculations (---) and (+++). Details are given in Jensen, 1979. The alternate theoretical curves, which use a screening parameter, are both flatter than the curve obtained here; and both give a higher mobility at high carrier concentrations. The shape of the solid curve is determined by τ rather than by the effective mass correction for degenerate samples, which is taken into account in Fig. 6 and which improves the agreement with experimental results at high concentrations.

TABLE VII

MOBILITIES CALCULATED FOR InSb USING THE APPROXIMATE
ANALYTICAL FORMULA[a]

a	$g^I/10^{-11}$ (eV^2 cm)	$n_e = N_i$ (cm^{-3})	$1/\tau_0^I \times 10^{-12}$ (sec^{-1})
0.06	1.293	1.369×10^{16}	.0811
0.1	2.941	3.113×10^{16}	.1844
0.2	9.475	1.003×10^{17}	.5941
0.3	19.64	2.079×10^{17}	1.231
0.4	33.79	3.577×10^{17}	2.119
0.5	52.36	5.543×10^{17}	3.285

a	$1/\tau_b^I$ (sec^{-1})	$e\tau_b^I/m_n$ (cm^2/V sec)	$e\tau_b^I/m_n\gamma(a)$ (cm^2/V sec)
0.06	0.7894×10^{12}	1.538×10^5	1.374×10^5
0.1	0.9334×10^{12}	1.301×10^5	1.084×10^5
0.2	1.119×10^{12}	1.085×10^5	7.750×10^4
0.3	1.223×10^{12}	9.930×10^4	6.206×10^4
0.4	1.297×10^{12}	9.363×10^4	5.202×10^4
0.5	1.355×10^{12}	8.962×10^4	4.481×10^4

[a] At 77 K; $m_n/m = 0.0145$. Mobilities calculated for InSb using Eq. (106) and the experimental parameters in Table 1.

TABLE VIII

Temperature Dependence of the Mobility for Degenerate GaAs

$a = E_f/G$	(at 77 K) $n_e \times 10^{-18}$ cm^{-3}	(sec^{-1}) $1/\tau^I$ (300 K)	(sec^{-1}) $1/\tau^I$ (77 K)	(sec^{-1}) $1/\tau^I$ (4 K)
0.01	.536	—	9.19×10^{12}	1.93×10^{13}
0.05	1.98	1.02×10^{13}	1.44×10^{13}	2.45×10^{13}
0.07	3.10	1.14×10^{13}	1.56×10^{13}	2.56×10^{13}
0.09	4.45	1.21×10^{13}	1.63×10^{13}	2.64×10^{13}
0.13	8.01	1.32×10^{13}	1.73×10^{13}	2.74×10^{13}
0.24	2.25×10	1.51×10^{13}	1.91×10^{13}	2.92×10^{13}
0.45	7.25×10	1.69×10^{13}	2.08×10^{13}	3.09×10^{13}
0.50	8.92×10	1.72×10^{13}	2.11×10^{13}	3.12×10^{13}

$a = E_f/G$	μ_T (300 K)	μ^I (300 K)	μ^I (77 K)	μ^I (4 K)
0.01	3900	—	2680	1290
0.05	2200	2210	1580	932
0.07	1900	1910	1410	859
0.09	1800	1740	1310	809
0.13	1500	1480	1150	728
0.24	1200	1110	890	582
0.50	780	719	595	403

[a] The temperature dependence of the mobility in highly doped but uncompensated GaAs. All mobilities are in cm^2/V · sec. All quantities with superscript I are impurity scattering rates and mobilities that are calculated using the approximate analytical expression for the impurity scattering rate given in Eqs. (111) and (113). The total mobility, which includes polar and impurity scattering and is calculated using numerical integration, is given at $T = 300$ K for comparison. The analytical expression is a better approximation at lower temperatures, as it requires $K_0 T/E_f \ll 1$. All mobilities are calculated using the effective mass at the Fermi level as given by the Kane theory, and it is assumed that $n_e V_c =$ constant.

TABLE IX

An Estimate of the Electron Scattering Rate and Mobility for the Ternary Compound $Ga_{0.47}In_{0.53}As$[a]

$n_e = N_i$ (cm^{-3})	$1/\tau$ (sec^{-1})	μ (cm^2/V sec)
7×10^{15}	2.30×10^{12}	1.86×10^4
1×10^{16}	2.35×10^{12}	1.82×10^4
2×10^{16}	2.40×10^{12}	1.78×10^4
3×10^{16}	2.50×10^{12}	1.71×10^4
5×10^{16}	2.61×10^{12}	1.64×10^4
6×10^{16}	2.7×10^{12}	1.59×10^4
1.25×10^{17}	2.81×10^{12}	1.52×10^4
3.00×10^{17}	3.68×10^{12}	1.16×10^4

[a] Scattering rates and mobilities calculated with effective mass $m_n/m = 0.041$ (Pearsall et al., 1978).

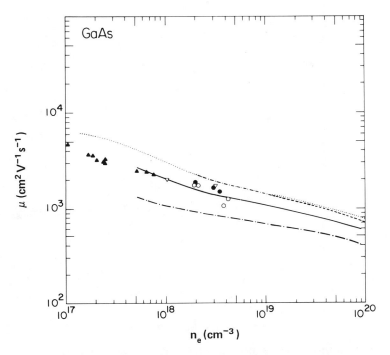

FIG. 8   Theoretical results for the impurity scattering mobility for GaAs, as calculated from the analytical expression for three temperatures, are compared with the experimental results of Poth *et al.*(1978), at 77 K and with the result for the total mobility as calculated by numerical integration at 300 K. The analytical expression alone gives the order of magnitude of the experimental results at 77 K. Numerical results are given in Table VIII. 300 K total, $\cdots$; 300 K total, $\cdots$; 300 K impurity, ---; 77 K impurity, solid line; 77 K experimental, $\bigcirc \bullet \triangledown \blacktriangle$; 4 K impurity, $- \cdot -$.

using the experimentally determined value of $m_n = 0.041\,m$ (Pearsall *et al.*, 1978). Results are shown in Fig. 9 and are compared with experimental data for low concentrations for which the effective mass correction is expected to be small. The agreement is reasonable considering the simple approximation and suggests that results obtained here may possibly be extended to ternary compounds for which alloy scattering is not important.

In summary, the polar optical mode and impurity scattering rates that at low frequencies give the quasi-classical Drude limit of the quantum theory have been derived and calculated for a variety of compounds. When used in the Drude–Zener theory at high frequencies, these rates give the necessary quantum-mechanical extension.

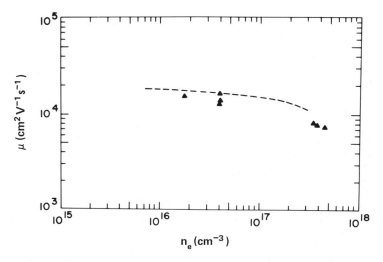

FIG. 9   The mobility versus carrier concentration for $Ga_{0.47}In_{0.53}As$. Experimental results are taken from Pearsall *et al.*(1978) as is the effective mass of $m_n = 0.041$ m, which is assumed constant. Experiment, ▲; theory, ---.

Appendix A.   Theory

The derivation of the high-frequency optical conductivity from the equation of motion of the quantum density matrix, which is discussed in detail in Jensen, (1975) will be summarized here.

We consider a system of dynamically independent electrons in interaction with a scattering system of optical or acoustical mode phonons and an equilibrium radiation field. A perturbing radiation field is applied. The total system is then described by a quantum density matrix R with the equation of motion (Argyres, 1961; Jensen, 1975; Kohn and Luttinger, 1957; Price, 1966).

$$R = R^0 + p, \tag{A1}$$

$$i\hbar\dot{p} = [H, R], \tag{A2}$$

where R^0 is the equilibrium density matrix and p is the deviation from equilibrium. The total Hamiltonian of the system H is given by

$$H = H^0 + H' + H^{er}, \tag{A3}$$

where H^0 is the sum of the Hamiltonians of the noninteracting subsystems consisting of electrons, phonons, and photons. The scattering interaction of the electrons with the phonon field is denoted by H', and the interaction of the electrons with the quantum-radiation field by

$$H^{er} = -(e/mc)(\mathbf{A} \cdot \mathbf{P}), \tag{A4}$$

where $\mathbf{A}$ is the vector potential of the radiation field and $\mathbf{P}$ is the momentum of the electron system. The Coulomb gauge is used, which implies that $\mathbf{A}$ and $\mathbf{E}$ are transverse where

$$\mathbf{E} = -\frac{1}{c}\dot{\mathbf{A}}. \tag{A5}$$

The form of H' depends on the scattering mechanism. Various cases are considered in Jensen, 1973.

We work in the number representation and take matrix elements of p between eigenstates $|N)$ and $|N')$ of H^0. We define

$$\begin{align}
(N'|p|N) &= p_{n'v'\gamma,nv\gamma} = p_{N'N}, \\
(N|p|N) &= p_N, \tag{A6}
\end{align}$$

where

$$|N) = |n)|v)|\gamma), \qquad H^0|N) = E_N|N), \tag{A7}$$
$$E_N = E_n + E_v + E_\gamma \qquad E_{N'N} = E_{N'} - E_N,$$

and $|n)$, $|v)$, and $|\gamma)$ are the wave functions of the subsystems consisting of electrons, phonons, and photons, respectively, and E_n, E_v, and E_γ are the corresponding energies.

The following generalized Boltzmann equation can be derived for the Fourier component $p_{nv}^{\pm q} = (nv|p^{\pm q}|nv)$ (Jensen, 1975):

$$i(\mp w - is)\, p_{nv}^{\pm q} - (i/\hbar)C_{nv}^{\pm q} = \sum_{n'v'} \Delta_{n'v',nv}^{\pm w}S_{nv,n'v'}(p_{n'v'}^{\pm q} - p_{nv}^{\pm q}), \tag{A8}$$

where $q = nw/c$ is the photon wave vector and

$$S_{nv,n'v'} = (2\pi/\hbar)H'_{nv,n'v'}H'_{n'v',nv}, \tag{A9}$$

$$\Delta_{n'v',nv}^{\pm w} = \frac{1}{2\pi i}\left[\frac{1}{E_{n'v',nv} \mp \hbar w - i\hbar s} + \frac{1}{E_{nv,n'v'} \mp \hbar w - i\hbar s}\right], \tag{A10}$$

$$E_{n'v',nv} = E_{n'} + E_{v'} - E_n - E_v. \tag{A11}$$

In Eq. (A10) one now uses the relation

$$1/(y - is) = P(1/y) + i\pi\delta(y), \tag{A12}$$

where the principal part $P(1/y)$ does not contribute to the Boltzmann equation, and the delta function ensures energy conservation among the subsystems. The operator C is defined by

$$C = [R_d, H^{er}], \tag{A13}$$

where R_d is the diagonal part of R, and H^{er} is associated with the driving force $e\mathbf{E}$. Equation (A8) is analogous to Eq. (26) of Price, (1966) which is shown to reduce to the quasi-classical Boltzmann transport equation in the limit of small w and q. When the right-hand side of Eq. (A8) is small compared with the terms on the left, which corresponds to the limit $w\tau \gg 1$, where τ is the electron scattering time, one can expand Eq. (A8) as a power series in $1/w$. One obtains to order $1/w^2$, letting $s \rightarrow 0$,

$$p_{nv}^{(2)\pm q} = \frac{i\hbar}{(\hbar w)^2} \sum_{n'v'} S_{nv,n'v'} \Delta_{n'v',nv}^{\pm}(C_{nv}^{\pm} - C_{n'v'}^{\pm}). \tag{A14}$$

Assuming statistical independence of the subsystems, which implies

$$(N|R|N) = R_n R_v R_\gamma, \tag{A15}$$

where

$$\begin{aligned} R_n &= (n|R(n)|n), \\ R_v &= (v|R(v)|v), \\ R_\gamma &= (\gamma|R(\gamma)|\gamma), \end{aligned} \tag{A16}$$

one has, on multiplying Eq. (A14) through by

$$n_k = (N|\hat{N}_k|N) = (n|\hat{N}_k|n), \tag{A17}$$

where $\hat{N}_k$ is the number operator for an electron in the state denoted by wave vector k, and taking the trace over $|n)|v)$, the following result:

$$\sum_{nv} n_k p_{nv}^{(2)} = R(\gamma)g_k^q. \tag{A18}$$

The quantity g_k^q is the deviation from equilibrium of the Fourier component of the electron distribution function. The electron current $\mathbf{j}^q$ is given by the expression

$$\mathbf{J}^q = \sum_k e\mathbf{v}_k g_k^q. \tag{A19}$$

On inserting (A14) into (A18) to obtain g_k^q, and subsequently using g_k^q in (A19) to calculate $\mathbf{J}^q$, one finds

$$\mathbf{J}^q = \sigma(-w):\mathbf{E}(-w) + \sigma(w):\mathbf{E}(w). \tag{A20}$$

The conductivities $\sigma(\mp w)$, which are tensor quantities, are given by Jensen (1975) as

$$
\left.\begin{array}{c}\sigma(-w)\\\sigma(w)\end{array}\right\} = (e^2/w^2 V) \sum_{\substack{kk'\\Q=|k'-k|}} (S_{kk'}/\hbar w) f_{k'}^0 (1 - f_k^0)(e^{\beta\hbar w} - 1)
$$

$$
\times [(n_Q^0 + 1)\delta(e_{k'k} - \hbar w^+) + n_Q^0 \delta(e_{k'k} - \hbar w^-)]
$$

$$
\times (\mathbf{v}_k - \mathbf{v}_{k'}) \begin{cases} v_k \\ -v_{k'} \end{cases} \tag{A21}
$$

where

$$
S_{kk'} = (2\pi/\hbar)|V_{k'k}|^2 \rightarrow \tfrac{1}{2} \sum_{s_i s_f} (2\pi/\hbar)|V_{k'k}|^2 \tag{A22}
$$

and $V_{k'k}$ is essentially the Fourier transform of the scattering interaction $H'(\mathbf{r})$ between electronic states characterized by wave vectors k' and k. The average over initial and sum over final electron spin states in (A22) is made in the case of spin degeneracy. V is the crystal volume.

In (A21), n_Q^0 is the equilibrium occupation number at temperature $T = 1/k_0\beta$ for phonons of wave vector Q, f_k^0 is the equilibrium electron Fermi distribution function, $\mathbf{v}_k$ is the velocity of an electron in the Bloch state characterized by wave vector k and energy e_k. The energies of the photon and phonon are denoted by $\hbar w$ and $\hbar w_Q$, respectively, and

$$
e_{k'k} = e_{k'} - e_k, \tag{A23}
$$

$$
\hbar w^\pm = \hbar w \pm \hbar w_Q. \tag{A24}
$$

The expressions for $\sigma(\mp w)$ in Eq. (A21) are tensor quantities. To obtain a scalar optical conductivity σ for comparison with the scalar absorption coefficient α, one takes the trace over $|\gamma\rangle$ of $\mathbf{J} \cdot \mathbf{E}$ as follows.

$$
\tfrac{1}{2}\langle \mathbf{J} \cdot \mathbf{E} + \mathbf{E} \cdot \mathbf{J} \rangle_q = \tfrac{1}{2} \sum_\gamma R_\gamma \langle \gamma | \mathbf{J} \cdot \mathbf{E} + \mathbf{E} \cdot \mathbf{J} | \gamma \rangle = \sigma \langle E^2 \rangle_q, \tag{A25}
$$

where

$$
\langle E^2 \rangle_q = \text{tr}\{R(\gamma)(\mathbf{E} \cdot \mathbf{E})\}. \tag{A26}
$$

The coefficient of $\langle E^2 \rangle_q$ is identified as the generalized optical conductivity. On summing over photon polarization vectors for an unpolarized $\mathbf{E}$ and averaging over the direction of the photon wave vector $\mathbf{q}$ (or equivalently, for a cubic crystal, over the direction of $\mathbf{v}_k - \mathbf{v}_{k'}$), it may be written

$$\sigma = \frac{e^2}{3w^2V} \sum_{\substack{kk' \\ Q=|k'-k|}} \frac{S_{kk'}}{\hbar w} |\mathbf{v}_k - \mathbf{v}_{k'}|^2 f_{k'}^0 (1 - f_k^0)(e^{\beta \hbar w} - 1)$$

$$\times [(n_q^0 + 1)\delta(e_{k'k} - \hbar w^+) + n_Q^0 \delta(e_{k'k} - \hbar w^-)]$$

$$= (c/4\pi)n\alpha. \tag{A27}$$

On solving Eq. (A27) for $n\alpha$, where n is the real part of the refractive index and α is the absorption coefficient, one obtains the exact quantum-mechanical expression for $n\alpha$ which has been calculated using the methods of time-dependent perturbation theory (Jensen, 1973).

Using the principle of detailed balance and the properties of the delta function, it can be shown that

$$(n_Q^0 + 1)f_{k'}^0 (1 - f_k^0)(e^{\beta \hbar w} - 1)\delta(e_{k'k} - \hbar w^+)$$
$$= F_+(w, w_Q)(f_k^0 - f_{k'}^0)\delta(e_{k'k} - \hbar w^+), \tag{A28}$$

$$n_Q^0 f_{k'}^0 (1 - f_k^0)(e^{\beta \hbar w} - 1)\delta(e_{k'k} - \hbar w^-)$$
$$= F_-(w, w_Q)(f_k^0 - f_{k'}^0)\delta(e_{k'k} - \hbar w^-), \tag{A29}$$

$$(n_Q^0 + {}_1^0)f_{k'}^0 (1 - f_k^0)(e^{\beta \hbar w} - 1)\delta(e_{k'k} - \hbar w^{\pm})$$
$$= (n_Q^0 + {}_1^0)f_k^0(1 - f_{k'}^0)(1 - e^{-\beta \hbar w})\delta(e_{k'k} - \hbar w^{\pm}), \tag{A30}$$

where

$$F_{\pm}(w, w_Q) = \frac{\sinh(\hbar w/2k_0T)}{2 \sinh(\hbar w_Q/2k_0T) \sinh(\hbar w^{\pm}/2k_0T)}. \tag{A31}$$

It is instructive to write the expressions in Eq. (A31) as follows:

$$F_+(w, w_Q) = \frac{(1 - e^{-\beta \hbar w})}{(e^{\beta \hbar w_Q} - 1)(1 - e^{-\beta \hbar w}e^{-\beta \hbar w_Q})}$$
$$= n_Q^0(1 - e^{-\beta \hbar w})/(1 - e^{-\beta \hbar w^+}) \to n_q^0 \to 0, \qquad T \to 0, \tag{A32}$$

$$F_-(w, w_Q) = (1 - e^{-\beta \hbar w})/(1 - e^{-\beta \hbar w_Q})(1 - e^{-\beta \hbar w}e^{\beta \hbar w_Q})$$
$$= (n_Q^0 + 1)(1 - e^{-\beta \hbar w})/(1 - e^{-\beta \hbar w}e^{\beta \hbar w_Q})$$
$$\to (n_Q^0 + 1) \to 1, \qquad T \to 0, \hbar w > \hbar w_Q. \tag{A33}$$

Thus although the contribution to the optical conductivity from photon absorption accompanied by phonon absorption is surpressed at low temperatures, photon absorption accompanied by phonon emission can still occur if it is not precluded by "freezing out" of the free carriers. It can occur as long as the photon energy is greater than the phonon energy and the transition is energetically permissible. For this contribution, the temperature dependent-factor reduces to a constant. The functions $F_{\pm}(w, w_Q)$ also reduce to $(n_Q^0 + {}^0_1)$ when $\hbar w \gg k_0 T, \hbar w > \hbar w_Q$ and when $k_0 T \gg \hbar w > \hbar w_Q$, as seen from Eqs. (A32) and (A33).

Therefore one can write, from Eq. (A27),

$$n\alpha = \frac{4\pi\sigma}{c} = \frac{4\pi e^2}{w^2 c} \sum_{\substack{k'k \\ Q=|k'-k|}} \frac{S_{kk'}}{V\hbar w}$$

$$\times \frac{1}{3} |v_k - v_{k'}|^2 (f_k^0 - f_{k'}^0) \cdot [F_+(w, w_Q)\delta(e_{k'k} - \hbar w^+)$$

$$+ F_-(w, w_Q)\delta(e_{k'k} - \hbar w^-)], \tag{A34}$$

which gives a convenient form for numerical calculations and also exhibits the equivalence with the quantum result (Jensen, 1973).

Appendix B. Summary of Theoretical Results

The derivation of the high-frequency optical conductivity from the generalized Boltzmann equation, summarized in Appendix A, gives the same result as the quantum-mechanical treatment that uses the methods of time-dependent perturbation theory (Jensen, 1973). The expression obtained is valid over the entire frequency range below the fundamental absorption edge. For elastic-scattering mechanisms, such as impurity or deformation potential scattering, the generalized optical conductivity given in Eq. (A27) reduces to the classical Drude result in the limit $\hbar w \ll k_0 T, E_f$ for $w\tau_{\pm} \gg 1$ and to the high-frequency quantum result when $\hbar w \gg k_0 T, E_f$ and $w\tau_{\pm} \gg 1$. In the intermediate range $\hbar w \sim k_0 T, E_f$, the conductivity can be expressed in terms of a generalized relaxation time that is related to the cross section per electron and becomes frequency dependent at high photon energies.

Polar optical mode scattering and impurity scattering are expected to be the most important scattering mechanisms in polar materials such as GaAs, InP, InSb, ZnSe, and CdTe. For polar scattering, which is not elastic, one does not obtain the classical Drude expression in the low-frequency limit unless the phonon energy is small compared with $k_0 T$. This condition does not hold at room temperature, and one finds, instead, the optical conductivity and generalized relaxation time given in Eqs. (52) and (53), respectively. As shown in Sections IV and V, they can be written in the following form, where the subscripts $+$ and $-$ correspond to photon absorption accompanied by phonon absorption and phonon emission, respectively, and where the total optical conductivity due to polar scattering is the sum of the two terms:

$$\sigma_{\pm} = (w_p^2/4\pi w^2)(1/\tau_{\pm}), \qquad w\tau_{\pm} \gg 1, \tag{B1}$$

$$1/\tau_{\pm} = \langle 1/\tau_{\pm} \rangle = (1/\tau_p)(F_{\pm}(w, w_Q)/X)(I_{\pm}/12n_eV_c). \tag{B2}$$

The conductivity and generalized relaxation time are defined in terms of experimental parameters for a given compound

$$w_p^2 = 4\pi n_e e^2/m_n, \qquad X = \hbar w/G, \tag{B3}$$

where G is the band gap energy and m_n the effective mass at $k = 0$. $F_{\pm}(w, w_Q)$ is a function of temperature and frequency, given in Eq. (A31), and $I_{\pm}/n_eV_c$ is a dimensionless integral that is frequency independent at low w and becomes independent of the carrier concentration n_e and a function only of frequency at high photon energies.

The constants $1/V_c$ and $1/\tau_p$ are given by

$$1/V_c = 2/\pi^2 \lambda_c^3, \qquad \lambda_c = \hbar/m_n\alpha_0 c, \qquad \alpha_0 = (G/2m_nc^2)^{1/2},$$
$$1/\tau_p = g/\pi\hbar^2\alpha_0 c \quad (\text{sec}^{-1}), \tag{B4}$$
$$g = 2\pi e^2\hbar w_Q(e_\infty^{-1} - e_0^{-1})eV^2 \quad \text{cm},$$

where $\hbar w_Q$ is the energy of the longitudinal optical phonon and e_0 and e_∞ are the static and high-frequency lattice dielectric constants, respectively. The generalized relaxation time in Eq. (B2) is related to the cross section per electron $\Sigma_{\pm}$, where n is the real part of the refractive index, as

$$n\Sigma_{\pm} = (w_p^2/n_ew^2)(1/c\tau_{\pm}) = n\Sigma_0(F_{\pm}(w, w_Q)/X^3)(I_{\pm}/n_eV_c), \tag{B5}$$

$$n\Sigma_0 = (1/12n_e)(w_p^2/w_g^2)(1/c\tau_p) = (e^2/4\pi\hbar c)\phi_p, \tag{B6}$$

$$\phi_p = (8\pi/3)R_p^2, \qquad w_g = G/\hbar, \qquad R_p^2 = \hbar\alpha_0 cg/G^3. \tag{B7}$$

For the case of impurity scattering, the optical conductivity assumes the form

$$\sigma^I = (1/4\pi)(w_p^2/w^2)(1/\tau^I), \qquad w\tau^I \gg 1, \tag{B8}$$

where $1/\tau^I$ is the generalized relaxation rate appropriate to impurity scattering

$$1/\tau^I = \langle 1/\tau^I \rangle = (1/\tau_0^I)(I_+^I/6n_eV_cX). \tag{B9}$$

The constant $1/\tau_0^I$ is given in terms of experimental parameters by

$$1/\tau_0^I = g^I/\pi\hbar^2\alpha_0 c, \qquad g^I = (8N_iV_c)(e^2/e_0)^2/\lambda_c, \qquad (B10)$$

where N_i is the number of charged impurities per unit volume.

The conductivity for impurity scattering in Eq. (B8) reduces to the classical Drude expression in the limit that the photon energy becomes small compared with the electron energy, which is of the order E_f, or k_0T, for degenerate or nondegenerate materials, respectively. In the high-energy limit, when $\hbar w \gg E_f$ or k_0T, the generalized relaxation time τ^I becomes independent of the electron concentration and energy and is a function of frequency only.

The relaxation rate $1/\tau^I$ can be related to the scattering cross section per electron Σ^I for impurity scattering as follows:

$$n\Sigma^I = (w_p^2/n_e w^2 c)\,1/\tau^I = (n\Sigma_0^I/X^3)(I_+^I/2n_eV_c), \qquad (B11)$$

$$n\Sigma_0^I = (e^2/4\pi\hbar c)\phi_I = (1/12n_e)(w_p^2/w_g^2 c)\,1/\tau_0^I, \qquad (B12)$$

$$\phi_I = (8\pi/3)R_I^2, \qquad R_I^2 = (\hbar\alpha_0 cg^I/G^3),$$
$$w\tau^I \gg 1. \qquad (B13)$$

The total conductivity σ, which is the sum of the conductivities due to polar optical mode and impurity scattering, is proportional to the total effective relaxation rate $1/\tau$ and the total absorption coefficient α where

$$4\pi\sigma = 4\pi(\sigma_+ + \sigma_- + \sigma^I) = (w_p^2/w^2)(1/\tau) = cn\alpha, \quad w\tau \gg 1, \quad (B14)$$

$$1/\tau = 1/\tau_+ + 1/\tau_- + 1/\tau^I. \qquad (B15)$$

Equation (B14) gives the high-frequency limit in which $w\tau \gg 1$. The transition to the low-frequency regime, in which $w\tau \lesssim 1$, is given, as usual, by

$$\sigma = \sigma_0/w^2\tau^2\eta = \sigma_0/(1 + w^2\tau^2) \rightarrow \begin{cases} \sigma_0, & w\tau \ll 1, \\ \sigma_0/w^2\tau^2, & w\tau \gg 1, \end{cases} \quad (B16)$$

where

$$\eta = 1 + (w\tau)^{-2}, \qquad (B17)$$

$$\sigma_0 = n_e e^2\tau/m_n, \qquad k_0T \gg E_f, \qquad (B18)$$

$$\sigma_0 = n_e e^2\tau/m^*, \qquad k_0T \ll E_f, \qquad (B19)$$

and $m^* = m_n(1 + 2E_f/G)$ is the effective mass at the Fermi level according to the Kane theory of the band structure (Kane, 1957).

Appendix C.   Angle-Averaged Matrix Element for
Polar Scattering

The overlap wave function factor $G(k, k'_\pm, y)$ of the Kane theory, given in Eq. (18), is discussed in Haga and Kimura (1963) and in Jensen (1973). A short summary of important results used in this paper follows.

The function has the general form

$$G(k, k'_\pm, y) = A_\pm + B_\pm y + C_\pm y^2, \tag{C1}$$

where the coefficients $A_\pm$, $B_\pm$, and $C_\pm$ are functions of the dimensionless variables u, $u_\pm$ defined in Eq. (35). In the small spin orbit splitting limit one has

$$\begin{aligned}
A_\pm &= (1 + u)(1 + u_\pm)/\gamma\gamma_\pm, \\
B_\pm &= \beta\beta_\pm/2, \\
C_\pm &= uu_\pm/\gamma\gamma_\pm,
\end{aligned} \tag{C2}$$

where β and γ are defined in Eq. (49). Using the above expression for $G(k, k'_\pm, y)$ in Eq. (33) and subsequently in Eq. (36), one obtains the following expression for the dimensionless quantity $d(u, u_\pm)$:

$$d(u, u_\pm) = \tfrac{1}{2} \int dy\, D(u, u_\pm, y) = Q_1(u, u_\pm) + Q_2(u, u_\pm), \tag{C3}$$

where

$$Q_1(u, u_\pm) = (1/4\gamma_\pm^2 W_\pm)\, 2d_0^\mp \ln(1 + W_\pm)/|1 - W_\pm|, \tag{C4}$$

and

$$\begin{aligned}
Q_2(u, u_\pm) = (1/4\gamma_\pm^2 W_\pm)[d_1^\mp \Delta^\pm(2) + (d_2^\mp/2)\Delta^\pm(4) \\
+ (d_3^\mp/3)\Delta^\pm(6)].
\end{aligned} \tag{C5}$$

In Eqs. (C4) and (C5), the quantity $W_\pm$ is given by

$$W_\pm = k/k'_\pm = \beta\gamma/\beta_\pm\gamma_\pm = [u(1 + u)/u_\pm(1 + u_\pm)]^{1/2}. \tag{C6}$$

The function $\Delta^\pm(N)$ is defined as follows, where N is an integer:

$$\Delta^\pm(N) = (1 + W_\pm)^N - (1 - W_\pm)^N, \qquad N = 2, 4, 6. \tag{C7}$$

Also

$$\begin{aligned}
d_0^\mp &= J_1^\mp J_2^\mp, \\
d_1^\mp &= J_2^\mp J_4^\mp - J_1^\mp J_3^\mp, \\
d_2^\mp &= J_1^\mp J_5^\mp - J_3^\mp J_4^\mp, \\
d_3^\mp &= J_4^\mp J_5^\mp,
\end{aligned} \tag{C8}$$

where

$$J_1^\pm = m_\pm - (r_\pm/s_\pm)n_\pm,$$
$$J_2^\pm = A_\pm + (r_\pm/s_\pm)B_\pm + (r_\pm/s_\pm)^2 C_\pm,$$
$$J_3^\pm = (1/s_\pm)B_\pm + 2(r_\pm/s_\pm^2)C_\pm, \tag{C9}$$
$$J_4^\pm = (n_\pm/s_\pm),$$
$$J_5^\pm = C_\pm/s_\pm^2,$$

and

$$m_\pm = 1 + (v_k/v_{k'_\pm})^2 = 1 + (\beta/\beta_\pm)^2,$$
$$n_\pm = 2(v_k/v_{k'_\pm}) = 2(\beta/\beta_\pm),$$
$$r_\pm = 1 + (k/k'_\pm)^2 = 1 + W_\pm^2, \tag{C10}$$
$$s_\pm = 2W_\pm.$$

Hence, recalling

$$\beta = 2(u(1+u))^{1/2}/(1+2u), \qquad \gamma = 1 + 2u, \tag{C11}$$

$$\beta_\pm = 2(u_\pm(1+u_\pm))^{1/2}/(1+2u_\pm), \qquad \gamma_\pm = 1 + 2u_\pm,$$
$$u_\pm = u + X^\pm, \qquad u = e_k/G, \qquad X^\pm = (\hbar w \pm \hbar w_Q)/G, \tag{C12}$$

it is seen that the angle-averaged quantity $d(u, u_+)$ in the relaxation time for polar scattering in Eq. (56) is given in terms of the band structure as a function of frequency and electron energy $e_k = Gu$. The integration over u, the dimensionless energy variable, can now be performed numerically to obtain the effective electron relaxation time and hence the generalized optical conductivity in Eq. (62).

Appendix D. Angle Averaged Matrix Element for Impurity Scattering

The factor $d^I(u, u_+)$ in the generalized relaxation time for impurity scattering in Eq. (81) is given as

$$d^I(u, u_+) = \frac{1}{W_+\beta_+^2\gamma_+^4}\left[-d_0^\pm\Delta^+(-2)\right.$$
$$\left. + 2d_1^\pm \ln\frac{(1+W_+)}{|1-W_+|} + d_2^\pm\Delta^+(2) + \frac{d_3^\pm}{2}\Delta^+(4)\right]. \tag{D1}$$

The notation is the same as in Appendix C with the modification that in Eq. (D1), one has for any function of u and u_+,

$$d_i^\pm(u, u_+) = d_i(u, u_+)|_{u_+=u+X}, \qquad i = 0, 1, 2, 3, \tag{D2}$$

because impurity scattering is elastic and hence $\hbar w^\pm \to \hbar w$.

Comparison with the cross section for impurity scattering discussed in Appendix D in Jensen (1973) can be facilitated by writing Eq. (D1) in the form

$$d^I(u, u_+) = \frac{2}{\beta_+^2 \gamma_+^4} \left[\frac{d_1^+}{s_+} \ln \frac{(r_+ + s_+)}{|r_+ - s_+|} + 2 \left(\frac{d_0^+}{r_+^2 - s_+^2} + d_2^+ + r_+ d_3^+ \right) \right]$$
$$= 2Q^I(u, u_+)/\beta_+^2 \gamma_+^4, \tag{D3}$$

where $Q^I(u, u_+)$ is defined by the quantity in square brackets in Eq. (D3). The cross section for an electron of energy $e_k = Gu$ is given by

$$n\Sigma^I(u, u_+) = \frac{n\Sigma_0^l}{X^3} h(u_+) \, d^I(u, u_+) = \frac{n\Sigma_0^l}{X^3} \frac{2Q^I(u, u_+)}{h(u_+)}. \tag{D4}$$

This is the result given in Eq. D-2 Appendix D of Jensen (1973) where the sum over initial spin states has been taken. For the case of degeneracy, the effective electrons that participate in the absorption are those in the Fermi surface. For $k_0 T \le \hbar w \ll E_f$ the absorption coefficient is given, using Eqs. (85), (37), and (38), by

$$\alpha = \tfrac{1}{2}(1/V_c)\int h(u) \, du \, \Sigma^I(u, u_+)[f^0(u) - f^0(u_+)]$$
$$\cong \tfrac{1}{2}(X/V_c)\int h(u) \, du \, \Sigma^I(u, u_+)\delta(u - a)$$
$$= \tfrac{1}{2}(Xh(a)/V_c)\Sigma^I(a, a_+), \tag{D5}$$

where $(Xh(a)/V_c)$ is the effective electron density. Using Eq. (D4) one may write

$$n\alpha = 4\pi\sigma/c = [Xh(a)/V_c](n\Sigma_0^l/X^3)[Q^I(a, a_+)/h(a_+)]$$
$$\cong (n\Sigma_0^l/X^3)[XQ^I(a, a_+)/V_c]$$
$$= (w_p^2/w^2 c)(Q^I(a, a_+)/12n_e V_c \tau_0^l), \tag{D6}$$

where Eq. (93) has been used to eliminate $n\Sigma_0^l/X^3$. It is seen from Eq. (D6) that the generalized expression for the impurity scattering rate is given by

$$1/\tau^I(a, a_+) = (1/12n_e V_c \tau_0^l)Q^I(a, a_+),$$

$$Q^I(a, a_+) = \left[\frac{d_1^+}{s_+} \ln \frac{r_+ + s_+}{|r_+ - s_+|} + 2 \left(\frac{d_0^+}{r_+^2 - s_+^2} + d_2^+ + r_+ d_3^+ \right) \right] \tag{D7}$$
$$\cong \left(\ln \frac{4a}{X} + \frac{1}{2} \right) \left(1 - \frac{\delta_c}{2 \ln 4a/X + 1} \right), \qquad k_0 T \le \hbar w \ll E_f,$$

where

$$\delta_c = 2\{1 - [1 - a/\gamma(a)]^2\}. \tag{D8}$$

REFERENCES

Argyres, P. N. (1961). *J. Phys. Chem.* **19**, 66.

Bardeen, J., and Shockley, W. (1950). *Phys. Rev.* **80**, 72.

Chang, Chin-An, Ludke, R., Chang, L. L., and Esaki, L. (1977). *Appl. Phys. Lett.* **31**, (11), 759.

Dingle, R., Stormer, H. L., Gossard, A. C., and Wiegmann, W. (1978). *Inst. Phys. Conf. Ser.* **45**, 250.

Dumke, W. P. (1961). *Phys. Rev.* **124**, 1813.

Dumke, W. P., Lorenz, M. R., and Pettit, G. D. (1970). *Phys. Rev. B* **1**, 4668.

Dutt, B. V., Al-Delaimi, M., and Spitzer, W. G. (1976). *J. Appl. Phys.* **47**, 565.

Dutt, B. V., Kim, O., and Spitzer, W. G. (1977). *J. Appl. Phys.* **48**, 2110.

Ehrenreich, H. (1959). *J. Phys. Chem.* **9**, 129.

Fan, H. F. (1967). In "Semiconductors and Semimetals" (R. K. Willardson and A. C. Beer, eds.), Vol. 3. Academic Press, New York.

Haga, E., and Kimura, H. J. (1963). *J. Phys. Soc. Japan* **18**, 777.

Haga, E., and Kimura, H. J. (1964). *J. Phys. Soc. Japan* **19**, 471, 658, 1596.

Harman, T. C., Goering, H. L., and Beer, A. C. (1956). *Phys. Rev.* **104**, 1562.

Holm, R. T., Gibson, J. W., and Palik, E. D. (1977). *J. Appl. Phys.* **48**, 212.

Iwasa, S., Balslev, I., and Burstein, E. (1964). "Proceedings International Conference Physics Semiconductors, Paris 1964." Dunod, Paris, p. 1077.

Jensen, B. (1973). *Ann. Phys.* **80**, 284.

Jensen, B. (1975). *Ann. Phys.* **95**, 229.

Jensen, B. (1978). *Phys. Status Solidi* **86**, 291.

Jensen, B. (1979). *J. Appl. Phys.* **50**, 5800.

Kane, E. O. (1957). *J. Phys. Chem.* **1**, 249.

Kohn, W., and Luttinger, J. M. (1957). *Phys. Rev.* **108**, 590.

Kung, J. K., and Spitzer, W. G. (1974). *Electrochem. Soc.* **121**, 1482.

Littlejohn, M. A., Sadler, R. A., Glisson, T. H., and Hauser, J. R. (1978). *Inst. Phys. Conf. Ser.* **45**, 242.

Long, D. (1968). "Energy Bands in Semiconductors." Wiley, New York.

Madelung, O. (1964). "Physics of the III-V Compounds." Wiley, New York.

Matz, D. (1967). *Phys. Rev.* **168**, 843.

Palik, E. D., and Holm, R. T. (1979). "Nondestructive Evaluation of Semiconductor Materials and Devices" (J. N. Zemel, ed.), Chapter 7. Plenum, New York.

Pankove, J. I. (1971). "Optical Processes in Semiconductors." Prentice–Hall, Englewood Cliffs, New Jersey.

Pearsall, T. P. *et al.* (1978). *Inst. Phys. Conf. Ser.* **45**, 94.

Perkowitz, S., and Thorland, R. H. (1974). *Phys. Rev. B* **9**, 545.

Price, P. J. (1966). *IBM J. Res. Dev.* **10**, 395.

Poth, H., Bruch, H., Heyen, M., and Balk, P. (1978). *J. Appl. Phys.* **49**, 285.

Rode, D. L. (1971). *Phys. Rev. B* **10**, 3287.

Spitzer, W. G., and Whelan, J. W. (1959). *Phys. Rev.* **114**, 59.

CHAPTER 5

Review of Recent Improvements in Pyroelectric Detectors

A. Hadni

Laboratoire Infrarouge Lointain
Université de Nancy I
Nancy, France

I. Introduction

Since 1963, pyroelectric detectors have been steadily improved. In this introduction we shall briefly give the state of the art concerning single-element detectors, pyroelectric vidicons, and detector arrays. We shall then look at some peculiar developments that are less known: epitaxial face-electrodes detectors, edge-electrode detectors, new pyroelectrics, and better use of pyroelectric materials.

173

A.   SINGLE-ELEMENT DETECTORS

1.   Detectivity D^*

The best detectivity for the face-electrode configuration has been claimed by the Mullard Company and cited by Putley (1980a) in its Fig. 3.3 with D^* (290 K, 6.5 Hz) = $3.7 \times 10^9/W$ cm $Hz^{1/2}$ for a research sample of alanine-doped triglycine sulphate 10 μm thick and 1.5×1.5 mm in area. This detectivity value is within one order of magnitude of

$$D^*_{ideal} (290 \text{ K}) = 2 \times 10^{10}/W \text{ cm Hz}^{1/2}. \tag{1}$$

It slightly surpasses that of a Golay detector, which up to now was the best one for thermal detection [D^* (7 mm^2, 6.5 Hz) = 3.6×10^9]. All the experimental values are corrected in order to compare with D^*_{ideal}, which does not take into account any loss due to the window needed to protect the detector and which assumes a perfect sine-wave illumination.

2.   The Use of D^* Can Be Misleading for Small-Area Detectors

The use of normalized detectivity $D^* = D\sqrt{A}$ is justified only when D^* is independent of target area A, in which case detectivity $D = D^*/\sqrt{A}$. Now $D = \mathcal{R}/B$, $\mathcal{R}$ being the responsivity, B the noise, and (assuming $B \simeq B_j \propto R^{-1/2}$) B_j the Johnson noise, and R the real part of the detector impedance. We have $B_j \propto A^{-1/2}$ and $\mathcal{R} \propto D^*/A$. This is also easily found by computing the voltage signal directly (cf. Hadni, 1980):

$$S = \frac{\alpha T \Lambda \Omega \, \Delta L_M \, \pi \omega R}{2\mathcal{G}(1 + \omega^2\tau^2)^{1/2}(1 + \omega^2\tau'^2)^{1/2}}. \tag{2}$$

It is proportional to the product $A \times R$, and assuming that $R_I = r_0(d/L1)$ for a type I (face-electrode) detector, it is seen that S is independent of target area $A = L1$. Now $\mathcal{R} = S/\phi$ and $\mathcal{R} \propto 1/A$. Direct evidence can be given if we look at a large-area face-electrode detector uniformly illuminated. Let us decompose it into a number of small-area elemental detectors. Each one will give the same signal voltage, while the corresponding incident flux will decrease proportionnaly to the elemental area so that responsivity $\mathcal{R} \propto 1/A$.

This result is verified (Fig. 1) for $A > A_0 = 0.5$ mm^2.

For small target areas ($A < A_0$) the capacitance of the detector itself becomes smaller than the stray capacitance in a standard preamplifier. With $A_0 = 0.5 \times 10^{-6}$ m^2, a thickness $d = 50$ μm, and $\varepsilon_R = 30$, we obtain an order of magnitude for the stray capacitance: $C_S \simeq 3$ pF. Assuming again a uniform illumination e, we have a given charge $\Delta Q = \pi \, \Delta T \times A$ on the

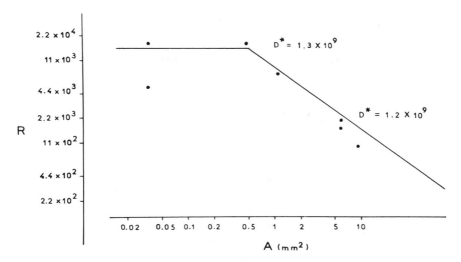

FIG. 1 Responsivity R of a face-electrode TGS pyroelectric detector versus target area A showing that $R \propto 1/A$ for values of A down to 0.5 mm². For smaller values of A, R is limited to a constant value of around 15×10^3 V/W because of stray capacitance $C_S \simeq 3$ pF. Normalized detectivity D^* is kept constant for $A > 0.5$ mm², D^* (300 K; 12.5 Hz; A > 0.5 mm²) $\simeq 1.3 \times 10^9$ W⁻¹ cm Hz¹ᐟ², a value slightly smaller than the best one claimed by Mullard: D^* (300 K; 6.5 Hz; A = 2.25 mm²) $= 3.7 \times 10^9$.

electrodes, and the signal voltage across them is written

$$S = \Delta Q/(C + C_S) \simeq \pi A \, \Delta T/C_S,$$

and the responsivity $\mathcal{R} = S/\phi$, i.e., $\mathcal{R} = \pi \, \Delta T/C_S \cdot e = \text{const.}$ with respect to A.

In conclusion, for $A > A_0$, D^* is constant with respect to A, while $D = D^*/\sqrt{A}$ and $R \propto 1/A$ are increasing when A decreases. However, for $A < A_0$, D^* becomes proportional to A, D to $\sqrt{A}$, and they are decreasing with A, while R is now a constant. A_0 is the smaller detector area that gives a capacitance higher than the stray capacitance due to connections and a preamplifier.

Let us notice that while Johnson and temperature noise are proportional to $A^{-1/2}$, amplifier noise is made of two terms that are respectively proportional to and independent of A (see Fig. 5 in Putley, 1970). When Johnson and temperature noise are negligible compared with amplifier noise, D^* is again dependent on A and hence meaningless. This is (for instance) the case for high frequencies ($B_J \propto \omega^{-1/2}$ becomes negligible).

3. Detectivity Decreases When Frequency Is Increased above an Optimum Frequency

Looking at Eq. (2) giving the signal and assuming purely dielectric losses, we write the resistivity as $r_0 = (\varepsilon_0 \varepsilon'' \omega)^{-1}$, so that the detectivity

$$D_{I,II} \propto \omega^{1/2}/(1 + \omega^2 \tau^2)^{1/2}$$

and $dD/d\omega = 0$ for $\omega_{opt} = 1/\tau$, with $\tau = \mathcal{C}/\mathcal{G}$. Then

$$D_{I,II}^* = \alpha T \pi/4\mathcal{G}^{1/2} (2)^{1/2} (kt\,\Delta f)^{1/2} (\varepsilon_0 \varepsilon'')^{1/2} C'^{1/2}, \qquad \omega = 1/\tau. \qquad (3)$$

To increase D^* ($\omega = 1/\tau$), the loss per unit area $\mathcal{G}$ must be as small as possible. Now $\mathcal{G} = \lambda'/e' + 4\sigma T_0^3$ is reduced to a minimum when $\lambda'/e' = 0$ (pure radiative losses).

In conclusion, for any pyroelectric detector there is an optimum frequency. For that frequency the detectivity is best when losses are smaller (i.e., purely radiative), and it can reach the background limit, as in Eq. (1). This optimum frequency is usually very low, because the losses are purely radiative. The best way to increase it to a practical value is to make very thin pyroelectric plates. For TGS with $e = 0.8\ \mu m$, Putley (1980b) did find $f_{opt} = 0.5$ Hz. Above the optimum frequency, D^* will decrease but not so rapidly as for classical thermal detectors. The main reason is that the internal resistance r_0 of the crystal is inversely proportional to the frequency, and thus Johnson noise is decreasing as $\omega^{-1/2}$, whereas it is a constant for either a bolometer or a thermocouple (see Section VI.C.1).

B. PYROELECTRIC VIDICONS

The practical feasibility of a pyroelectric vidicon was shown a few years after that of a single-element detector. The Pyricon now gives thermal images of very good quality (Figs. 2–4) with a number of applications to passive detection, fire detection, night-flying helicopters, and medical care. The definition of D^* is difficult, but Singer (1977) has claimed $D^* = 10^8/W$ cm $Hz^{1/2}$ with a DTGFB retina, after a digital treatment of the signal. With a reticulated DTGS target and optics opened to $f/1$, a 5 line-pair per millimeter spatial resolution can be achieved with a 0.5-K thermal resolution on an object at 300 K (Nelson, 1976). The AGA or the TRT camera, which employs a mechanical sweeping device and a cooled semiconductor detector (InSG or HgCdTe, respectively), is still more sensitive with a 0.1-K thermal resolution, but improvements are to be expected for the Pyricon, which is cheaper and more convenient.

C. PYROELECTRIC DETECTOR ARRAYS

Still more recent, pyroelectric detector arrays are now commercially available from Spiricon with 128 $LiTaO_3$ elements and 100 μm separation

FIG. 2   Pyroelectric vidicon image of a dressed man, courtesy of Thomson–CSF (St. Egrève, France). It is a color image with 5°C temperature difference from violet (28°C) to yellow (33°C).

(Roundy, 1979; Spiricon, 1981). They are made of a monolithic LiTaO$_3$ array, coupled to an integrated circuit array of Si self-scanning FET multiplexing switches for periodic signal readout. They achieve a NEP of 2×10^{-7} W in chopped CW operation at 200 Hz, a 75-V/W responsivity, and a 10^4 dynamic range. The NEP is expected to decrease to 10^{-8} W/Hz$^{1/2}$ with an electronic integrator under development. Reset noise is the dominant noise factor [$N_R = (kT/C)^{1/2}$ in volts, where C is the detector and amplifier

FIG. 3 Close-up of a Renault 5 taken with an RSRE pyroelectric vidicon camera using an English electric tube with a square reticulated DTGS target (Warner *et al.*, 1981) and a Rank Irtal 3 lens (Putley, 1981b).

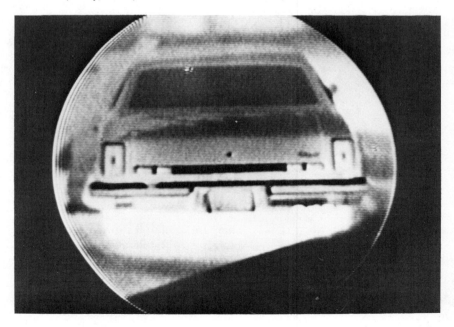

FIG. 4 Pyroelectric vidicon image of an American car from North American Philips (Briarcliff Manor, New York).

capacitance]. The thermal cross talk is reduced to about 10% at 200 Hz (i.e., by focusing a spot into one element, 10% of the response is seen on each adjacent element). With an $f/1$ optics the expected thermal resolution at 300 K is $\Delta T = 0.8°$ with integration. With pulsed radiation, from subnanosecond to millisecond, the detectors integrate the signal. The sensitivity is around 3×10^4 V/J.

Larger elements (1 mm $\times$ 0.9 mm) are available from Molectron (1981), with 1 mm spacing and up to 128 elements.

More recently American Philips has studied a 32×32 TGS element solid-state array and is expecting an NEP $= 1.25 \times 10^{-10}$ W/Hz$^{1/2}$ for a 125×125 μm element that should detect 0.02°C at room temperature (Carlson et al.,1981).

II. Epitaxial Face-Electrode Detectors

It is possible (Hadni and Thomas, 1981) to grow an epitaxial layer of TGS on a gold electrode, with a thickness smaller than 1 μm, and thus to make a thin pyroelectric detector (Fig. 5). Thin pyroelectric detectors are needed for two reasons. First, the simple theory leading to Eq. (2) assumes the whole pyro-electric plate to be at a uniform temperature. This is valid only if the thermal diffusion length Λ is larger than the sample thickness d. Let us recall that $\Lambda = (2\chi/\omega)^{1/2}$ is the distance at which the amplitude of the thermal wave is divided by e and the phase is retarded by one radian. It is obvious that for $d > \Lambda$, only a surface layer $d \simeq \Lambda$ will give a significant contribution to the pyroelectric signal, and the voltage responsivity S will show a value much smaller than given by Eq. (2). The values of Λ for TGS at different frequencies f are $f = 20$ Hz, $\Lambda = 64$ μm; $f = 200$ Hz, $\Lambda = 19.2$ μm; $f = 20 \times 10^4$ Hz, $\Lambda = 0.64$ μm.

Second, thin pyroelectric detectors are needed because the reduction in thickness should increase the responsivity according to Eq. (2). However, the detector is fixed to a TGS single crystal and not in vacuum, which increases thermal losses. The improvement is expected only for frequencies

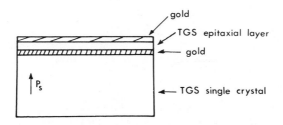

FIG. 5 The scheme of an epitaxial pyroelectric detector.

higher than $1/\tau$, when responsivity becomes independent of thermal losses $\mathcal{G}$. Preliminary results (Claudel and Hadni, 1976) have indeed shown that for $f > 200$ Hz, responsivity becomes higher with a 0.8-μm-thick TGS crystal than with a 20-μm-thick crystal.

A. SPECIAL PROPERTIES OF EPITAXIAL LAYERS

With epitaxial detectors we have found that the dielectric properties of thin ferroelectric crystals are often quite different from those of the bulk material. Figure 6 shows that the dielectric constant is drastically reduced. In Fig. 7 it is seen that for a thin sample the effect of a dc field is to reduce slightly the dielectric constant whereas for a thick sample the effect is considerable. This could lead to the assumption that a thin layer is subject to

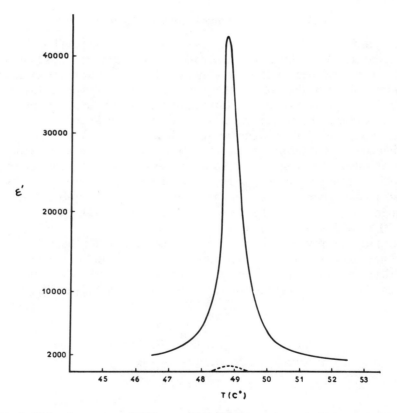

FIG. 6   Dielectric constant of TGS versus T. Full line; crystal sample thickness $e = 95 \ \mu$m; dotted line; crystal sample thickness $e = 1.3 \ \mu$m. The value of the maximum is reduced from 42,000 to 550.

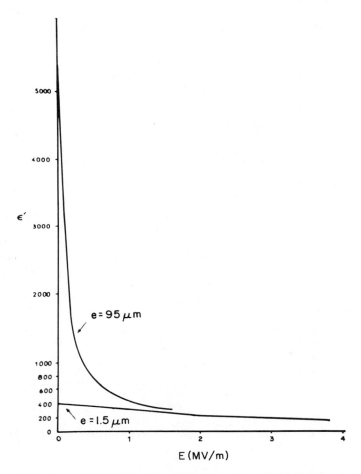

FIG. 7   Dielectric constant of TGS at $T = 50°C$ versus a dc applied field E showing a drastic reduction for the thickest plate and a small linear one for the thinnest. For $E > 1.6$ MV/m the dielectric constants have similar values.

an high internal dc field. The magnitude of the dc field is of the MV/m order, and the ratio of the surface layer e_s to the dielectric constant ε_s is $e_s/\varepsilon_s \simeq 16$ Å. In Fig. 8 we see that at room temperature the increased values of coercive field are two orders of magnitude higher than for thick samples, and they remain high at the Curie temperature. Figure 9 shows that spontaneous polarization is still high at room temperature, but the Curie temperature is raised substantially above 49°C. We could thus expect to make epitaxial detectors with a smooth response with respect to temperature, a

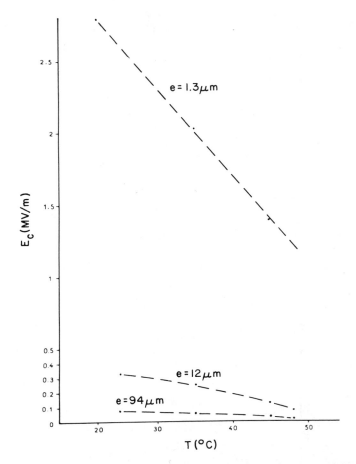

FIG. 8 Coercive field E_c versus temperature T for crystal plates of different thicknesses: $e = 94\ \mu m$, $e = 12\ \mu m$, $e = 1.3\ \mu m$.

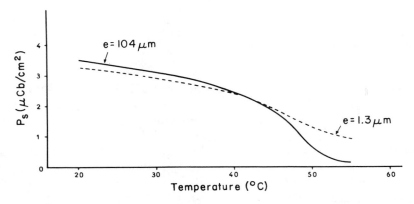

FIG. 9 Spontaneous polarization versus T for two crystal plates: $e = 104\ \mu m$, $e = 1.3\ \mu m$.

larger temperature range, and also with interesting detectivities at high frequency. These peculiar properties of thin ferroelectric crystals are explained by surface layers for which a detailed model has been proposed (Hadni *et al.*, 1983).

B. RESPONSIVITY OF EPITAXIAL FACE-ELECTRODE DETECTORS

1. *TGS*

In good accordance with expectation, Fig. 10 gives the low-frequency responsivity R of an epitaxial face-electrode detector with respect to T with a smooth peak at 38°C and significant values up to 60°C. Noise is smallest at 49°C, and detectivity is a maximum at 40°C with D (TGS 40°C; 12.5 Hz) = $2.2 \times 10^9/\text{W Hz}^{1/2}$.

2. *TGSe*

Figure 11 gives R with respect to T for TGSe with the highest value at room temperature and significant values up to 40°C, much above the Curie temperature ($T_c = 22°C$). Noise is a minimum at 27°C, and detectivity is a maximum at 23°C.

Figure 11 corresponds to temperatures above the Curie temperature, and the change of temperature does not give rise to the specific noise that impedes rapid measurements below T_c. It has thus been easy to decrease temperature after an experiment up to 40°C. There is a temperature hysteresis, and R is smaller for decreasing temperatures.

Figure 12 gives R with respect to frequency for the same detector used at

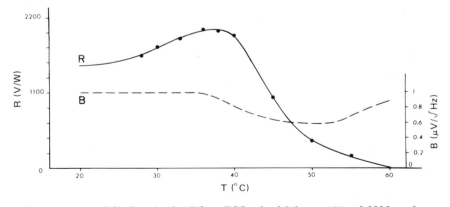

FIG. 10   Responsivity R and noise B for a TGS epitaxial detector ($A = 0.0095$ mm², $e = 1.3$ μm, $f = 12.5$ Hz).

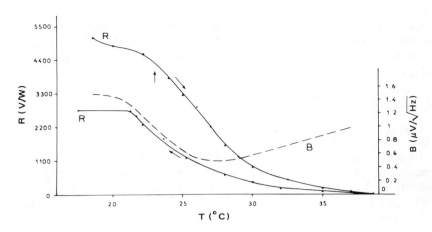

FIG. 11   A TGSe epitaxial detector grown on a TGS bulk crystal ($A = 0.01$ mm^2, $e = 1$ μm, $f = 12.5$ Hz) versus T. Noise is a minimum for $T \simeq 27°$C $> T_c = 22°$C. Thermal hysteresis is observed for R.

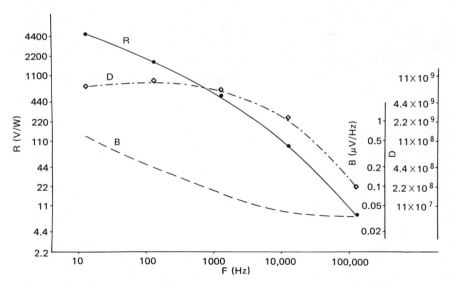

FIG. 12   Responsivity R, noise B, and detectivity D (per W Hz$^{1/2}$) versus frequency f for a TGSe detector (0.01 mm^2, 1 μm, 19°C, 12.5 Hz). A flat response is shown from 12.5 to 1250 Hz ($D \simeq 8.8 \times 10^9$/W Hz$^{1/2}$) and still a valuable detectivity at 1.25×10^5 Hz ($D \simeq 2.2 \times 10^8$/W Hz$^{1/2}$).

constant temperature (19°C). At 12.5 Hz a NEP slightly better than that given previously is observed,

$$NEP(TGSe\ 19°C,\ 12.5\ Hz) = 1.1 \times 10^{-10}\ W/Hz^{1/2},$$

or

$$D(TGSe\ 19°C,\ 12.5\ Hz) = 8.8 \times 10^{9}/W\ Hz^{1/2}.$$

It is remarkable that the NEP is kept nearly constant up to 10^3 Hz, and it is six times better than the best value from Mullard at this frequency.

At higher frequencies, detectivity is lower but still interesting:

$$D(TGSe,\ 10^4\ Hz) = 3.2 \times 10^{9}/W\ Hz^{1/2},$$

or

$$NEP(TGSe,\ 10^4\ Hz) = 3.10^{-10}\ W/Hz^{1/2};$$

$$D(TGSe,\ 10^5\ Hz) = 2.2 \times 10^{8}/W\ Hz^{1/2},$$

or

$$NEP(TGSe,\ 10^5\ Hz) = 4.5 \times 10^{-9}\ W/Hz^{1/2}.$$

At that frequency the thermal diffusion length in TGSe is Λ (TGSe) $\simeq$

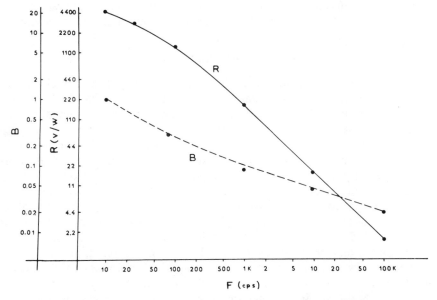

FIG. 13   Responsivity R and noise B for a bulk TGS detector ($A = 0.04\ mm^2$, $e = 20\ \mu m$) for $f = 12.5$ Hz, versus frequency; B in $\mu V/\sqrt{Hz}$.

0.9 μm, which is close to the value of the detector thickness. This could explain why Putley (1970) in his Fig. 12 could only get $D = (10^5$ Hz$) = 10^7/W$ Hz$^{1/2}$ with a bulk detector ($A = 3 \times 10^{-2}$ cm^2, $e = 10 \mu$m), a value 22 times smaller. Figure 13 gives R and B with respect to frequency for a TGS bulk detector ($e = 20 \mu$m); and indeed, at 10^5 Hz, B has the same order of magnitude as before ($B \simeq 0.02 \mu$V$/\sqrt{\text{Hz}}$) whereas R is smaller (4.4 V/W instead of 30 V/W).

C. CONCLUSIONS ON EPITAXIAL DETECTORS

Epitaxial detectors give a flat detectivity from 10 to 1250 Hz that is the best observed up to now at the latter frequency,

$$D(\text{TGSe}, A = 0.01 \text{ mm}^2, e = 1 \mu\text{m}, 19°\text{C}, 12.5 \rightarrow 1250 \text{ Hz})$$
$$= 8.8 \times 10^9/W \text{ Hz}^{1/2}. \quad (4)$$

From 1250 to 125,000 Hz, detectivity is decreasing but is still, at least, one order of magnitude better than for bulk detectors.

This improvement can be used in a number of applications, i.e., Fourier spectroscopy, night observation, etc.

More important than any technical application are the peculiar properties of thin ferroelectric films, which will be discussed elsewhere. Understanding these films will probably lead to new applications. We have already used the displacement of the Curie temperature toward higher temperature that makes the pyroelectric detectors usable over a wider range of temperatures.

III. Edge-Electrode Detectors

Edge-electrode detectors are made according to Fig. 14a. The electrodes are still perpendicular to the polar axis, but the radiation is incident on a face parallel to the axis. The main difference is that now radiation is not absorbed by the electrodes but generally by the crystal volume itself. This may be useful for very-high-frequency operation in which for face-electrode detectors some delay is unavoidable in transporting into the crystal the heat generated at the electrodes (i.e., 1 msec for gold-black electrodes).

Another difference is that impedance of an edge-electrode detector with the same target and dimensions as a face-electrode detector is much higher, with a capacitance smaller than 1 pF and a resistance higher than 10^{12} ohms at 10 Hz. This allows use of a higher load resistance, which could be an advantage at high frequencies (Putley, 1970). The problem, however, is to reduce the circuit capacitance to a value smaller than that of the detector itself. We shall see that it is now possible to reduce stray capacitance to 0.1 pF, and edge-electrode detectors seem to be the best with which to look at the shape of very short, intense radiation pulses. The charge resistor must be kept equal to the crystal resistance and is one or two orders of magnitude higher for edge-electrode than for face-electrode detectors. This may be a difficulty at very low frequencies where this resistance is considerable, but at very high frequencies this is an advantage. Most studies on very short pulses have used edge-electrode detectors:

1. Byer and Roundy (1972) used a SBN (strontium–barium–niobate) edge-electrode detector to look at CO_2 laser pulses 0.1 μsec long.

2. Auston and Glass (1972) have obtained 10-psec pyroelectric pulses, 250 V high, with a $LiTaO_3 : Cu^{2+}$ edge-electrode detector illuminated by a 10-psec infrared pulse at 1.06 μm, cited by Lines and Glass (1977), p. 603.

3. Glass and Auston (1972) have shown the possibility of excited-state polarization in an edge-electrode $LiNbO_3 : Cr$ at 173 K, cited by Lines and Glass (1977), p. 451.

4. However, Wood *et al.* (1970) have used a SBN face-electrode detector to look at a mode-locked TE CO_2 laser giving 1-MW pulses, 2 nsec long.

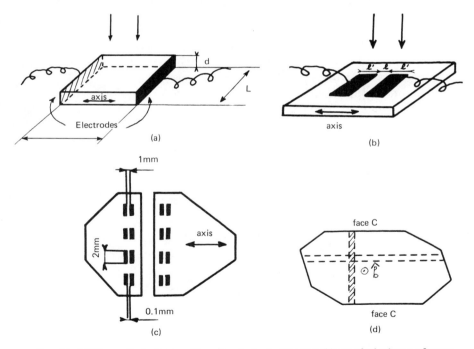

FIG. 14 (a) Theoretical scheme of an edge-electrode detector, (b) practical scheme of a type II detector, and (c) practical scheme of a series of type II detectors. (d) There are an infinity of orientations possible for a plate parallel to the monoclinic b axis; we have generally chose plates perpendicular to the C face (crossed area).

An approach to edge-electrode detectors is given in Fig. 14b. Such detectors (let us call them type II detectors) are very easy to make by printing two electrodes on a crystal face parallel to the polar axis. For high-frequency operation it is not necessary to cut a plate; the bulk crystal can be used. A series of detectors can be made in this way on any face of the crystal if it is parallel to the polar axis (Fig. 14c). However, for TGS we have generally chosen a face perpendicular to the largest natural face, i.e., the C face (Fig. 14d).

A. SCHEME OF A TYPE II DETECTOR

Type II detectors are closely related to edge-electrode detectors because the target is not an electrode and is parallel to the polar axis. However, we have to assume field lines perpendicular to the electrodes (Fig. 15), and the simplest model is to draw circles centered in the middle of the target. In fact, TGS is monoclinic with a higher dielectric constant along the b axis, and the field lines computed from the dielectric constant tensor are ellipses (ellipticity $= \sqrt{\varepsilon_1/\varepsilon_2}$). This has not been considered here, but certainly the field lines are closer to the surface.

In the isotropic model we can compute the capacitance of the type II detector represented in Fig. 15 to be $C = (\varepsilon_0 \varepsilon_R A)/e$, where $\varepsilon_0 \simeq 10^{-11}$; $A \simeq 2 \times 10^{-6}$ m²; $e \simeq 3.14 \times 10^{-3}$ m; hence $C \simeq 0.6 \times 10^{-14}\, \varepsilon_R$. We have been able to measure C with respect to T with a General Electric bridge (Fig. 16). The maximum value is only a few picofarads, and the main problem in such measurements is to reduce the circuit capacitance to values much under 1 pF. From Fig. 16 we have at $T = 23°C$, $C \simeq 0.1$ pF, and hence $\varepsilon_R \simeq 30$; at

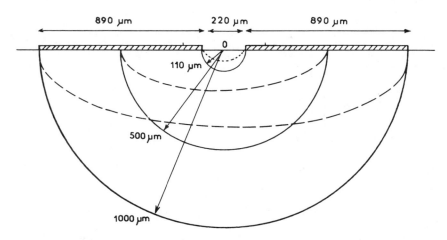

FIG. 15   Cross section of a type II detector showing the electric field lines for the case of an isotropic material (solid lines) and an anisotropic crystal (dotted lines).

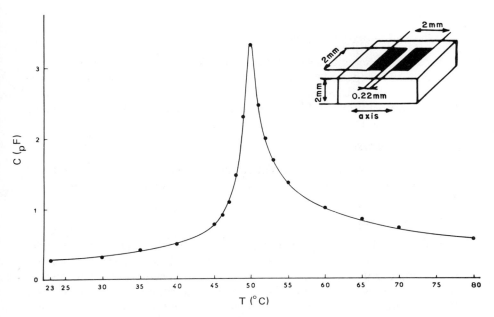

FIG. 16   Capacitance of a type II detector versus T (bulk TGS crystal).

$T = 50°C$, $C \simeq 3.2$ pF, and hence $\varepsilon_R \simeq 550$. This last value cannot be considered too small, because the field lines close to the electrodes lie in a direction for which the dielectric constant is not very sensitive to temperature. An exact calculation could be made, but it is still possible to say that the model in Fig. 15 appears reasonable as a crude first approximation.

The properties of such a pyroelectric detector can be qualitatively predicted from the model:

1.   The pyroelectric response occurs from elements that are deep inside crystal but certainly not so deep as in the isotropic model. The anisotropy of the dielectric constant, especially high close to the Curie temperature, is a strong factor in displacing the field lines in a direction close to the crystal surface.

2.   Infrared radiation from 5 to 100 μm is absorbed by a surface layer 3 to 5 μm thick that should give a null contribution to the pyroelectric response in the isotropic model but in fact gives a definite but very small signal. Heat has to propagate into the crystal to give the pyroelectric signal.

3.   Alteration of the target area by deposition of any insulating material or by any mechanical or chemical etching should not modify the responsivity, because it occurs mainly from the inside of the crystal.

B. EXPERIMENTAL DATA

1. *Reduction of Stray Capacitance*

Because the capacitance of a type II detector is between 0.1 and 4 pF, stray capacitance has to be reduced to less than 0.1 pF, which is not easy. The first measurements of responsivity were very poor (i.e., 88 V/W), but they have been improved up to 34,000 V/W by reducing stray capacitance and also stray light which can give false signals (see Section II.B.4).

2. *Optimum Geometry*

In Fig. 17 it is seen that the distance l between electrodes must be reduced to $l = 100$ μm, and probably $l = 50$ μm would be better. On an other hand, L has to be large (i.e., $L = 5$ mm).

3. *Crystal Temperature*

In Fig. 17 it is also seen that an optimum temperature is especially critical for small values of l. This value has to be reached very slowly to avoid microscopic domain switchings which lead to specific noise and also to prevent hysteresis effects which can be very important. Figure 18 shows a typical temperature hysteresis observed in an alanine-doped TGS crystal when temperature is increased at 1°C/mn. Similar hysteresis behavior in face-electrode detectors has been explained by Poprawski and Mroz (1981).

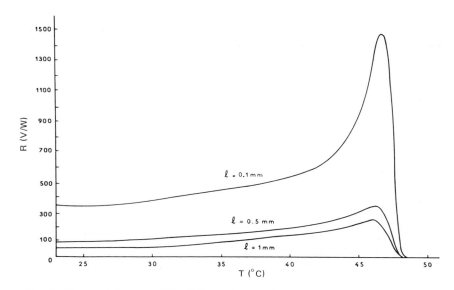

FIG. 17 Responsivity versus T for different distances l between electrodes, for a TGS type II detector.

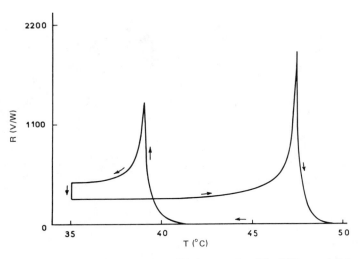

FIG. 18   Hysteresis in the curve $R(T)$ when T is increased rapidly (1°C per min), for a TGS type II detector.

4.   Domain Texture, Additional Signals

It is possible to look at the domain texture of a type II detector, as with a type I detector, using the pyroelectric probe technique (Ungar *et al.*, 1981). However, for type II it is more difficult, because the signal may be smaller and because the pyroelectric target, which is not an electrode, does not absorb the He–Ne laser radiation used for the inspection. Nevertheless, signals have been observed on the target area close to the electrodes. Probably the electrode edges are thinner than the bulk and can absorb the laser radiation. In Fig. 19a the two white lines indicate that the target is monodomain. This is not generally the rule, and domains of opposite polarization are often observed, usually crossing the target area from one electrode to the other but sometimes localized close to one electrode. Figure 19b is a schematic of the electrodes that explains Fig. 19a which is a photograph of the pyroelectric inspection of the target. By comparing the two, one can see that pyroelectric signals are obtained by illuminating the target area close to the electrodes or the external parts of the electrodes. This can be qualitatively understood if we consider a monodomain illuminated by a spot located on the inner part of an electrode and on the outer part. In both cases the heating gives rise to an electric dipole of the same orientation. In one case it will attract positive charges to the electrode and in the other, negative charges, hence giving pyroelectric signals of opposite signs.

In Fig. 19c different parts of the crystal are cross-hatched to show they also give signals when illuminated. To avoid these additional signals only the

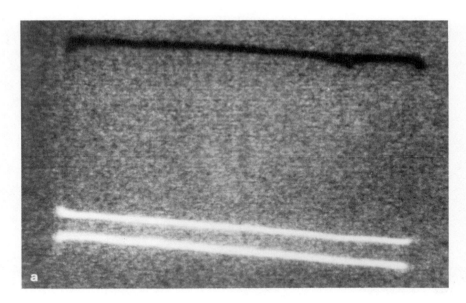

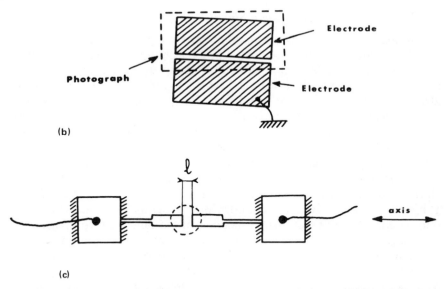

FIG. 19 (a) Pyroelectric map of a type II detector taken with an He–Ne laser and showing signals only close to the electrode. The target lies between the two white signal lines. (b) Scheme showing the limits of the above photograph for target and electrodes. (c) Scheme showing how the electrodes are connected to the two largest metal areas where silver-paste contacts are made. All cross-hatched areas give pyroelectric signals when illuminated.

target should be illuminated, and all connexions must be isolated from the crystal by a thick layer of insulating varnish. From Fig. 20 it is possible to expect useful electrodes limited to a $190 \times 55 \ \mu$m area.

5. Sensitivity Variations Across the Target

The sensitivity variation along the target can be studied by sweeping a heating spot on a line perpendicular to the electrodes. We have used a blue laser that can give a small spot (2-μm diameter), modulated at different frequencies to get different thermal diffusion lengths, and a blackbody that can illuminate a thin slit at low frequencies.

a. *Blue Laser Source Modulated at a Low Frequency* ($F = 90 \ Hz$). Figure 21a shows no significant response on the target (TGS does not absorb in the blue). The response appears at the limit of the electrodes. Less expected is the fact that responsivity does not disappear when the spot is on the electrodes and is still 10% of the maximum, at 150 μm inside the bismuth electrodes. A significant negative signal is observed at the outer part of the electrodes, much smaller on the grounded electrode than on the other one. In any case the light is absorbed in different parts of the detector surface, and at the low frequency considered the thermal diffusion length is around 30 μm, long enough to give a thermal signal at the field lines. To allow some absorption on the target itself (and to increase that on the electrodes), the whole detector is covered by a thick layer of Mn-phthalocyanine (Strimer *et al.*, 1981). There is now a large signal when the target is illuminated, with a maximum in the middle as long as the electrodes are separated by less than 230 μm (Fig. 21b and c). The increase of sensitivity at the limit of the electrodes is now hardly visible, but the signal is still visible at 300 μm inside the electrodes. Hence we draw three conclusions:

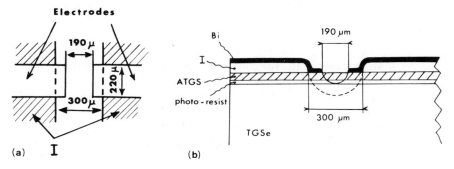

FIG. 20   (a) Upper view of the electrodes with I, an insulating varnish, separating the connections from the crystal surface. (b) Cross section of the electrodes for an epitaxial TGS detector grown on a TGSe bulk single crystal.

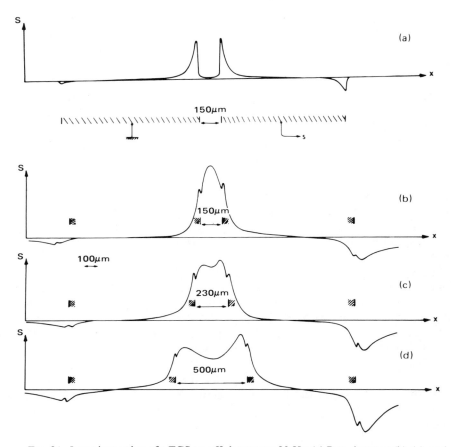

FIG. 21 Laser inspection of a TGS type II detector at 90 Hz. (a) Bare detector. (b), (c), and (d) The whole detector is covered with Mn-phthalocyanine of thickness $l = 100$, 250, and 500 μm, respectively.

1. The signal is generally maximum at the center of the target when it is covered by a suitable layer that both absorbs and electrically insulates. This is true only when l is small; and for $l = 500$ μm (Fig. 21d) it is close to the electrodes that the signal is maximum.

2. There is some absorbtion of radiation in the Bi electrodes that gives positive signals when the spot is on the target sides and negative ones outside.

3. The integrated signal $\mathcal{S} = \int_{-\Delta x/2}^{+0x/2} s\, dx$ gives the response corresponding to uniform illumination of a rectangle of width Δx. Responsivity $S/\phi \propto \mathcal{S}/\Delta x$ can thus be made maximum by using a thin rectangular slit centered in the middle of the target and placed at a very short distance in front of it.

b. *Blue Laser Source Modulated at Different Frequencies up to $f = 2 \times 10^5$ Hz.* In Fig. 22 we again have signal s with respect to displacement x of a blue laser spot on a line perpendicular to the electrode edges. The type II pyroelectric detector has not been coated. Three chopping frequencies are used. At very low frequency (9 Hz) a signal is now observed on all the Bi electrodes. At 90 Hz we have a signal similar to the one previously observed (Fig. 21a). At 90 kHz the signal is limited to the electrode edges. The absorption by the metallic electrodes is difficult to understand. The effect of frequency is caused by the decrease in thermal diffusion length when the frequency increases: $l = 95 \, \mu m$ ($f = 9$ Hz); $l = 30 \, \mu m$ ($f = 90$ Hz); $l = 0.95 \, \mu m$ ($f = 9 \times 10^4$ Hz). The temperature change for the highest frequency used is confined to just under the electrodes where the field lines are perpendicular to polarization, and the polarization change has no significant component on the field lines.

c. *Blackbody Source Modulated at Low Frequency (12.5 Hz).* The scheme of the experiment is given on the right-hand side of Fig. 23. A type II detector is made of two electrodes, $L = 2$ mm, $l' = 0.9$ mm, spaced by $l = 0.2$ mm. The detector is separated from a blackbody source at 500 K by a screen having a slit 2 mm long and 40 μm wide. The slit is illuminated at a 12.5 Hz frequency by an aperture $\phi = 20$ mm in the blackbody source

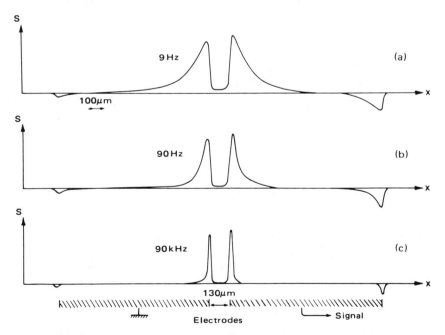

FIG. 22 Laser inspection of a bare TGS type II detector at different frequencies.

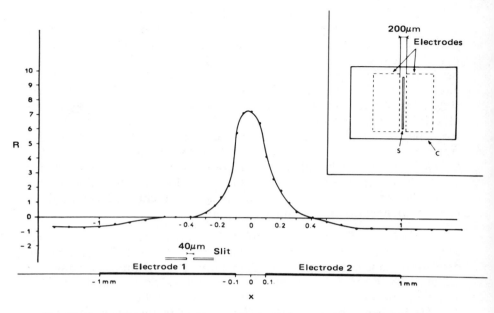

FIG. 23 Responsivity versus position x of a thin slit placed across the target and electrodes of a TGS type II detector. The slit is illuminated by a blackbody at 500 K. There is a positive signal from both the target and the first part of the electrodes. A small negative signal is observed at the outer side of the electrodes.

located at 25 cm. Absorption occurs in a thin layer (2 to 5 μm thick) at the surface of the TGS crystal, and heat is propagated into the bulk with a thermal diffusion length $\Lambda \simeq 80\ \mu$m. However, the results are the same as in section a above: the signal is maximum when the slit is located just between electrodes, and there is still a signal when the electrodes are illuminated (10% of the maximum, at 150 μm inside the electrodes).

6. Spectral Sensitivity

It has been shown (Gerbaux et al., 1978) that a TGS plate 1 μm thick has a average transmission coefficient of around 60% in the spectral range 3 to 100 μm. We can thus expect that absorption in the target is limited to a surface layer 3 to 5 μm thick and that responsivity is not significantly different for the radiation of a black body at 590 K ($\lambda_M \simeq 6\ \mu$m) and that of a CO_2 laser ($\lambda \simeq 10\ \mu$m). In fact we obtained $R(46°C, 10\ Hz, blackbody\ 590\ K) = 2112$ V/W and $R(46°C, 10\ Hz, \lambda = 10\ \mu m) = 1500$ V/W.

7. Chopping Frequency

Figure 24 gives responsivity R, noise B, and detectivity D with respect to frequency f for a type II TGS detector at 46°C with a high load resistance

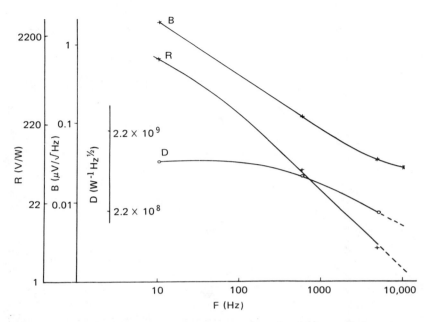

FIG. 24   Responsivity R, noise B, and detectivity D of a TGS type II detector versus frequency.

$(10^{10}$ ohms$)$. It is seen that $R \propto 1/f$, which shows that $f > 1/\tau'$, because of the high value of τ'. This is confirmed by the triangular shape of the signal.

8.   *Type II Epitaxial Detectors*

From the Fig. 15 model it is expected that a thin layer deposited on the target should not significantly modify the pyroelectric properties. This has been verified.

Figure 25 shows that epitaxial layers parallel to the b axis can be grown by the GETS method. Figure 26 gives the responsivity with respect to T of a type II detector made with a bulk TGS crystal covered by a thin deuterated TGSe layer. Only one peak is observed at 48°C, close to the TGS Curie temperature $(T_c = 49°C)$. A small signal is observed down to room temperature, but there is no anomaly at 30°C where the pyroelectric coefficient of TGSe–D is maximum. This is a confirmation that only the inside of the crystal is inspected by the electrodes of a type II detector. The same result is obtained with a type II detector made with a bulk TGSe crystal covered with a thin TGS oriented layer. In Fig. 27 we observe a remarkable responsivity at 23°C (34,000 V/W) corresponding to the bulk TGSe and a broad maximum with only 50 V/W at 46°C due to a small contribution of the TGS surface layer.

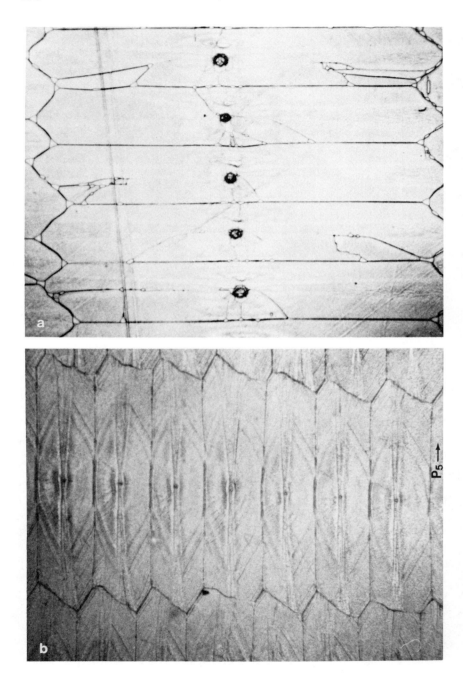

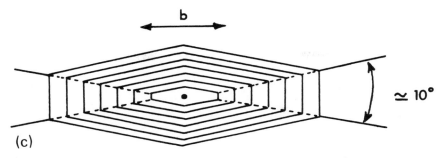

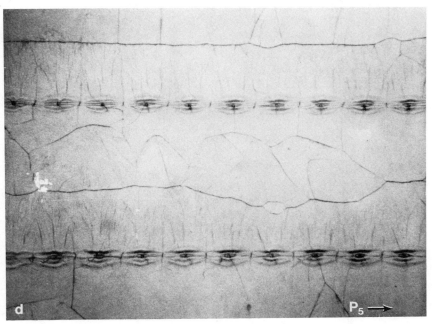

FIG. 25   (a) Flat oriented TGS single crystals, $80 \times 540 \ \mu m$, made in a closed box at 80°C. (b) Flat oriented TGS single crystals, $200 \times 800 \ \mu m$, made in a closed box with saturated water vapor. The line containing the holes is perpendicular to the b axis of the bulk crystal. (c) Tentative explanation of the fringes observed in Fig. 19b. They seem to be ridges corresponding to a greater thickness. (d) In this case the lines containing the holes are parallel to the b axis. Crystal growth is still perpendicular to the lines; but because the velocity is slower, the crystals issued from the seed holes on two adjacent lines do not meet directly.

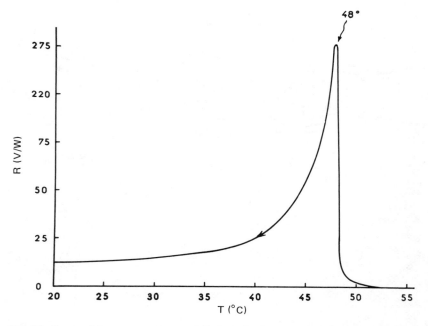

FIG. 26 Responsivity versus temperature for an epitaxial TGSe type II detector grown on a bulk TGS single crystal. There is only one maximum corresponding to the bulk.

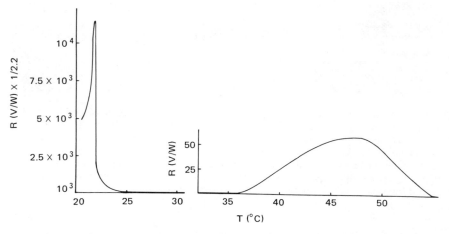

FIG. 27 Responsivity versus temperature for an epitaxial TGS type II detector grown on a bulk TGSe single crystal. There is a high maximim corresponding to the bulk at 23°C and a much smaller one at 46°C.

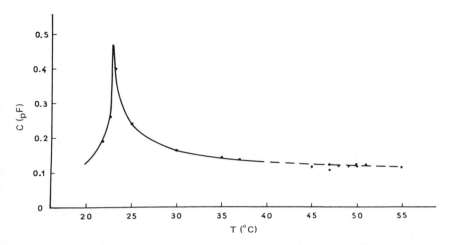

FIG. 28 Capacitance C versus temperature T for a TGS epitaxial type II detector grown on a TGSe bulk single crystal. There is only one peak at 23°C, corresponding to the bulk TGSe.

The dielectic constant of the TGS epitaxial detector grown on a bulk TGSe detector has been studied with respect to T (Fig. 28). There is a peak at 23°C (only 0.5 pF) but no anomaly at 49°C. Assuming $A = 220 \times 55 \ \mu m^2$ and $e = 3.14 \times 122.5 \ \mu m$, we obtain $C = 0.31 \times 10^{-16} \ \varepsilon_R$; for $T = 23°C$ we have measured $C = 0.5 \times 10^{-12} \ F$, hence $\varepsilon_R \simeq 1.6 \times 10^4$; for $T = 50°C$ we have measured $C = 0.12 \times 10^{-12} \ F$, hence $\varepsilon_R = 0.4 \times 10^4$. The 50°C value seems too large; the measured capacitance cannot be ascribed entirely to the electrodes as they are defined in Fig. 20. There is a stray capacitance C_S, the order of magnitude is $C_S \simeq 0.1$ pF. It is due to the connections, to the preamplifier FET, and to the parts of the electrodes that are assumed to be disconnected by the insulating layer of varnish (see Fig. 20b).

9.   The Effect of Water Adsorption

When a TGS type II detector is in a box connected to either a container dried with P_2O_5 or a tube filled with water at room temperature, the responsivity can be modified by a factor 3.2 (Fig. 29). It appears that for a critical water vapor pressure there is a maximum responsivity. For the room temperature saturated vapor pressure, a minimum is observed whereas the phase retardation of the signal is modified by more than 100°. Such a water vapor pressure probably leads to enough water molecules adsorbed to increase significantly the surface conductivity and reduce the responsivity.

The critical value of the water vapor pressure that makes the responsivity maximum either may introduce some absorption on the electrodes (from

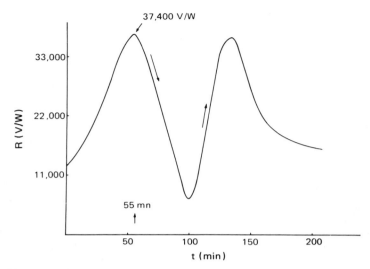

FIG. 29   Responsivity versus time for a TGSe type II detector, 12.5 Hz, in communication either with a liquid water container or a P_2O_5 drying box.

Section 5, the increase should be 50% at most) or, more probably, may increase the spatial extension of the electrodes, and the reduction of l should be responsible for the increase of responsivity. It has been shown spectroscopically that this increase is independent of the wavelength and the first assumption can be rejected.

10.   Response to Short Infrared Pulses

In these cases the radiation is still absorbed in a layer 1 to 5 μm thick, but now, because the pulse is short ($\simeq 1$ μsec), the thermal diffusion length is negligible. The main electric field lines are not reached by the heat pulse. However a substantial signal is observed. It has been explained by a pure piezoelectric effect (Ungar and Hadni, 1983).

   a.   *Far Infrared Laser Pulse* ($\lambda = 118$ μm). On a TGS plate cut parallel to the monoclinic axis, several type II detectors have been made according to Fig. 14c, except that all detectors now have the same distance l between electrodes. The upper detector in the picture receives a laser pulse from a water vapor laser ($\lambda = 118$ μm). The signal is displayed on a cathode ray oscilloscope. It has been observed that the other detectors receive a delayed signal. The delay has not been measured, but the attenuation is small enough (only one order of magnitude at a 10 mm distance) to make such a determination possible with a box-car integrator.

b. *CO_2 Laser Pulse ($\lambda \simeq 10 \mu m$)*. Q-switching with a rotating mirror may allow us to obtain pulses 0.5 μsec wide that can be timed with 0.1 μsec accuracy. The propagation of the pulse could be studied accurately. Up to now we can only say that propagation velocity is too high for a thermal wave, for which $v = \sqrt{\chi\omega}$, $v \simeq 3$ mm/0.1 msec, with f = 5 MHz and $\chi = 26 \times 10^{-4}$ cm²/sec. It is probably the case of an elastic surface wave (Ungar and Hadni, 1983).

For such experiments the CO_2 laser brings two advantages over the pulsed H_2O laser. It is a CW laser, and the pulses from the rotating mirror are not accompanied by electric disturbances; and because the wavelength is shorter, the radiation is more easily focused in the center of the small target of a type II detector.

C. CONCLUSIONS ON TYPE II DETECTORS

Type II detectors have given the highest responsivities observed up to now with TGSe, i.e., 37,400 V/W, in special conditions. This number is not definite because of experimental difficulties; we can only say that type II responsivity is especially high close to the Curie temperature. Type II detectors are also the easiest ones to make.

They can be improved by deviating the field lines closer to the surface, to a distance smaller than the thermal diffusion length. The electrodes can be made closer together etc...

Applications are numerous: high-resolution spectroscopy, burglar alarms, and heterodyne detection, to name a few.

IV. New Ferroelectrics

For pyroelectric detection only a few ferroelectrics have given a high responsivity: TGS, $LiTaO_3$, and SBN (strontium-barium-niobate). The most useful factor of merit is $M_V = \pi/C_p\varepsilon'$, because it is proportional to the voltage responsivity and also to the detectivity of a pyroelectric vidicon. There is also $M_D = \pi/C_p\varepsilon''^{1/2}$ which is proportional to the detectivity D^* of a single-element detector. M_D is the more difficult to discuss because ε'' depends on the care taken in preparing the crystal, and the values of ε'' differ widely in the literature. Sometimes $M_i = \pi/C_p$ is also used; it is proportional to the current responsivity. Table 16.1, p. 570, in "Lines and Glass" (1977) gives some data concerning TGS, $LiTaO_3$, SBN, PLZT, and PVF_2.

One challenge is to find new materials that could give better factors of merit. About 2000 pyroelectric crystals have been studied up to now. New ones can be found by looking at the many organic and inorganic crystals that were first synthesized in the late 1890s and that are known to belong to one of the ten crystallographic classes that are pyroelectric. The experimental problem is to know the size of the pyroelectric coefficient and the different factors of merit. A promising crystal has been studied recently, ammonium phosphotellurate (Gauthier, 1981). It is grown from a water solution, belongs to the monoclinic m crystallographic class, and has properties very close to TGS: $T_c = 48°C$, P_s (20°C) = 2.1 μCb/cm, with the polarization direction located in a plane perpendicular to the monoclinic axis where it is not significantly varying. It also has M_V values comparable to those of TGS: M_V(PTA) = 2550 V/cm J (i.e., 4000 for TGS). It seems that significant improvements cannot be expected for proper ferroelectrics because ε' is increases with π. It has been shown that for improper ferroelectrics such as $LiNH_4SO_4$, boracites, etc., ε' has a very small anomaly at the Curie temperature and π can be considerably increased. Boracites are considered to be the most promising, but they are difficult to grow, especially those that give a high value to M_V, i.e., Ni–Br and Ni–I boracites (CuCl boracites are easy to prepare but have a much smaller M_V).

Up to now TGS and its derivatives are the only ones convenient for high-performance pyroelectric vidicons. TGS is still used in France ($\varepsilon = 40$) whereas DTGS seems to be used more in Great Britain because of its lower dielectric constant and because it is easy to reticulate. DTGFB is preferred by American Philips because of a still lower dielectric constant at room temperature ($\varepsilon = 13$), but it might be difficult to etch by ion milling and to control the etching.

V.   Better Use of Pyroelectric Materials

We have seen two new possibilities, the use of epitaxial pyroelectric layers and type II detectors that have given high responsivities. We shall now look at two stimulating recent proposals, from Shaulov and from Putley, and we shall see that the one from Shaulov may give the key to explaining the high responsivity of type II detectors.

A.   OBLIQUE CUTS

Shaulov et al. (1981) investigated a DTGS flat parallel-faced plate cut normal to a direction that was 74° from the pyroelectric axis. They showed both experimentally and theoretically that the factor of merit M_V can be three times higher than with a plate cut normal to the pyroelectric axis.

The explanation relies on the fact that $M_V \propto \pi/\varepsilon'$, and although π is a maximum for a plate cut perpendicular to the polar axis, ε' is also a maximum, and such a cut is not optimum, especially close to the Curie temperature where the anisotropy of the dielectric susceptibility is especially high.

The results of this analysis were examined in DTGFB. The dielectric constant with respect to T is given in Fig. 30a for the three principal directions, and for the optimum polar angle in Fig. 30b. The gain over normal incidence can be as high as 7. Experimentally the factor of merit in DTGFB at a normal cut and in a cut perpendicular to a direction at 74° from the pyroelectric axis is given with respect to T in Fig. 30c. For the oblique cut there is a maximum at 64°C in good accordance with Fig. 30b. The improvement on the normal cut is a factor of three.

Let us note the analogy between Fig. 30c (oblique cut) and Fig. 17 concerning a type II detector. In both cases the increase of sensitivity with respect to T is due to the increase in anisotropy. The great responsivity of type II detectors occurs mostly because field lines in the most part are not parallel to the polar axis.

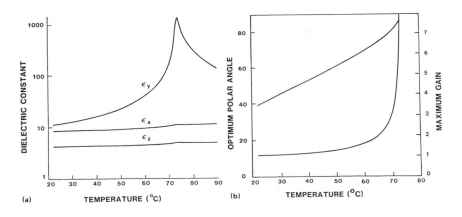

(a)   (b)

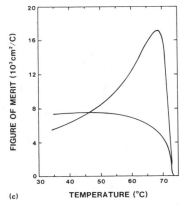

FIG. 30   (a) The anisotropy of the DTGFB dielectric constant versus temperature for a sample $e = 0.1$ mm thick. (b) Optimum cut angle θ (upper curve) and maximum gain (lower curve) obtained with that cut over a normal cut, versus T. (c) Computed factor of merit of DTGFB versus T for a normal cut plate (lower curve) and an oblique cut (upper curve). (From Shaulov, 1981.)

(c)

B. RADIATION SHIELDING

Recently Putley (1981) has claimed that the temperature noise equivalent power $NEP_{ideal} = (16\ k\sigma T^5)^{1/2}A^{1/2}$ W/Hz$^{1/2}$, which at $T = 290$ K is written $NEP_{ideal} = 5 \times 10^{-11}A^{1/2}$ W/Hz$^{1/2}$, is not really a limit. He has proposed a radiation shielding that in the simplest case is a reflective sphere at temperature T_2 (Fig. 31a) with a emissive coefficient ε_2 and has shown that now

$$NEP = NEP_{ideal} \times (\varepsilon_2/\varepsilon_1)^{1/2}. \qquad (5)$$

For a black detector $\varepsilon_1 \simeq 1$ and for a good infrared reflector $\varepsilon_2 \simeq 10^{-2}$, which suggests that an order of magnitude improvement in detectivity could be obtained. Now if the space between the outer and inner surface is filled with highly reflective but thermally isolated shields (Fig. 31b), the emissivity ε_2 may be replaced by $\varepsilon' = \varepsilon_2/(n + 1)$, where n is the number of shields. With $n = 30$, $\varepsilon' \simeq 3 \times 10^{-4}$, and the detectivity should be 50 times better than for the ideal limit, given by Eq. (5).

In fact, a polished light pipe is used to admit signal within a specified field of view, and the detector will also receive the component of background fluctuation associated with the field of view. With an $f/4$ aperture, we expect $D^* \simeq 1.6 \times 10^{11}$/W cm Hz$^{1/2}$ and a gain of 8.

The proposal is reminiscent of the use of cold apertures to reduce the field of view, but here the screens are at room temperature. The use of superinsulators, which are well known for cryogenic insulation, is a new suggestion for detectors and could also reduce microphonics.

C. HETERODYNE DETECTION

There are few publications on the use of pyroelectrics in heterodyne detection. Baynham *et al.* (1974) obtained a heterodyne NEP $\simeq 5 \times 10^{-14}$

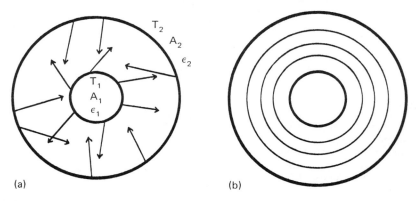

(a)   (b)

FIG. 31   Thermal equilibrium between a center black sphere ($\varepsilon_1 \simeq 1$) at temperature T_1 and (a) a metal sphere ($\varepsilon_2 \simeq 10^{-2}$) at temperature T_2 and (b) a series of concentric spheres. (From Putley, 1981.)

W/Hz$^{1/2}$ (10^6 Hz$^{1/2}$ IF) with a 4×10^{-2} W CO_2 local oscillator. With the progress in video pyroelectric detection at high frequencies and laser stabilization, this problem ought to be reconsidered.

VI.   Comparison with Quantum Detectors

A.   COMPARISON WITH SPECIFIC QUANTUM DETECTORS

1.   *Mercury Cadmium Telluride at 10 μm*

The best quantum detector with which to look at the room-temperature scenery today is the mercury cadmium telluride detector (MCT). It has to be cooled to 80 K and can be made with a target area ranging from 4×10^{-6} to 4×10^{-2} cm^2. The time constant is short ($\tau \simeq 10^{-6}$ sec), the responsivity is high (10^4 V/W), the impedance is small (R $\simeq 500$ ohms), and with a 20° field of view the specific detectivity D_λ^* at the wavelength giving the maximum sensitivity is $D_\lambda^* \simeq 10^{10}$ W^{-1} cm Hz$^{1/2}$ up to 10^6 Hz, which is only a factor of two under D_{ideal}^* for a thermal detector, and a factor of three under that of $D_{\lambda M}^*$ for a quantum detector. Now $D_\lambda = D_\lambda^*/\sqrt{A}$; let us choose $A = (2 \times 10^{-2}$ cm)^2 to make a comparison with the TGSe face-electrode detector; then $D_\lambda(4 \times 10^{-4}$ cm$^2) = 5 \times 10^{11}$/W Hz$^{1/2}$.

Comparing the MCT to a thermal detector operated at room temperature, we have to increase the field of view from 20° ($f/3$) to 90° ($f/0.5$) and introduce a factor of 4.5 to reduce D_λ; $D_\lambda(4 \times 10^{-4}$ cm^2; 180°) $\simeq 10^{11}$/W Hz$^{1/2}$ and NEP$_\lambda$(4×10^{-4} cm^2; 90°, 10 μm, 10^6 Hz) $\simeq 10^{-11}$ W/Hz$^{1/2}$. Moreover, NEP (blackbody) > NEP$_\lambda$, and only the NEP (blackbody 300 K) is suitable for predicting the performance of the detector looking at room-temperature scenery. We can roughly estimate NEP (blackbody) $\simeq 2$ NEP ($\lambda = 10 \mu$), and then finally NEP (blackbody; MCT 80° K; 4×10^{-4} cm^2; 90°; 10^6 Hz) $\simeq 0.2 \times 10^{-10}$ W Hz$^{1/2}$. Let us recall that with a TGSe epitaxial detector we obtained five times larger NEP at 1250 Hz and that such a factor might be recovered by the use of an oblique cut.

In conclusion, to look at room temperature scenery, the epitaxial TGSe detector is comparable to MCT for chopping frequencies up to 10^3 Hz, which is not surprising if we remember that the curve of the ideal $D_{\lambda M}^*$ for a photon detector looking at a 180° background at room temperature is close to the ideal D* of a thermal detector for 5 μm < λ < 60 μm. This is completely different for a background at 3 K, and Richards (1983) has presented a review on extremely sensitive photoconductors for astronomy.

2.   *Doped Germanium Detectors for the 10–60-μm Range*

The quantum detectors for this frequency range are doped germanium detectors operated at 5 K. The detectivity $D_{\lambda M}^*$ and the time constants are comparable to those of MCT operated at 10 μM, both looking at a 290-K background. The comparison with the new pyroelectric detectors is thus still in favor of the pyroelectrics for $f < 10^3$ Hz. We give these detectors with their wavelength range: GeCu, $1-30\ \mu$m; GeHg, $8-12\ \mu$m, and GeZn, $20-40\ \mu$m.

3. Far-Infrared ($\lambda > 60\ \mu m$) Quantum Detectors

For this frequency range the detectors most used are doped germanium detectors and the so-called electronic bolometer, which is made of very pure InSb. In both cases, for room temperature background the detectivity may be one order of magnitude higher than for MCT, and the time constant may be of μsec order. Doped germanium detectors, i.e. Ge:Ga, Ge:Bi and Ge:Sb, are usable at 2 K in small spectral ranges from 60 to 120 μm. InSb detectors at 2 K (Putley, 1964) make use of the small coupling of the conductivity electrons and lattice at low temperature. The NEP is around 10^{-11} W/Hz$^{1/2}$, and the detector resistance is small enough so that it is easy to incorporate it in circuits that will not damp its short natural time constant ($\tau \simeq 10^{-6}$ sec).

Both detectors have to be used at 2 K, but they are better than pyroelectric detectors even for looking at a room-temperature background.

B. PYROELECTRIC DETECTORS CAN INTRINSICALLY BE AMONG THE FASTEST DETECTORS†

1. Classical Treatment (Thermal Detector)

We have seen in Sections II.B and II.C that pyroelectric detectors have useful capabilities at frequencies far above 10^6 Hz. The ultimate frequency limitation of a pyroelectric detector is determined by the rate at which the absorbed radiation thermalizes in the crystal. Let us consider three possibilities for absorption.

(a) Absorbing blacks. The radiation is absorbed on the target and the response is limited by a thermal diffusion time to the crystal: gold balck, 10^{-3} sec; thin metallic electrodes, 10^{-9} sec.

(b) The crystal itself (edge-electrode detectors, type II detectors). The radiation is generally absorbed directly into an infrared absorption band of the crystal itself, and the relaxation time is determined by the coupling of these phonons with the anharmonic lattice modes. This is expected to be in the 10^{-10} sec range at room temperature.

(c) Electronic transition: The radiation may excite an electronic transition in doped materials (e.g., LiNbO$_3$:Cu^{2+}); then the pyroelectric response is limited by a nonradiative electronic relaxation time.

† See Lines and Glass (1977).

2. Special Cases Where the Pyroelectric Can Work As a Quantum Detector

In both previous cases, the electronic polarization, or the time average of the unit-cell polarization (which is not zero because of anharmonicity) may also contribute to the response, which is then intrinsically shorter than 10^{-12} sec. In these cases, the response occurs from a quantum transition, respectively electronic or molecular, and is observed before thermalization has been achieved. We are dealing with a quantum detector.

C. EXTERNAL LIMITATIONS

1. Current Responsivity

The current responsivity is independent of the electrical characteristics of the crystal and amplifier and so is independent of frequency. However, for measurements, the current must not be short-circuited inside the crystal, i.e., the load impedance must be lower than that of the crystal. The crystal impedance is written $|Z| = \rho'(1 + \omega^2\tau'^2)^{-1/2}$, ρ' being the crystal internal resistance, $\tau' = \rho'C$, with C the capacitance. For high frequencies, $Z_{HF} \simeq \rho'/\omega\tau'$, with $\rho' = r_0 d/A$ and $C = \varepsilon_0\varepsilon_R A/d$ for a Type I detector; hence $\tau' = r_0\varepsilon_0\varepsilon_R$.

Now let us assume pure dielectric losses with $r_0 \simeq (\varepsilon_0\varepsilon''\omega)^{-1}$, $\omega\tau' = \varepsilon_R/\varepsilon''$, and $Z_{HF} = d/A\varepsilon_0\varepsilon_R\omega$. We see that at high frequencies Z can be small. For instance with $d = 10^{-6}$ m; $A = 10^{-6}$ m^2; $\varepsilon_R = 10$, we have $Z_{HF} = 10^{10}/\omega$; for $\omega = 10^6$ rad/sec, $Z_{HF} = 10^4$ ohms.

To measure the full pyroelectric current at any frequency of a given frequency band, the bias resistance ρ must be smaller than the smaller value of Z, i.e., the one corresponding to the highest limit of the frequency band. In these conditions $V = Zi$ is proportional to $1/\omega_{max}$. These are good conditions for observing the shape of a radiation pulse. We accept a priori the loss of a given amount of sensitivity, and then the responsivity is independent of ω (see Fig. II.30 in Wolfe and Zissis, 1978). We cannot speak of a response time.

2. Voltage Responsivity

If we want to get the highest responsivity, we have to use a bias resister ρ higher than the highest value of $|Z|$ in a given frequency band, i.e., ρ is chosen equal to the crystal resistance value ρ' at the lower limit of the frequency band to be studied. In that case $\mathcal{R} \propto \rho/\sqrt{1 + \omega^2\tau'^2}$ and $\tau' = \varepsilon_0r_0\varepsilon'$ is effectively the time constant of the detector, with $r_0 = 1/\varepsilon_0\varepsilon''\,\omega_{min} = $ const.

VII. Conclusion

Two kinds of considerations have been reviewed: There are *practical considerations,* which make pyroelectric detectors comparable to quantum detectors for frequencies up to 10^3 Hz and background at room temperature.

    1.   Plates cut at a large angle from the pyroelectric axis may have a responsivity multipled by a factor three.

    2.   Type II detectors also show a high responsivity close to the Curie temperature.

    3.   Epitaxial detectors give an NEP as small as 10^{-10} W at modulation frequencies up to 10^3 Hz.

    4.   Array detectors with up to 128 detectors have shown a NEP $\simeq 10^{-8}$ W/Hz$^{1/2}$ in a self-scanning operation at a 200-Hz frequency.

And there are *theoretical considerations,* which show that pyroelectric detectors can achieve theoretical limits as well as can quantum detectors, at least for the 10-μm range.

    1.   The intrinsic response of a pyroelectric detector may be as fast as that of a quantum detector.

    2.   Time constants are introduced by electric parameters as in the case of quantum detectors. In both cases they are the limiting factors.

    3.   In some cases the pyroelectric detector can be considered as a quantum detector.

REFERENCES

Auston, D. H., and Glass, A. M. (1972). *Appl. Phys. Lett.* **20,** 398.
Baynham, A. C., Elliott, C. T., Shaw, N., and Wilson, D. J. (1974). "Digest Conference Submillimeter Waves, Atlanta," p. 153. MIT Press, Cambridge, Massachusetts.
Byer, R. L., and Roundy, C. B. (1972). *Ferroelectrics* **3,** 333.
Carlson, A., Wittig, K., Singer, B., and Silver, A. (1981). *SPIE* **267,** 86.
Claudel, J., and Hadni, A. (1976). Electro-Optics/Laser International 76 UK, p. 149.
Gauthier, S. (1981). Propriétés ferroélectriques du phospho-tellurate d'ammonium, Thesis, University of Grenoble.
Gerbaux, X., Waldschmidt, J. M. and Hadni, A. (1978). *Appl. Opt.* **17,** 1616.
Glass, A. M., and Auston, D. H. (1972). *Opt. Commun.* **5,** 45.
Hadni, A. (1980). *In* "Infrared and Millimeter Waves" (K. J. Button, ed.), *Vol. 3.* Academic Press, New York.
Hadni, A., and Thomas, R. (1981). *Thin Solid Films* **81,** 247.
Hadni, A., Thomas, R., Unger, S. and Gerbaux, X. (1983). "Ferroelectrics," in press.
Lines, M. E., and Glass, A. M. (1977). "Principles and Applications of Ferroelectrics and Related Materials." Oxford Univ. Press, London and New York.
Molectron Corporation, (1981), Application note. Sunnyvale, California.
Nelson, D. F. and Lax, M. (1976). *Phys. Rev. B* **14,** 1759.

Poprawski, R., and Mróz, J. (1981). *Ferroelectrics* **33**, 17.

Putley, E. H. (1964). *Phys. Stat. Sol.* **6**, 571.

Putley, E. H. (1970). *In* "Semiconductors and Semimetals" Vol. 5, Chapter 6. Academic Press, New York.

Putley, E. H. (1980a). *In* "Topics in Applied Physics." Springer-Verlag, Berlin and New York.

Putley, E. H. (1980b). *Infrared Phys.* **20**, 149.

Putley, E. H. (1981a). *Infrared Phys.* **21.**

Putley, E. H. (1981b). *Ferroelectrics* **33**, 20.

Richards, P. (1983). "Proceedings of the 7th International Conference on Infrared and Millimeter Waves" (Prof. Coulon, Ed.). Université de Marseille, France.

Roundy, C. B. (1979). *Appl. Opt.* **18**, 943.

Shaulov, A., Smith, W. A., and Rao, N. V. (1981). *Appl. Phys. Lett.* **39**, 180.

Singer, B. (1977). *Adv. Image Pickup Display* **13**(3), 1–82.

Spiricon (1981). *Application note AN-10.* Logan, Utah.

Strimer, P., Gerbaux, X., Hadni, A., and Souel, T. (1981). *Infrared Phys.* **21**, 37.

Ungar, S. and Hadni, A. (1983). "Infrared and Millimeter Waves, in press.

Ungar, S., Hadni, A., Thomas, R., and Strimer, P. (1981). *Ferroelectrics* **33**, 43.

Vogt, H., Würfel, P., Hetzler, U., and Ruppel, W. (1981a). *Ferroelectrics* **33**, 243.

Vogt, H., Zepf, H. P., Würfel, P., and Ruppel, W. (1981b). *Ferroelectrics* **33**, 53.

Warner, D. J., Pedder, D. J., Moody, I. S., and Burrage, J. (1981). *Ferroelectrics* **33**, 249.

Wolfe, W. L., and Zissis, G. J. (1978). "The Infrared Handbook." Office of Naval Research, Arlington, Virginia.

Wood, O. R., Abrams, R. L., and Bridges, T. J. (1970), *Appl. Phys. Lett.* **17**, 376.

CHAPTER 6

Cyclotron and Zeeman Transitions in Photoexcited Semiconductors at Far Infrared

T. Ohyama and E. Otsuka

Department of Physics
College of General Education
Osaka University
Toyonaka, Osaka, Japan

I. Introduction

This is the second part of a discussion of off-thermal equilibrium cyclotron resonance. In the first part (Vol. 3, Chap. 7, hereafter referred to as article I), the author dealt mostly with the hot-electron system in n-type InSb and the exciton system in Ge. In the first part, the existing system of the free or loosely bound donor electrons was energized by a pulsed electric field to yield asymmetric and broadened cyclotron resonance lines. The degree of asymmetry and the extent of line broadening depend on the applied electric field and also on the relative geometry of the electric and magnetic fields. Electron temperature was determined as a function of the applied electric field, both at the geometry of $E \perp B$ and at $E \parallel B$, where E and B are the static electric and magnetic fields, respectively, applied to the InSb sample. Occurrence of the so-called cyclotron emission from the hot-electron system was also observed. The second part of article I discussed some dynamics concerning the coexisting systems of excitons and electron–hole drops (EHD) in Ge. Argument was further extended to the magnetospectroscopy as well as to the quantum oscillations. However, the recent rapid development in the study of EHD, including that within the authors' own group (Nakata and Otsuka, 1981), has forced us to modify some of the conclusions drawn in article I. The whole summary, or review, of the physical natures of EHD envisaged from the far infrared perhaps should wait another year or two even for Ge. The rapid technical progress now underway for exploring further details also holds us back from entering onto that subject here (Nakata and Otsuka, 1982).

What we employ in this article is a kind of cross-disciplinary approach that is complementary to that of article I and that introduces a rather new feature of semiconductor physics. On the one hand we put aside electrical excitation and keep InSb, while on the other hand we keep photoexcitation but put aside Ge for the moment. In other words, we have carried out cyclotron resonance in photoexcited InSb; then we go over to GaAs. Photoexcitation has indeed been the prerequisite for cyclotron resonance experiments with Ge or Si, but strange as it seems, it has seldom been used for observing cyclotron resonance in InSb or any other III–V intermetallic semiconductor compound. For n-type InSb, the carriers needed for observing cyclotron resonance are already present at liquid-helium temperatures even without any kind of excitation. For p-type InSb, people used to wait until the ambient temperature rose above the threshold at which holes bound to acceptors start to dissociate (Button *et al.*, 1968; Ranvaud *et al.*, 1979).

Use of photoexcitation for intermetallic semiconductor compounds, specifically for InSb and GaAs, lays open a new feature in cyclotron resonance.

Naturally the photoexcited carriers are of transient existence. Instead of steady-state illumination, use has been made of the light pulse. This enables us to take advantage of the time-resolution method. In other words, one can pursue the transient behavior of the photoexcited carriers. Their time-dependent characteristics have been found to be full of interesting variety from the dynamical point of view—perhaps even more interesting than in the dynamics of the exciton system in Ge.

Photoexcitation produces both types of carriers, namely, electrons and holes. Accordingly, observation of cyclotron resonance for either type of carrier, in principle, should be independent of the type of the semiconductor, whether n or p. But it has in fact been easier to observe electron cyclotron resonance in p-type than in n-type material. Of course such a statement is dangerous for general application, but we will see an example of the case in a later section. It is also possible to make a joint observation of electron and hole cyclotron resonance signals by means of photoexcitation. This fact, simple as it seems, brings forth unexpected help for understanding the band structure of InSb.

Photoexcited cyclotron resonance of GaAs is similar to that of InSb to a certain extent. It is accompanied by impurity Zeeman transition. Yet the aspect is different from the InSb case in many subtle respects. Information obtained from the time-resolved examination of the photoexcited carrier dynamics in GaAs is unique in the study of transient phenomena occurring within this material. Again, both n- and p-type materials have been investigated. The experimental result is rather sensitive to the method of growing crystals. The far-infrared magnetooptical measurement in time-resolution, accordingly, is useful for characterizing the grown material. Because GaAs has been attracting great attention for applications the experimental method introduced here would serve as a crucial test of the material applied to special purposes.

Though we will avoid further discussion of photoexcited Ge, we do introduce some new results from photoexcited Si (Ohyama, 1981a), including the first observation of the excitonic Zeeman line, a new reliable determination of the work function of EHD, etc. Exciton scattering by phonons and impurities is envisaged for the first time, to our knowledge, from the approach by far-infrared laser magnetospectroscopy.

II. Experimental Arrangements

A. PHOTOEXCITATION

By *photoexcitation* we mean excitation by intrinsic illumination, in other words, creation of an electron–hole pair by a single photon of band-gap light. Excitation such as ejecting the electrons (holes) bound at the donor (acceptor) impurity sites into the conduction (valence) band continuum by long wavelength light (the mechanism of a far-infrared detector in most cases) will not be included. Neither will we consider the two-photon process, which creates electron–hole pairs by less-than-the-gap photon energy. We mean only the more-than-the-gap photon energy. This excitation energy is produced mainly by a xenon flash lamp which is a general-purpose excitation producer. Because of the wide range of emission spectra produced by this lamp, the band gaps of most semiconductors lie within the reach of intrinsic excitation, though the intensity of a particular emission line is naturally much weaker than that of one created by an appropriate laser device. If the role of intrinsic illumination is simply to produce electron–hole pairs, there is no obvious advantage in using a laser such as argon-ion, krypton-ion, YAG, optically pumped dye, etc., in many strong-excitation experiments. Electron–hole drops in Ge, either ordinary small drops or strain-confined large drops, are readily produced by a xenon lamp (EG&G FX108). Most of the power, however, is dissipated without effectively producing electron–hole pairs in our semiconductor sample.

The xenon light, emitted isotropically, is collected with the help of a concave reflector and guided to one end of the self-focused concentric double light guides (SELFOC) produced by Nippon Sheet Glass Co. Ltd. The light power is measured to be 17 mW/cm^2 near the entrance to SELFOC. The cross section and side view of SELFOC are given in Fig. 1. It is slightly tapered, but typical diameters are 2.5 mmϕ (i.d.) and 3.0 mmϕ (o.d.). The inner part of the guide rod is made of barium silicate glass containing excess Ba atoms, and the outer part is made of sodium silicate glass containing excess Na atoms. So that the presence of this rod will not prevent the passage of the FIR beam, SELFOC is situated along the outer side of the FIR light pipe and is introduced, after bending, into the pipe immediately above the sample holder (Fig. 2). The greatest merit of SELFOC is that there is no observable loss of beam despite rather acute bending. The SELFOC transmits the band-gap light (400 nm to 2 μm) to the surface of our sample with a loss of less than 20% per meter. The power actually hitting the sample is $\sim$ 3 mW/cm^2, which is frequently too much for making meaningful measurements, and a considerable reduction of power has to be imposed before the sample absorbs the radiation. Another merit of SELFOC is that it can cut off the far-infrared part of the blackbody radiation from random sources outside at room temperature. The same shielding can also be achieved with a fused-quartz tube, but there is too much loss of band-gap light transmission after bending.

Pulsed photoexcitation has a two-fold advantage. It minimizes the temperature rise of the sample, and it is synchronized with the pulsing of the FIR laser. The typical frequency of operation has been 10 Hz. Because the width

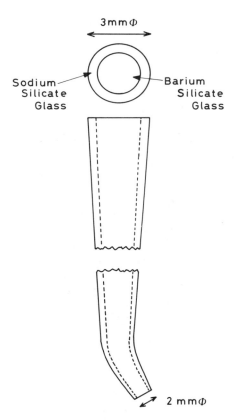

FIG. 1 Cross section and side view of the SELFOC.

of the light flash is ~ 1 μsec the duty ratio is only 10^{-5}, which very effectively prevents the sample temperature from rising. Nevertheless a temperature rise was measured in the case of Ge, both directly by means of a superconducting bolometer and indirectly by watching the spectral behavior of EHD. It has been concluded that for our standard size Ge sample ($3 \times 3 \times 1$ mm³) the rise is ~ 0.1 K during 1 μsec immediately after the photopulse at maximum intensity: then the sample cools off rapidly. The amount of rise can change if one inserts a different kind and size of sample. In no case, however, does the maximum rise exceed 1 K. Because as a rule a meaningful measurement has been made more than 1 μsec after the light flash, and frequently under reduced intensity of excitation, sample heating can practically be ignored. This consideration is particularly important when one deals with the electron temperature, which should be defined independently of the lattice temperature.

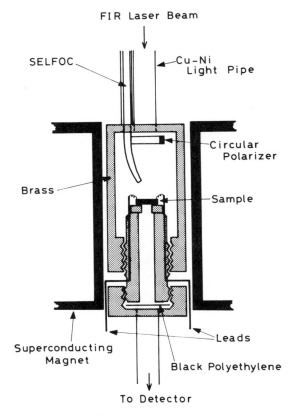

FIG. 2   Detailed sketch of the sample holder.

B.   FIR Laser System

The main aspect of the cyclotron resonance experiment is the far-infrared laser source. This particular series of experiments has mostly used a discharge-type laser rather than an optically pumped unit. The latter is certainly more useful for spectroscopic purposes because of its ability to produce many lasing wavelengths, but the maintenance of high stability, etc., cannot necessarily be had at all wavelengths. In particular, when dealing with transient phenomena it is desirable to be able to pay attention exclusively to the time-dependent behavior of the sample itself rather than to the laser operation. In our experience the operation of the discharge-type laser requires less labor in many respects, but we must be content with a fixed laser wavelength. Of course, we can secure more than one wavelength if necessary, even in the discharge-type laser, simply by changing the vapor in the cavity or by adjusting the tuning. As a matter of fact, five different wavelengths were obtained from a single discharge-type laser: 84 μm (3571 GHz), 163 μm (1840 GHz), and 172 μm (1744 GHz) from D_2O vapor; and 119 μm (2521 GHz) and 220 μm (1364 GHz) from H_2O vapor.

The general features of the discharge-type laser are more or less conventional; it is driven by a dc 8-kV power supply: the air-cooled glass tube has a length of 2 m and a diameter of 80 mmϕ: it has a concave mirror with a long focal length (2 m) and a plane mirror and so on. A few new things are worth mentioning. The D_2O or H_2O vapor contained within the cavity has been mixed with He gas by mole ratio of 1 : 1 with the total gas pressure being kept at 0.1 Torr. Addition of He gas lengthens the discharge time to nearly 100 μsec and the corresponding lasting time to $\sim$ 50 μsec. The laser oscillation becomes more stable, and the output power is increased by 50% at each lasing wavelength.

Occasionally a handy grating monochromator with a focal length of 10 cm was inserted between the FIR laser output and our cryostat. The blaze wavelength and groove density of the grating are 112.5 μm and 7.9 mm^{-1}, respectively. With a slit opening of 4 mm, a resolution of $\lambda/\Delta\lambda \gtrsim 50$ is obtained. Because the insertion of the monochromator reduces the FIR laser beam intensity by a factor of 5, its general use should be avoided. Incidentally, the insertion led us to the discovery of a new lasing line from D_2O vapor (Ohyama, 1981b) having a wavelength of 163 μm that has not been reported elsewhere to our knowledge.

Only for the supplementary purpose of dealing with the exciton system in Si, the wavelengths 96.5 μm (3109 GHz) from CH_3OH and 103 μm (2913 GHz) from CH_3OD have been employed with the help of CO_2 laser pumping.

The standard operation of a discharge-type laser is described everywhere and does not deserve a detailed discussion. The only new thing here is its use in combination with pulsed photoexcitation. In order to make a synchronized combination with the photoexcitation repeated at 10 Hz, the FIR laser discharge is triggered at 20 Hz. The synchronization therefore is at every other pulse, or else every other FIR laser pulse (nearly 50 μsec wide) is not accompanied by photoexcitation. By such operations one can extract the exact contribution of the photoexcited carriers to cyclotron resonance.

C. FIR DETECTORS

The FIR detectors are all homemade. We frequently tried doped Ge detectors, the dopant depending on the FIR wavelength employed. Otherwise we used n-type InSb accompanied by a superconducting solenoid, the so-called Putley detector; this detector has a general-purpose character at the sacrifice of sensitivity. Only for a special case was use made of n-type GaAs. All these detectors have both merits and drawbacks.

In the experiments noted in this article, best use has been made of the

Putley unit, at the expense of response time and sensitivity. The basic material is an *n*-type InSb rectangular piece ($2 \times 4 \times 0.3$ mm³, $N_D - N_A = 2 \times 10^{14}$ cm⁻³). It was polished by emery and alumina (0.1 μmϕ) and etched by CP–4 solution diluted 50% by water. Electrodes in ohmic contact were achieved with In. The material had a resistivity of 0.03 $\Omega \cdot$ cm at 300 K, with 0.23 $\Omega \cdot$ cm (B = 0) and 230 $\Omega \cdot$ cm (B = 1 T) at 4.2 K. The bias was adjusted by a 1.5-V battery with the help of a 100-kΩ variable resistor in series. The responsivity, which is tunable with the help of cyclotron–resonance-assisted photoconductance, was optimum between 150 and 200 μm in radiation wavelength. The sensitivity could certainly be increased by raising resistivity, either by making the material thinner or by choosing a smaller $N_D - N_A$. But the capacitance–resistance time constant actually hampered its free operation, especially at lower temperatures, say at 1.5 K.

The associated superconducting solenoid has an outside diameter of 22 mmϕ, a length of 80 mm, and a bore diameter of 9 mmϕ. The winding is 1800 turns of Nb–Ti single-core wire with an o.d. of 0.33 mmϕ. The unit is illustrated in Fig. 3. By passing current through the heater wire, one keeps

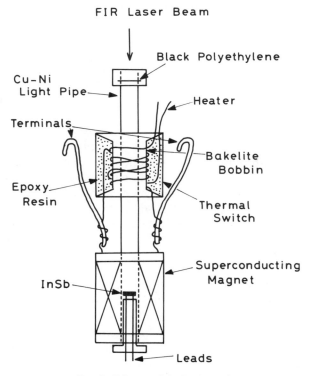

FIG. 3 Scheme of the Putley unit.

the temperature of the coil above T_c. The coil can be energized up to 1.5 T in one minute. At the appropriate magnetic field strength, the heater current is shut off. After a few minutes the energization current is brought down to zero. Thus we complete the persistent-current magnet. The time variation of the magnetic field measured by a GaAs Hall element is given in Fig. 4. The decay rate of the field strength is less than 0.01% per hour. At the optimum working condition the sensitivity of this detector unit is 500 times lower than our optimum working sensitivity for Sb-doped Ge. An advantage, however, is the faster response time, which in our experience is less than 0.1 μsec at the detector. This is suited for time-resolution analysis of a fast transient phenomenon. In actual measurements one has to allow for a somewhat slower response because of the net capacitance – resistance effect of the circuit and the bandwidth of the amplifier. All in all, the effective risetime has been confirmed to be less than 0.5 μsec.

For observing cyclotron emission a higher sensitivity is required, because the emission signal is usually rather weak. Thus, typically, use was made of Sb-doped Ge ($3 \times 3 \times 0.5$ mm³, $N_D - N_A = 1.4 \times 10^{14}$ cm⁻³). The electrode was soldered with a 90%Pb – 10%Sb alloy. The material had a resistivity of 5 $\Omega \cdot$ cm at 300 K and more than $5 \times 10^5 \Omega \cdot$ cm in operation at 4.2 K. The bias was applied again with the help of a 1.5-V battery, but we found the variable resistor in series should be ~ 2 MΩ. Though the response time of the Ge piece itself was 0.1 μsec, the long RC time constant due to the high

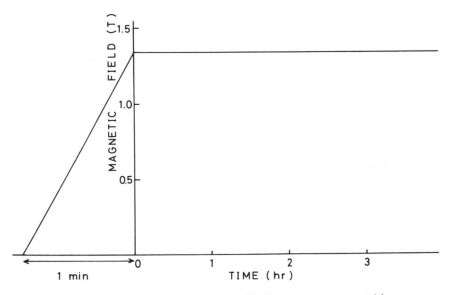

FIG. 4 Decay rate of the field strength generated by the permanent current-driven magnet. The origin of the horizontal axis corresponds to the time when the thermal switch is shut off.

resistance made the apparent response time of the entire detector system as long as $\sim 10\,\mu$sec. For this detector material, compensation was very small. To see the shift of maximum photoresponse in the magnetic field another detector (As-doped Ge with $N_D = 5 \times 10^{13}$ cm^{-3}) was used. In previous experiments many other detectors were used to analyze the cyclotron emissions generated by electrical excitation, because it was of primary importance to confirm the origin of emission for the first time. But this time, because we expected little more than simple confirmation of the emission phenomena, we used only two kinds of detectors.

D.   OTHER TECHNICAL ADVANTAGES OR FACILITIES

1.   Differential Method (Optical Modulation)

Synchronized combination of FIR laser pulses with photoexcitation pulses was mentioned in section B. Without photoexcitation we observe the cyclotron resonance in thermal equilibrium, which has been treated many times in the past. Let us denote the intensity of transmitted FIR beam without photoexcitation by I_0 and that with photoexcitation by I_L. The increment in absorption coefficient $\Delta\alpha$ upon photoexcitation can be expressed

$$d\Delta\alpha = \ln(I_0/I_L), \tag{1}$$

where d is the thickness of the sample. By means of a twin boxcar system, the quantity $\ln(I_0/I_L)$ can be derived. Thus one can obtain $\Delta\alpha$, the genuine contribution of photoexcited carriers to the cyclotron resonance.

2.   Circular Polarizer

When it is necessary to distinguish electron and hole signals, it is helpful to use a circular polarizer. We have prepared one for the wavelength of $119\,\mu$m. The polarization achieved is 90%. As we shall see in section III, hole signals in InSb are identified with the help of this polarizer. Once the signals are identified at 119 μm, their identification at other wavelengths is quite straightforward; the use of the polarizer at each wavelength is not essential.

3.   Auxiliary Means of Photoexcitation

Though a xenon flash lamp has mainly been used for excitation, we also have employed a tungsten lamp, a YAG laser, and an Ar$^+$ laser system. These were used only in observing the cyclotron emission from photoexcited carriers in InSb. The tungsten lamp is a 24-V 150-W movie projector bulb. Its radiation was mechanically chopped at 30 Hz to yield chopped light of ~ 1 msec duration. The YAG laser (Coherent, Model 60), 5 W if operated in the continuous wave (CW) mode, is also put on trial. In

Q-switching it produces a peak power of 600 W at a repetition rate of 1000 Hz, 0.6 W on the average, with a pulse width of 200 nsec.

4. External Magnetic Field

The static external magnetic field was produced by a small vertical superconducting solenoid, homemade equipment wound with Nb–Ti wires. The solenoid has a bore diameter of 20 mmϕ, the height and outer-diameter being 70 mm and 60 mmϕ, respectively. The highest field available was $\gtrsim 5$ T at 40 A. Nearly seven experimental hours were obtained at 4.2 K and five hours at 1.7 K with 2.2 l of liquid helium.

III. Hot Carrier Experiments in InSb

A. SUMMARY OF THE PAST

The subject of hot carriers — or to be more exact, hot electrons — in InSb is not new. The association of cyclotron resonance with hot electrons has been made more than once, even within the authors' group (Kobayashi and Otsuka, 1974; Matsuda and Otsuka, 1979). In all cases, however, electrons contributing to cyclotron resonance were already present prior to electrical excitation. The donor impurity levels are so shallow that in n-type material a large portion of the donor electrons are released into the conduction band even at liquid-helium temperatures. We have, accordingly, no difficulty in seeing the electron cyclotron resonance line at a practical thermal equilibrium. (In a strict sense, the system is not in thermal equilibrium, because we are perturbing it with FIR laser irradiations.) Upon application of an electric field, donor electrons are ejected into the conduction band, thus increasing the intensity of the electron resonance. One should keep in mind that another absorption peak exists that we call, for convenience, *impurity cyclotron resonance* (ICR). This should be distinguished from the genuine cyclotron resonance of conduction electrons (CCR). The ICR absorption peak arises from the donor bound electrons, which will be swallowed by the conduction band after the transition. In the absence of a magnetic field this transition reduces to the conventional $1s \rightarrow 2p$ type transition of a hydrogenlike impurity. Under a "strong" magnetic field satisfying the condition

$$\hbar\omega_c/2 \gg R_y^*, \tag{3.1}$$

the conventional idea for such a transition is no longer valid. Here ω_c is the cyclotron frequency for electrons and R_y^* is the effective Rydberg, or the binding energy, of the donor electron in the absence of a magnetic field. The new situation is that even the bound electron has the character of a "free" electron, because the zero-point energy $\hbar\omega_c/2$ of the electron is much larger than the binding energy of the donor impurity. In other words, the donor bound electron can be regarded as a kind of hindered harmonic oscillator. This is the reason why the name "cyclotron resonance" is attached even to the impurity electron absorption. The ICR line appears side by side with the CCR line, the former lying in the lower magnetic field. The ICR line disappears when the ambient temperature is raised or when an electric field is applied to the InSb sample.

The CCR line, or the conduction electron cyclotron resonance, also offers a tool for studying the hot-electron system. Due to the relatively narrow energy gap of InSb (0.24 eV at 4.2 K), the bottom of the conduction band deviates somewhat from a parabola. This deviation causes the splitting of different cyclotron transitions; in other words, we observe more than one cyclotron transition, which we may denote C_1, C_2, ... from the lower energy side. The appearance and the intensity of the second or third line, C_2 or C_3, depends on whether the temperature of the electron system is determined by the lattice or by the applied electric field. The electron temperature under the application of an electric field has been determined either by measuring the relative intensity of C_1 and C_2, or by analyzing the line shape of the C_1 line. The electron temperature derived in the first way is called the *intersubband electron temperature,* whereas that determined in the second way is the *intrasubband electron temperature.* An outstanding feature of the earlier experiment with electrical excitation has been the geometry effect, i.e., the difference in distribution of the electron system between the geometries $E \perp B$ and $E \parallel B$, where E and B are applied electric and magnetic fields, respectively. For a given applied electric field the electron temperature is different for the two geometries. The difference is reflected in the line shape of C_1, the relative intensities C_1/C_2, etc. It is also reflected in the relative intensity of ICR. All these features are summarized in article I.

For InSb, the cyclotron resonance of holes has attracted independent attention. Apart from Ge and Si, InSb is perhaps the only material for which the structure of the valence band is reflected in cyclotron resonance by the appearance of the so-called quantum lines, i.e., the various cyclotron transitions arising from different energies between the Landau levels. The first irrefutable observation of quantum lines in the valence band of InSb was made by Button *et al.* (1968). Using *p*-type InSb, they observed the quantum lines due to the holes that were thermally released from the acceptor impurities. Because the energy levels of the acceptor are rather deep in InSb, the temperature had to be raised to nearly 20 K in order to release the holes. The second reference to hole cyclotron resonance can be found in Ranvaud *et al.* (1979). A new feature was added by the application of uniaxial stress, which led to a new determination of the Luttinger band parameters. Prior to the cyclotron resonance work of Ranvaud *et al.,* a pioneering determination of the band parameters had been made by Pidgeon and Brown (1966). These authors, however, did not perform cyclotron resonance experiments but saw the interband magnetooptical absorption, i.e., Landau transitions.

In all past experiments the free holes were investigated under thermal equilibrium. The difficulty of creating "hot" holes will again be experienced here. The main reason for this difficulty must be attributed to the heavier average mass of holes in comparison with the average mass of electrons.

B. SAMPLES

As mentioned in section I, it is not of vital importance for photoexcited cyclotron resonance whether one selects n- or p-type InSb, because both electrons and holes can be produced, at least in principle, in both types of material. So long as the crystal is pure, one can readily observe electron (or hole) cyclotron resonance in p-type (or n-type) material. As we shall see, however, sometimes it is convenient to select a specific type of material for a specific purpose.

We chose three kinds of samples. However, for different experimental purposes we did not always use the same piece of the same kind. For example, cyclotron emission was looked for in two different pieces of a sample for optical and electrical excitations, one without and the other with electrodes attached. Even in such cases care was taken to take the two pieces from areas in the same ingot as close as possible to each other. Still, with all this care, greatly different properties were sometimes observed in the two pieces, especially in the case of p-type material, in which a large compensation could be expected for a high-purity crystal.

The impurity characteristics of the samples employed are listed in Table I. We classify the samples A, B, and C, with C being p-type material. All the samples are shaped so as to be convenient for far-infrared transmission measurement. The plane facing the incident radiation is a square, 4×4 mm^2. Thickness varies between 0.15 and 0.50 mm. The flat $\langle 111 \rangle$ face has always been used.

C. GENERAL BEHAVIOR OF PHOTOEXCITED CYCLOTRON RESONANCE

The most essential feature of the photoexcitation experiment is that both electron and hole signals can be observed at the same time; however, this is possible only in the case of steady-state excitation. In the time-resolution experiment, as in the present case, the hole signals quickly die away. The electron signals, on the other hand, stay much longer. The overall intensity and duration of the signals further depend on the intensity of photoexcitation. As an example, cyclotron-resonance traces obtained at the FIR wavelength of 119 μm (H$_2$O) are shown in Fig. 5 for sample A (n-type). Electron

TABLE I

LIST OF THE InSb SAMPLES

Sample	Type	$N_D(/cm^3)$	$N_A(/cm^3)$
A	n	3.8×10^{14}	1.7×10^{14}
B	n	1.9×10^{14}	1.8×10^{14}
C	p	1.0×10^{14}	1.1×10^{14}

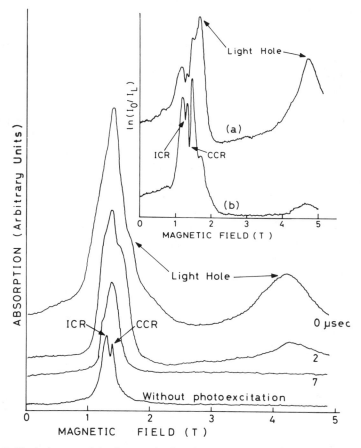

FIG. 5 Typical traces of the time-resolved cyclotron resonance of the photoexcited sample A for 119 μm, 4.2 K, $B \parallel \langle 111 \rangle$. For comparison, the signal without photoexcitation is also shown. The time zero is set at the top of the excitation pulse. The inset shows the signals for the differential method and a circular polarizer: (a) hole active mode and (b) electron active mode. In this method, ICR and CCR appear as dips.

signals are indicated as ICR (impurity cyclotron resonance) and CCR (conduction-electron cyclotron resonance). Two-hold signals are observable for this wavelength. They correspond to the transitions arising from the spin up and spin down states of the lowest Landau level for the "light hole." To identify the hole signals more clearly, use was made of a circular polarizer. Because holes appear only under photoexcitation, the differential method introduced in section IID turned out to be very effective.

The time-resolution procedure helps us grasp the general kinetics of photoexcited cyclotron resonance, as follows. With excessive band-gap photoexcitation many electrons and holes are produced. Part of the free carriers produced are lost in neutralizing the ionized donors and acceptors. All the impurities are thus made neutral. The remaining photocarriers give rise to the cyclotron-resonance signals in addition to those due to the carriers in thermal equilibrium. The free carriers recombine, however, within a very short time, after which the hole signals are no longer observable. Holes are still available at the neutral acceptor sites, but they cannot be set free at liquid-helium temperatures. Electrons, on the other hand, are observable at any time, which is no surprise, because even in thermal equilibrium at liquid-helium temperatures electrons readily leave the donor states. However, the overall signal intensity of the electron lines diminishes gradually after the disappearance of the hole signals. This is due essentially to the donor-to-acceptor electron transfer. Thus there are two stages in the energy-relaxation process of the electron system. If we measure the signal intensity of the C_1 line as a function of delay time, the time variation is found to be like that shown in Fig. 6. The initial rapid decay, which we shall call the first stage, is characterized by a time constant $\sim 1.5\ \mu\text{sec}$. The value of this time constant cannot be trusted, because the sampling is too small and because the apparent response time of our FIR detector system is of the order of 0.5 μsec. All we can guarantee from our experiment is that the actual recombination time is smaller than 1.5 μsec. For the second stage, on the other hand, we have enough sampling points. The observed time constant in the example of Fig. 6 is as long as 2.4 $m\text{sec}$. This time constant is nearly independent of the ambient temperature but strongly dependent on the sample; for sample C it is of the order of $\sim 10\ \mu\text{sec}$. It should further be noted that the second stage may not necessarily be characterized by a single exponential function of time. After a long time lapse, one arrives at the thermal-equilibrium signal trace, ICR and C_1 for n-type samples and nothing for p-type.

D.   ELECTRON CYCLOTRON RESONANCE

Photoexcited electrons are present in both n- and p-type materials. In n-type InSb electron signals are observed without photoexcitation, but in p-type material they are not. With photoexcitation, electron signals are of

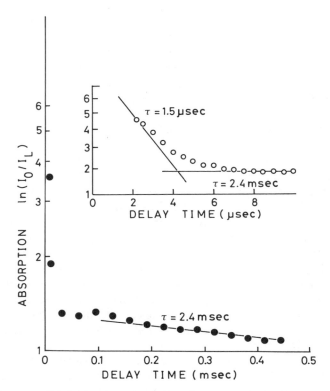

FIG. 6 Decay profile of the cyclotron resonance after photoexcitation for sample B, 119 μm, 4.2 K. The inset shows the initial fast decay more clearly.

both photoexcited and thermal-equilibrium origin in *n*-type material, whereas all signals are due to photoexcitation in the *p*-type material. In other words, no differential method is required for investigating photoelectrons in *p*-type InSb. Accordingly, the *p*-type material offers a better chance to understand the behavior of photoelectrons. Be that as it may, *n*-type InSb has its own characteristics, which should be examined briefly. Then emphasis will be put on *p*-type InSb.

1. n-*type InSb*

The electron cyclotron resonance in *n*-type InSb can directly be compared with that due to electrical excitation. A general view of the resonance behavior has already been given in Fig. 5, where hole resonance is also showing up. In the first stage, where excitation is so strong that lines overlap very little information other than the overall intensity of absorption can be extracted. In the second stage, each absorption line can be seen, and more information becomes available. We now take a serial trace of time-resolved cyclotron resonance signals, as in Fig. 5, using the same n-type sample A but with the excitation intensity reduced by an order of magnitude. By thus reducing the excitation intensity one can practically skip the first stage. A pattern of such a time-resolution series obtained at the FIR wavelength of 84 μm is given in Fig. 7. No hole signal is recognizable from the beginning, because the light is not strong enough to create the excess free carriers that give rise to hole signals even after the acceptors are neutralized. The off-thermal-equilibrium feature that comes right after photoexcitation can be seen both in the broader linewidth of C_1 and in the appearance of C_2. One finds something different from the behavior in the case of electrical excitation, which is that the relative intensity of the ICR line is always comparable to that of the C_1 line. This aspect of ICR is saying something new in the electron distribution function.

On closer examination of the C_1 line, one finds a shoulder on the lower magnetic field side. This is shown in Fig. 8 under stronger excitation. Two possible origins of this shoulder can be considered: one is that there are

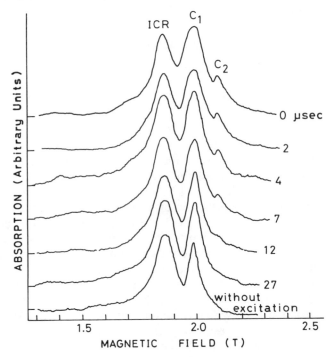

FIG. 7 Time-resolved cyclotron resonance absorption in sample A under the reduced photoexcitation (84 μm, 4.2 K, $B \parallel \langle 111 \rangle$), with an enlarged abscissa.

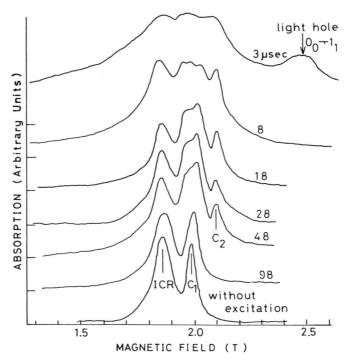

FIG. 8 Time-resolved cyclotron resonance absorption in sample A under the strong photoexcitation (84 μm, 4.2 K, $B \parallel \langle 111 \rangle$).

different kinds of impurities; the other is that it is due to transitions between the higher discrete levels of the same impurity system, which have been unidentified so far. The first possibility is unlikely, because such a shallow donor level as that in InSb cannot be expected to undergo an observable chemical shift. In other words, the effective-mass approximation for the donor electron system is very good. To discuss the second possibility, one has to recall the quantum-mechanical treatment of the impurity states in a strong magnetic field. Let us characterize the impurity levels, making use of the quantum numbers introduced by Hasegawa and Howard (1961). Then the commonly known ICR transition can be described as $000 \rightarrow 110$. According to Hasegawa and Howard, this is the only allowed transition between the discrete levels with infinite magnetic field strength. With a finite magnetic field, however, residual probability is still available for other transitions. One possibility would be $0\bar{1}0 \rightarrow 1\bar{1}0$, which was hinted at already in the work by Kobayashi and Otsuka (1974). A question arises, nevertheless. Because the shoulder hardly shows up for p-type material, a conclusive identification of this transition is still pending.

The time variation of the C_1 signal intensity is very slow in the second stage, as already shown in Fig. 6. On the other hand, the C_2 signal is rather short-lived. In Fig. 7, for reduced excitation, it is virtually unobservable at 27 μsec after the photopulse.

The contribution of the thermal-equilibrium signal relative to the entire signal is very large in the reduced-excitation experiment, which makes a quantitative analysis of the genuinely photo-excited carriers difficult. For that reason, emphasis will rather be on the data from p-type InSb.

2.   p-type InSb

The physical characteristics of the photoexcited electrons show up more distinctly in p-type material, because signals are all due to photoexcitation. We need not worry about the difference between equilibrium and nonequilibrium signals. In other words, no differential method is necessary.

As when we dealt with n-type material, we shall put aside the treatment of the first stage and focus our attention on the second stage. We take a series of time-resolved traces at reduced intensity of excitation, as for n-type material. Typical data are illustrated in Fig. 9. Holes are not seen from the beginning. The individual time variation of each signal is of interest. It is shown in Fig. 10 where the integrated intensity of each signal, which is proportional to the density of electrons, is plotted against delay time. The intensity of ICR gradually rises for the first several microseconds, indicating that it takes time to neutralize the donor impurities, which, being in p-type material, have been entirely ionized before photoexcitation. After reaching the maximum at 6.5 μsec in the data shown, the ICR intensity starts decaying. The C_1 signal, or the conduction-electron signal, starts decaying from the beginning, or within less than 1 μsec after the origin of the time scale, which is taken from the top of the photopulse. The C_1 signal does not represent the entire population of conduction electrons, because there exists an appreciable fraction of the C_2 signal at the early stage. The C_2 signal decays rather quickly in comparison with C_1 (see Fig. 10). In order to see the overall time process of the photo-excited electron system, we also plot in Fig. 10 the total intensity ICR + C_1 + C_2. It remains practically constant for the first several microseconds and then gradually starts decaying. The gradual decay, which is nearly exponential in time for this sample, is true for all the quantities, including C_1. The observed time constants are roughly as follows: 11.1 μsec for C_1, 2.1 μsec for C_2, 13.4 μsec for ICR and for ICR + C_1 + C_2, and 10.3 μsec for C_1 + C_2. It is evident that the time constant of the entire electron system is eventually governed by that of ICR.

One may be surprised to see such long-lived photoexcited carriers in a direct-gap semiconductor. If the recombination process were to be considered solely as the direct electron–hole interaction, the time constant would be much smaller. The real process, however, involves the impurity states,

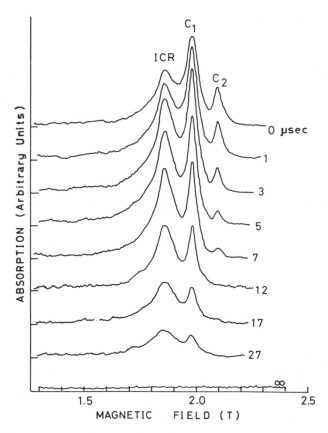

FIG. 9 Time-resolved cyclotron resonance absorption in sample C under the reduced photoexcitation (84 μm, 4.2 K, $B \parallel \langle 111 \rangle$). The lowest trace shows that nothing remains in the long run in p-type material.

which can accommodate electrons for a long time. There exists a kind of quasi-equilibrium between the electron system running freely within the conduction band and that anchored at the donor sites. Exchange of electrons between the donor levels and the conduction band frequently occurs, but it does not contribute to the energy relaxation of the electron system. Energy loss occurs only when an electron jumps down to the acceptor level where the electron–hole recombination finally takes place, thus leaving the acceptor ionized to make the entire system tend to thermal equilibrium. The frequency of the donor-to-acceptor electron transfer depends on concentration, distribution, and compensation of the impurities. The transfer of electrons may occur either by direct transfer (tunneling) or by the two-step

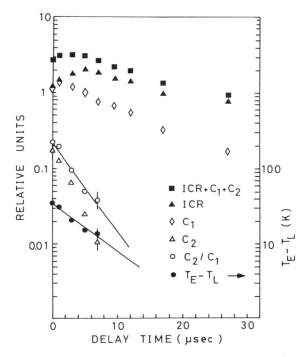

FIG. 10 Time variation of various quantities obtained from Fig. 9. Electron temperatures are also shown (see the scale on the right).

process (ionization of the donor followed by capture of the electron by the neutral acceptor). In the latter case, ionization of a neutral donor is expected to occur only by thermal agitation. As a matter of fact, our experience shows that, at least for the given sample, the apparent time constant for ICR does not show any appreciable temperature dependence. This indicates that the process occurs largely by the direct transfer of electrons from donor to acceptor, because the probability for thermal ionization of the donor electron is strongly temperature dependent. The transition probability for direct transfer will, on the other hand, be simply a function of the distance between the relevant donor and acceptor, i.e., a function of the form $\exp(-2r/r_0)$, where r is the distance and r_0 a constant.

An important feature of the time-resolved cyclotron resonance is the time variation of the electron temperature. The simplest and perhaps most intuitive definition of the electron temperature is given by the relation

$$C_2/C_1 = \exp[-(E_2 - E_1)/k_B T_E], \qquad (2)$$

where E_1 and E_2 are the energies at the minima of the Landau subbands 0^+

and 0^-, and T_E is the electron temperature. The notation C_2/C_1 is to be understood as giving the ratio of the integrated intensities of the two lines, which is equal to the electron population ratio n_-/n_+ between the two subbands. The electron temperature thus defined is called the intersubband electron temperature (Matsuda and Otsuka 1979), because it involves two different Landau subbands. Electron temperature can be defined in other ways, for example, by analyzing the line shape of the given resonance, specifically the C_1 line. This was also treated in our previous work, in which we called the temperature thus derived the intrasubband electron temperature. Furthermore, one can define electron temperature by examining the relative intensity of the ICR line. In the present case, however, we shall be content to discuss only one kind of temperature, i.e., the intersubband electron temperature as defined by Eq. (2), and call it simply *electron temperature*. We give in Fig. 10 the quantity C_2/C_1 as a function of time. The decay seems to be nearly exponential, and the time constant is found to be 3.0 μsec. From the ratio C_2/C_1 one can readily find the electron temperature T_E with the help of Eq. (2). We are interested in the manner in which the electron temperature decreases: in other words, we want to see if the classical Newton's law for cooling is valid for electron temperature. The reference temperature will of course be taken to be the lattice temperature, which we write T_L and assume equal to the bath temperature 4.2 K. We plot the difference $T_E - T_L$ as a function of delay time. The approximate linearity in the semilogarithmic plot shows the validity of the equation

$$-\frac{d(T_E - T_L)}{dt} = \frac{T_E - T_L}{\tau_c}. \tag{3}$$

This is Newton's law itself. Here τ_c is the time constant of cooling, $\sim 7\,\mu$sec, which is practically independent of temperature between 1.6 and 4.2 K. It is rather surprising to see the energy relaxation of electrons so slow. As a matter of fact, a qualitative estimate based on acoustical phonon emissions leads to a time constant of $\sim 0.1\,\mu$sec at 4.2 K. In other words, a bottleneck exists between the hot electron system and the lattice.

An independent interpretation, however, can arise here. In observing the intensities C_1 and C_2, we are dealing only with the electrons at 0^+ and 0^-. Transfer of electrons from 0^- to 0^+ states means the flipping of spins of the conduction electrons. The electron temperature T_E, accordingly, may also be regarded as the spin temperature. In other words, the cooling process we observe is the cooling of the spin temperature.

This situation tempts us to speak of our measurement of the spin relaxation. No measurement of spin–relaxation time is available for conduction electrons in InSb at this magnetic field strength. The direct observation of the electron populations at 0^- and 0^+, as carried out here, is certainly more convincing and more accurate in deriving the relaxation time than the conventional magnetic-resonance experiment. Thus we might as well connect the time constant τ_c, or rather the decay time of the ratio C_2/C_1, to the spin–relaxation time. Such an interpretation, attractive as it seems, needs caution. The spin system we are talking about is not closed. The electrons residing at 0^+ may drop down further to the impurity states. In the meantime, the electron population at 0^- may be recruited from the higher Landau subbands. Yet these effects would be quite negligible in the present case. As a matter of fact, a trial calculation of the spin-flipping time in terms of impurity scattering, for the impurity concentration of $10^{14}\,\text{cm}^{-3}$, yields ~ 3 μsec (Fujii, 1982), in good agreement with our experimental observation. Thus the spin flipping process appears to be a bottleneck in the energy relaxation of the electron system. With this spin-involving relaxation process in mind, let us go back for the moment to the original simple language of electron temperature.

The definition of electron temperature implicitly assumes a Boltzmann distribution of the electron system. This has not been directly confirmed. Derivation of T_E from an equation such as Eq. (2) can simply be done for any distribution function whatsoever, because it deals with only two subbands. The best way to see whether the distribution is Boltzmann or not is to measure the relative intensities of the higher cyclotron transitions, say, C_3 and, if possible, C_4. Unfortunately they are not strong enough to be observed. If the Boltzmann distribution is guaranteed, the relative intensities of cyclotron transitions should be along a straight line in the plot of energy versus intensity (Matsuda and Otsuka, 1979). In the case of electrical excitation, a nearly Boltzmann distribution was observed up to the intersubband electron temperature ~ 40 K for the geometry of $E \perp B$. Though the overall environmental situation is different in optical excitation, we assume for the moment that the electron distribution is nearly Boltzmann shortly after photoexcitation or under moderate excitation. In the case of strong excitation, the apparent relative intensity C_2/C_1 almost gives an impression of being $\lesssim 1$, which means that the electron temperature is infinite, a ridiculous situation.

Another nonequilibrium aspect of the electron system can be examined by looking at the linewidth and line shape of the cyclotron resonance. Measurement of the linewidth has been done both for C_1 and for C_2 in the same time-resolution series. We assume the Lorentzian line shape and derive the inverse relaxation time, or the collision frequency $1/\tau$, from the simple relation

$$\omega_c\tau = 2B_0/\Delta B, \tag{4}$$

where ω_c is the cyclotron frequency for electrons, B_0 the resonance field, and

ΔB the linewidth of the relevant cyclotron resonance. The $1/\tau$ value thus derived is plotted against time in Fig. 11 for both C_1 and C_2. It is seen that the linewidth reaches its saturation value shortly after the photopulse, at ~ 5 μsec of the delay time, where the electron system still has an electron temperature of ~ 20 K. The linewidth of C_2 is hard to follow beyond 5 μsec because of the fading signal intensity. The $1/\tau$ value at the apparent saturation is $(2.2 \pm 0.1) \times 10^{11}$/sec for C_1 and $(1.0 \pm 0.2) \times 10^{11}$/sec for C_2. It is of interest to see such a difference, a factor of 2, in collision frequencies between the electrons lying in the 0^+ subband and those lying in the 0^- subband. An interpretation that immediately arises is in terms of the spin-polarized electron scattering, which involves the polarized spin states of the donor impurities. The notion essentially assumes that electron–neutral-donor scatterings are predominantly contributing to the linewidth. If the donor electrons are practically all spin-polarized, the collision frequency will be quite different between 0^+ and 0^- electrons. Such a difference has indeed been observed very clearly for donor states in both Ge and Si by

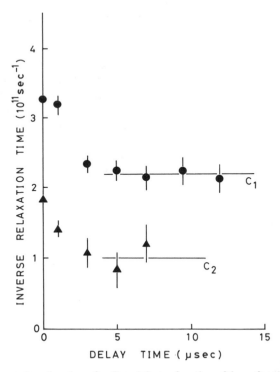

FIG. 11 Inverse relaxation times for C_1 and C_2 as a function of time after the photopulse (84 μm, 4.2 K). The horizontal lines in the figure show their saturated values.

Ohyama *et al.* (1970). The theory and observed experimental result for Ge and Si tell us that the singlet configuration gives a larger scattering cross section than does the triplet. A simple calculation predicts a factor of ~ 5 difference for a relatively low wave number of the incident electron, say for that with $ka_B^* \sim 0.2$, where a_B^* is the effective Bohr radius of the donor electron. If we interpret our experimental observation for InSb in terms of this mechanism, we have to conclude that the dominant donor-electron occupation is over the minus spin states. In other words, the g-factors of the donor electron and the conduction electron have opposite signs. A positive g-factor for the donor electron, however, is highly inconceivable considering the Roth formula (Roth *et al.*, 1959). The well-known negative g-value (~ -50) for electrons in InSb is caused primarily by the small effective mass. Even for the donor electron, assumption of equally small effective mass cannot be avoided in order to account for the large Bohr orbit or the small binding energy. The only chance for introducing a positive g-factor will thus be to build an entirely new expression for the donor electron, one different from the Roth formula. At this moment the authors are unable either to justify such a procedure or to find some way to succeed in it.

With the question of g-factor being put aside for the moment, we consider the donor impurities to be playing an important role in the energy relaxation of the photoexcited electron system. Because of the large Bohr orbit of the neutralized donor, electron–donor scattering will be quite frequent and could mask the electron–phonon interaction. In fact, the observed small temperature dependence of τ_c indicates the unimportance of the electron–phonon interaction. This argument holds its validity both for the purely orbital and for the spin-involved relaxation processes.

E. Hole Cyclotron Resonance

The hole cyclotron resonance in InSb has been determined only in p-type material and at temperatures near 20 K (Button *et al.*, 1968; Ranvaud *et al.*, 1979). The experiments were done without photoexcitation, and the carriers were those thermally provided by the acceptors. The acceptor levels are so deep that practically no free holes are available at liquid-helium temperatures.

Production of both electron and hole signals by band-gap photoexcitation is an old technique in cyclotron resonance. The reason that people did not use this technique for InSb in the past was simply to avoid possible confusion due to the intermingling of electron signals. To safely identify the quantum lines truly arising from the valence band, the choice is certainly understandable. But the electron signals are rather easy to identify, and one can also make full use of the circular polarizer mentioned in section IID. Moreover, the data on hole cyclotron resonance signals at liquid-helium temperature would be a contribution to semiconductor physics.

The purpose of this work, however, is not to obtain general information on the entire set of quantum lines, which would require an extremely high magnetic field, as high as 20 T even at the FIR wavelength of 337 μm. Instead, we will be satisfied with watching the behavior of the two quantum lines obtainable at low magnetic field. Hensel and Suzuki (1974) introduced a way of classifying the valence-band quantum transitions from cyclotron-resonance results for holes in Ge, and their method of classification is applicable to InSb. According to their quantum numbers for identifying the Landau levels, the two quantum lines that we observe here are described as the transitions $0_0 \rightarrow 1_1$ and $1_0 \rightarrow 2_1$. In semiclassical language these are the lowest energy cyclotron transitions of the "light hole," having two possible spin states. The transition $1_0 \rightarrow 2_1$ escapes observation below a magnetic field of 5 T at the FIR wavelength of 84 μm. For three other main wavelengths, 119, 172, and 220 μm, both of the lines are detectable. In Fig. 12 the two hole resonances are indicated by arrows.

One should recall that holes can be observed in InSb only if excess free carriers remain. In time-resolution series, they appear only in the first stage; therefore such observations should be done during excessive photoexcitation, and very quickly.

One thing worth mentioning is that, on the whole, observation of hole signals has been easier for n-type than for p-type material. Because we have tried only three kinds of InSb, it is dangerous to make a general statement. Perhaps the most important thing is not the type of material but the concentration of acceptors, a conjecture based on the assumption that the most dominant mechanism for determining the hole resonance linewidth is hole–impurity scattering. If the impurities are all neutralized by photocarriers, one will see hole–neutral-acceptor and hole–neutral-donor scatterings. These two scatterings are classified as electron–hydrogen ($e^- - H$) and positron–hydrogen ($e^+ - H$) scatterings, respectively (Otsuka et al., 1964, 1966). The former can be approximated by the Erginsoy formula

$$1/\tau = 20\hbar a_A^* N_A / m_h^*, \tag{5}$$

and the latter by

$$1/\tau = 3.4\hbar a_D^* N_D / m_h^*. \tag{6}$$

Here τ is the mean free time for the hole; a_A^* and a_D^* are the effective Bohr radii of the acceptor and donor, respectively; m_h^* is the effective mass of the incident hole; and N_A and N_D are the concentrations of fully neutralized acceptors and donors. Rigorously speaking, Eq. (6) is valid only for very low energy of the incident hole, i.e., at $T = 0$. At finite temperatures, the relative importance of Eq. (6) to Eq. (5) is reduced. One has to take into account that a_A^* is considerably smaller than a_D^* in the case of InSb. Thus it may not be

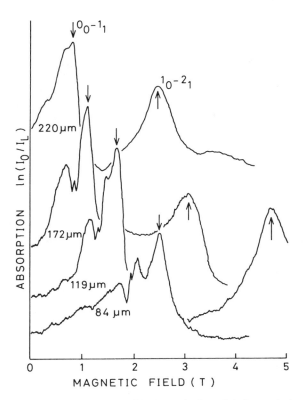

FIG. 12 Cyclotron resonance traces with the optical modulation technique (differential method) for wavelengths of 84, 119, 172, and 220 μm (sample A, 4.2 K, $B \parallel \langle 111 \rangle$). Two-hole quantum lines $0_0 \rightarrow 1_1$ and $1_0 \rightarrow 2_1$ are indicated with arrows. The $1_0 \rightarrow 2_1$ transition is out of scale for the wavelength of 84 μm.

concluded that hole–neutral-acceptor scatterings are more dominant than hole–neutral-donor scattering for the same donor or acceptor concentration. (If the above argument is applied to electrons, one can understand why it is easier to observe electron resonance in p-type than in n-type InSb.) Allowing for the possibility that part of the impurities are ionized, one has to compare hole–ionized-acceptor and hole–ionized-donor scattering. So long as one sticks to the Born approximation, as employed in the conventional theories of semiconductors (Brooks, 1951; Conwell and Weisskopf, 1950), there should arise no difference between these two types of scattering. However, there is experimental evidence that the Born approximation is not valid at the temperature with which we are concerned (Otsuka *et al.*, 1973). The calculations so far available indicate that hole–ionized-acceptor scattering will give rise to a larger cross-section than hole–ionized-donor scattering (Blatt, 1957; Otsuka, 1981). Thus in a crude sense one might as well expect the dominance of hole–acceptor (both neutral and ionized) scattering over hole–donor (both neutral and ionized) scattering. Perhaps the more conservative thing would be to expect better hole signals in the material with less impurities as a whole. All of the quantitative measurements for the hole resonance have been done with sample A.

The line shape of the hole resonance has been examined for the FIR wavelengths of 220 μm and 119 μm at the $1_0 \rightarrow 2_1$ transition line, because this line is completely free from the interference of the electron signals. It has been found that the line is very nearly Lorentzian. It is hard to see the variation of the linewidth with time, because the number of sampling points before the hole signal completely dies is too small. Nevertheless, repeated measurements indicate that the linewidth decreases by a few percent during the observation period. This is an indication that the hole system is also in the high-temperature state; in other words, we have hot holes. The relative heights of the two hole signals observed at the same delay time also indicate the maintainance of a considerably high hole temperature. Leaving the topic of hole temperature for the moment, one can roughly derive the collision frequency $1/\tau$ of holes from the nearly constant linewidth of the resonance, making use of Eq. (4). It is 1.8×10^{12}/sec for sample A, about an order of magnitude larger than that for electrons in sample C at the saturation region.

F. CYCLOTRON EMISSION

One of the most challenging features of the recent hot-electron topics is the phenomenon of cyclotron emission (Gornik, 1972; Kobayashi et al., 1973). Electrons that are elevated to the higher Landau subbands in the conduction band can lose energy by emitting photons having the cyclotron energy. The elevation of electrons to the higher Landau subbands has so far been achieved with the help of an electric field. We have seen here that electrons can be heated by the band-gap light. Naturally we also can expect the possibility of cyclotron emission from the optically heated carrier system.

Cyclotron emission has been clearly observed for sample B. Observation of the phenomenon has been favored by use of the mechanically chopped light of a tungsten lamp, use of a CW YAG laser, or use of an acoustooptical modulated Ar+ laser. Because the signal is much weaker than the absorption, the primary requirement of the detector is high sensitivity. We chose an Sb-doped Ge detector having a sensitivity two orders of magnitude higher than that of the Putley unit. The sole defect is its slow response, requiring nearly 10 μsec. Thus an optical excitation source that provides short pulses is not suitable. The mechanical chopping device employed here has produced pulses of 1 msec at a repetition of 30 Hz; thus the slowness of the detector response becomes immaterial. The emission signal obtained by tungsten-lamp excitation is shown in Fig. 13; for comparison, the emission signal obtained by electrical excitation, with a pulse width of 1 msec and repetition rate of 30 Hz at 30 V/cm is also shown. In this experiment one varies the magnetic field looking for the response of the detector. When the cyclotron energy emitted from the sample, which is tuned to the applied magnetic field, meets the maximum photoresponse of the detector, a peak appears. The width of the peak corresponds to the bandwidth of the detector. When we replace the detector with another one, say, As-doped Ge, the corresponding photoresponse peak appears at a slightly shifted position in the magnetic field (Kobayashi *et al.,* 1974; Otsuka, 1980). As one sees in Fig. 13,

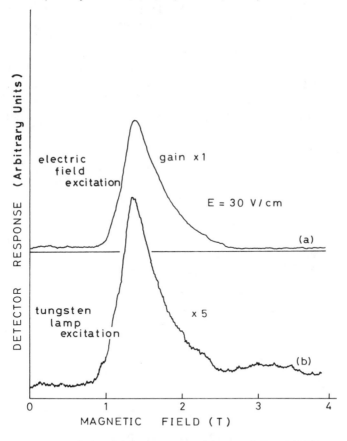

FIG. 13 Cyclotron emission signals are compared for sample B at 4.2 K for two different kinds of excitation with an Sb-doped Ge detector: (a) electric field excitation and (b) photoexcitation by tungsten lamp.

the emission signal obtained by optical excitation is considerably weaker than that obtained by electric excitation, at the best of our experimental skill. An emission signal of similar quality in optical excitation has also been obtained by using chopped lights from a CW-YAG laser. When we applied Q-switching to the YAG laser, no emission signal came out, probably because of the slow response time of the Sb-doped Ge detector. An intermediate case has been observed for xenon-lamp excitation. Its pulse width is ~ 1 μsec, certainly shorter than the response time of the detector but not so short as the Q-switched YAG laser pulse width (~ 200 nsec). A typical emission trace is shown in Fig. 14. Observation is made close to the top of the photopulse. The signal-to-noise ratio is poorer than with the tungsten lamp or the CW-YAG laser excitation. This is partly due to the responsivity of the detector. One thing of interest is that the emission signal seemingly has a long tail on the higher magnetic field side. A simple interpretation would be admitting the existence of high-energy cyclotron emission, either from electrons at higher Landau subbands or from electrons with higher values of k_z in the 1^+ subband, where k_z is the wave vector component along the magnetic field. Development of a quick and sensitive detector would open a new and unexpected feature of cyclotron or related emission from the hot-electron system. For the moment we have been unable to add any

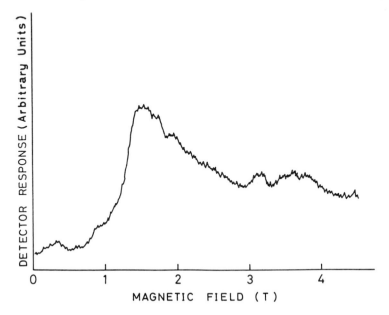

FIG. 14 Cyclotron emission signals by xenon-lamp excitation obtained with the same detector as in Fig. 13.

novel information with respect to cyclotron emission and so report a simple observation of emission by optical excitation.

IV. Carrier Dynamics in GaAs

A. MOTIVATION OF STUDY

Of all the III – V compound semiconductors, perhaps GaAs is most widely utilized for devices. One can find this material in semiconductor lasers, advanced integrated circuits, high-mobility transistors, etc. The effective operation of the devices is always achieved by controlling the nonequilibrium behavior of the carriers. The main probe for investigating nonequilibrium carriers, especially photoexcited carriers, has been restricted to recombination spectra. The first report of the optically hot electron system came from GaAs through the time-resolved luminescence study by Ulbrich (1973). The far-infrared magnetooptical absorption, including cyclotron resonance, has been regarded as a secondary, if not surplus, means of characterization. As we have seen in section III, however, this secondary method is bringing forth much new information on carrier dynamics in InSb. Why not in GaAs? It is true that more physical information in the far infrared has been obtained for InSb than for GaAs because of the ready accessibility of electron cyclotron resonance. Various delicate manipulations have been possible for InSb, because material purer than that for GaAs can be grown by ordinary crystal-pulling techniques, which enables us to have bulk, or thick, material. It is certainly more difficult to have GaAs crystals pulled out containing less than 10^{16} cm^{-3} impurities. Only as epitaxially grown layers can GaAs claim a purity as high as 10^{13} cm^{-3}. Yet even in this form, GaAs can provide us with plenty of physics from far-infrared experiments. A trial investigation of photoexcited carriers introduced here shows a new pattern for characterization. Many similar characteristics are observed for GaAs and InSb, and one new feature in particular through time resolution. This type of study has seldom been reported for GaAs. If this trial is successful, it will become a powerful tool for unveiling the behavior of the electron system in the rapidly developing new materials such as ternary or quarternary compounds, superlattices, and heterojunctions.

B. SAMPLES AND EXPERIMENTAL PROCEDURES

Three kinds of samples are used in this work. One is grown by the Ga – AsCl$_3$ – H$_2$ vapor-deposition method and the other two by the metallorganic chemical vapor deposition (MOCVD) method. Their impurity characteristics are given in Table II. Each sample is 5 μm thick and is

TABLE II

LIST OF THE GaAs SAMPLES

Sample	Type	$|N_D - N_A|$ ($10^{15}/cm^3$)	N_D ($10^{15}/cm^3$)	N_A ($10^{15}/cm^3$)	Mobility at 300 K (cm^2 v^{-1} sec^{-1})	Growth method
A	n	0.50	1.5	1.0	7000	Ga–AsCl$_3$–H$_2$
B	n	1.6	2.9	1.3	7560	MOCVD
C	p	0.86	4.3	5.2	310	MOCVD

deposited on a semi-insulating GaAs substrate containing Cr. The far-infrared transmission measurement is carried out with the sample sticking to the substrate. This procedure is based on the assumption that the substrate is making no contribution to the absorption spectra. In order to justify this assumption, we peeled off the epitaxial layer by chemical etching and then confirmed that no signals were coming out of the substrate. All the samples are rectangular, having dimensions of $4 \times 4 \times 0.5$ mm^3 including the substrate. For photoconductivity measurements and electrical excitation, two point probe electrodes are soldered onto the epilayer with indium metal. The far-infrared laser operation and the method of photoexcitation are much the same as were used with InSb. The differential method is also utilized to its maximum extent. The detector is again the InSb Putley unit because of its fast response, i.e., a time constant shorter than 0.5 μsec. Most of the experimental results presented here were obtained with the sample grown by the Ga–AsCl$_3$–H$_2$ vapor-deposition method.

C. ZEEMAN ABSORPTION AND CYCLOTRON RESONANCE UNDER PHOTOEXCITATION

A set of typical magnetoabsorption traces after xenon lamp illumination, with a delay time of 20 μsec, are shown in Fig. 15 for three different FIR laser wavelengths. The origin of the absorption, either impurity Zeeman or cyclotron resonance, is indicated. For the wavelength of 119 μm the cyclotron-resonance trace is out of scale. Without photoexcitation the cyclotron-resonance signal practically disappears for all wavelengths. Both signals vary with delay time after the photopulse. To make this feature more clear, a time-resolved series is given in Fig. 16 at a fixed FIR wavelength of 172 μm. The behavior of the Zeeman transition signal for the donor impurity is quite peculiar and complicated in comparison with the cyclotron resonance. No absorption signal comes from the acceptor because of its large binding energy.

Before entering into detailed discussion of this behavior, one might as well look at the magnetic-field dependence of transition energy, in other words, the variation of electron cyclotron mass with the FIR wavelength. We find that $m_c^* = (0.0685 \pm 0.0005)m_0$, $(0.0693 \pm 0.0005)m_0$, and $(0.0699 \pm 0.0005)m_0$ for 220, 172, and 119 μm, respectively. All these values are somewhat larger than the previously available one $(0.0667 \pm 0.0003)m_0$ (Chamberlain *et al.*, 1971), which has been obtained at a lower magnetic field corresponding to the wavelength of 337 μm. Probably the difference reflects the nonparabolicity of the conduction band. One may also note the value $0.06650m_0$ obtained by Fetterman *et al.* (1971) for 190–337 μm. The magnetic-field positions for the Zeeman absorption lines, on the other hand, are quite in accord with the calculation by Larsen (1968).

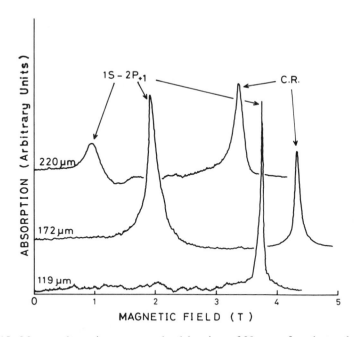

FIG. 15 Magnetoabsorption traces at the delay time of 20 μsec after photopulse for the wavelength of 119, 172, and 220 μm for sample A, 4.2 K. The signal indicated by $1s \rightarrow 2p_{+1}$ is the Zeeman absorption associated with the donor state. The abbreviation C. R. denotes the conduction electron cyclotron resonance arising from the photoexcited electrons. The Zeeman absorption is out of scale for the wavelength of 84 μm.

Now we go back to time variation. In order to envisage the behavior of the photoexcited carriers we use the differential method. The temperature is fixed at 1.8 K. The absorption, measured as $\ln(I_0/I_L)$, is given in Fig. 17 as a function of the magnetic field for various delay times. The quantities I_L and I_0, as before, are the transmitted FIR intensities with and without photoexcitation, respectively. The Zeeman line drastically decreases on intrinsic light illumination at the beginning, then recovers gradually, becoming even stronger than the absorption in the absence of photoexcitation. Looking at its behavior more closely, one finds a slight shift of resonance position with time. This is evident in the asymmetry in the signal. Broadening of the linewidth is also noticeable though the final linewidth seems to shrink again to some extent. The cyclotron-resonance behavior is much simpler. Its absorption intensity decays steadily with time except immediately after photoexcitation when the signal suddenly emerges from nothingness.

One can obtain the change in the electron population with time, for both the conduction band and the donor levels, from the absorption intensity,

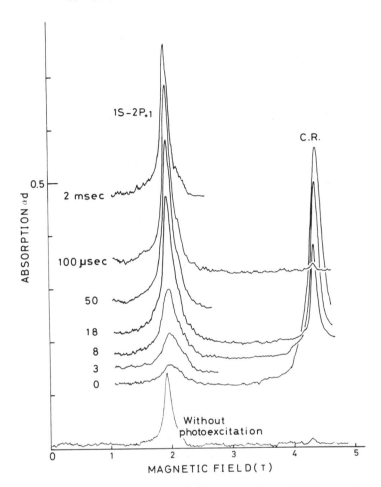

FIG. 16 The time-resolved cyclotron resonance absorption and the Zeeman absorption for sample A, 172 μm, 4.2 K. For comparison, the signal without photoexcitation is also shown.

i.e., the area under the absorption line. The absolute carrier densities are calibrated by the procedure described in section IVE. The result is plotted in Fig. 18 against delay time. It is found that though densities for both conduction electrons and donor electrons change with time the sum of the densities does not change appreciably for the first 50 μsec. Because the density of conduction electrons, obtained from the intensity of cyclotron resonance, steadily decreases at a time constant of ∼ 7 μsec, the total density can practically be represented by the donor electrons after 20 μsec. The total density, nearly constant in time, is still considerably higher than the final

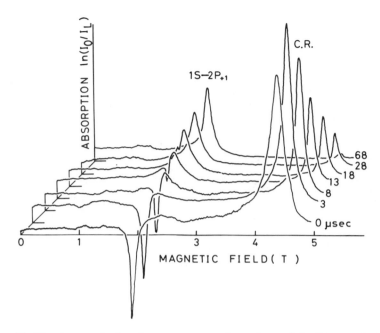

FIG. 17 The time-resolved cyclotron resonance absorption and the Zeeman absorption obtained with the optical modulation technique (differential method), for sample A, 172 μm, 1.8 K.

density, however. The final value of the donor electron density is indicated by a bar with $t = \infty$. The photoexcited electrons can live as long as 2 msec. Such behavior is similar to that of photoexcited electrons in n-InSb. But the lifetime of the conduction electrons relative to the donor electrons seems to be much shorter; in other words, a quasi-thermal equilibrium feature such as we saw in InSb between the conduction electrons and the donor systems is not directly observable, because the donors in GaAs have larger binding energies than those in InSb, which leads to the much smaller density of conduction electrons. On the other hand, the quasi-thermal equilibrium between the donor and the acceptor systems, which we might call the second-stage equilibrium, can last for quite a long time. The slow electron transfer from donor to acceptor may be described as for photoexcited GaP (Thomas *et al.*, 1965) by assuming a transition probability of the type

$$W(r) = W_{\max} \exp(-2r/a_B), \tag{7}$$

where r is the separation between the neutral donor and the neutral acceptor, a_B is the Bohr radius of the donor electron, and $W_{\max}$ is a constant factor.

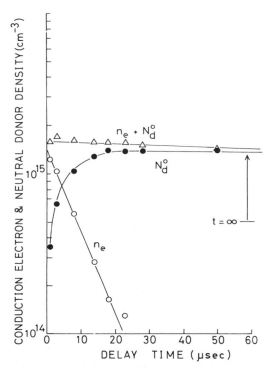

FIG. 18 Time variations of the conduction electron density, the neutral donor density and the sum of both after photoexcitation, obtained from Fig. 17, for sample A, 172 μm, 4.2 K. The level indicated by t = ∞ corresponds to the neutral-donor density at thermal equilibrium, i.e., without photoexcitation.

If one puts

$$W_{max} = 5 \times 10^7/\text{sec}, \tag{8}$$

the observed decay behavior of the donor absorption line can be fit by Eq. (7) with an appropriate ensemble average of r and the neutral impurity density obtained from the Zeeman-line signal in the time-resolved experiment (see Fig. 19).

D. DYNAMIC SCREENING OF THE DONOR POTENTIAL

The photoexcited carriers are expected to neutralize the impurity centers promptly. The acceptor levels are so deep, ~ 30 meV, that the centers, once they capture the photoexcited holes, will hardly release them into the valence band at liquid-helium temperatures. The donor levels, on the

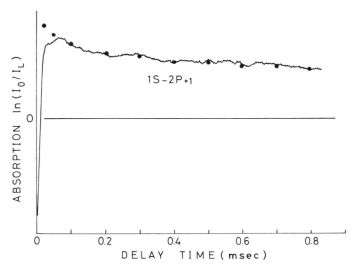

FIG. 19   Decay profile of the Zeeman absorption after photo-excitation, for sample A, 172 μm, 4.2 K. The negative absorption for the first 10 μsec arises from the decrease of the neutral impurities. The closed circles are from the calculation by Thomas *et al.* (1965) with $W_{max} = 5 \times 10^7$ sec^{-1} (see text).

contrary, are shallow enough to exchange electrons with the conduction band even at low temperatures. In the presence of a considerable density of free carriers, the capturing potential of an ionized donor center may be expressed in the form of a screened Coulomb potential

$$V(r) = -(e^2/\kappa r) \exp(-qr). \tag{9}$$

Krieger (1969) and Katana *et al.* (1970) have calculated the electron binding energy E_D of the center as a function of qa_B. Here q^{-1} is the screening length, which can be expressed as

$$q^{-1} = (\kappa k_B T/4\pi e^2 n_e)^{1/2} \tag{10}$$

according to the Debye-Hückel approximation, where n_e is the density of the conduction electrons. Taking $a_B = 96$ Å, the calculated binding energy is given in Fig. 20 as a function of n_e. For comparison, the experimental points obtained in our experiment are shown in the same figure. The procedure for deriving these experimental values is as follows:

$$e^- + D^+ \rightleftarrows D^0. \tag{11}$$

Then, because of the law of mass action, we have

$$n_e N_D^+/N_D^0 = K(T, E_D), \tag{12}$$

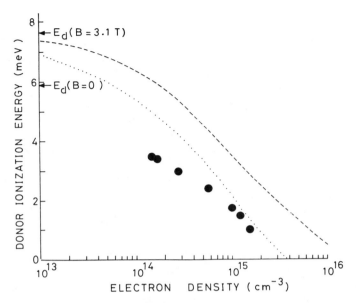

FIG. 20 Calculated binding energies by Krieger (1969) (dotted line) and Katana *et al.* (1970) (dashed line) as a function of the conduction electron density, at 4.2 K. For the binding energy at the low electron concentration limit, the value of 7.8 meV is taken with a magnetic field of 3.1 T. For comparison, experimental points obtained in our work are shown in the same figure.

where N_D^+ and N_D^0 are the densities of ionized and neutral donors, respectively, and $K(T,E_D)$ is the equilibrium constant. Statistics show that $K(T,E_D)$ has the form

$$K(T,E_D) = N_c \exp(-E_D/k_B T) \qquad (13)$$

with

$$N_c = (\tfrac{1}{2}\pi^2)(2\pi m_c^* k_B T/\hbar^2)^{1/2}(eB/c\hbar). \qquad (14)$$

Combination of the experimentally obtained values of n_e and N_D^0 yields the value of E_D. In this derivation, the magnetic field has been set 3.1 T, the mean value between the cyclotron resonance and the Zeeman transition. The apparent agreement between theory and experiment indicates the validity of the screening model. However, this agreement cannot explain all the features of the observed phenomena. Intuitively, the variation of the binding energy should also affect the Zeeman transition energy. But no appreciable shift of the peak position has been detected. The big difference between thermal and optical activation energy might be connected with this observation.

E. CARRIER EXCITATION BY ELECTRIC FIELD

Unlike InSb, no cyclotron-resonance signal for GaAs is observable in the dark. Application of an electric field, however, may induce impact ionization of neutral donors and thus yield the electron cyclotron resonance. For InSb, transfer of electrons from the donor levels to the conduction band on application of a pulsed electric field was observed very clearly (Kobayashi and Otsuka, 1974; Otsuka, 1980). The same procedure, or the pulsed electric field modulated cyclotron resonance (PEM-CR) technique, has been used for GaAs. The resultant traces for various electric field strengths are given in Fig. 21. Because the PEM-CR technique is a kind of differential method, only feeble signals emerge at low electric field. When the electric field strength is increased, its pulse width being 10 μsec and repetition rate 10 Hz, the Zeeman-transition signal starts going down and the cyclotron-resonance signal goes up. Differing from photoexcitation, the Zeeman signal never reverses itself but saturates its downward signal intensity in correspondance to the total ionization. The cyclotron-resonance signal at this stage gives the $N_D - N_A$ value, or 5.0×10^{14} cm^{-3} for sample A. This value sets the standard for determining the density of photoexcited electrons. Similarly, the density of electrons localized at the donor levels can be determined from the Zeeman transition intensity in the absence of optical or electrical excitation, which is the same as the saturated downward signal in the presence of electrical excitation. From the observed absorption coefficient and the known density of neutral donors one can estimate the absorption cross section σ_a for the $1s \rightarrow 2p_{+1}$ transition to be

$$\sigma_a = 1.2 \times 10^{-12} \text{ cm}^2. \tag{15}$$

F. LINEWIDTH ANALYSIS

As in the case of InSb, linewidth study of the resonant absorption in GaAs yields information on carrier dynamics and phase transition. Because we observe two main absorption lines, cyclotron resonance and Zeeman transition, we shall treat them separately.

1. Cyclotron Resonance

A quantitative linewidth measurement has been carried out for sample A at the FIR wavelength of 172 μm and at the fixed temperature of 4.2 K. In time-resolved measurements, the linewidth decreases considerably with time. Naturally, the intensity of resonance also decreases in accordance with the decrease of photoexcited carriers. The inverse relaxation time can be derived from the well-known simplified relationship

$$\omega\tau = 2B_r/\Delta B, \tag{16}$$

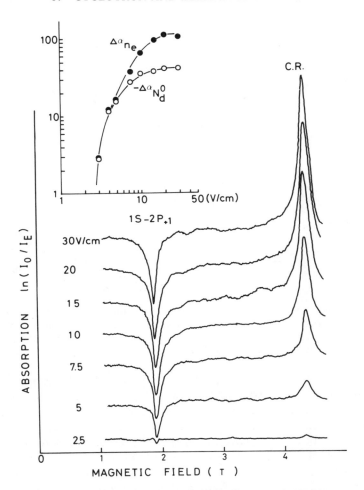

FIG. 21 Dependence of the pulsed electric field-modulated cyclotron resonance and the Zeeman absorption on the applied electric field for sample A, 172 μm, 4.2 K. The inset shows the change of the absorption as a function of the applied electric field.

where ω is the angular frequency of the FIR laser oscillation, B_r is the resonance field, and ΔB is the linewidth measured in the magnetic field. The analysis has been made at various delay times; in other words, $1/\tau$ is derived as a function of n_e. The result, given in Fig. 22 for samples A and C, shows that the inverse relaxation time can be expressed in the form

$$1/\tau = a + b\, n_e^\gamma, \tag{17}$$

where a and b are constants and γ a numerical exponent. From the result for

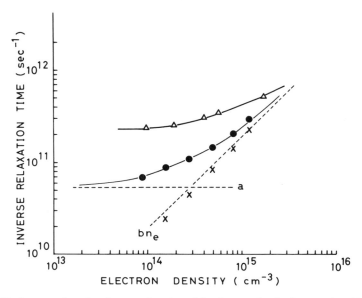

FIG. 22 Inverse relaxation time as a function of the electron density for sample A (● and X) and sample C (△) for 172 μm, 4.2 K. The horizontal line indicated by a and the oblique line indicated by bn_e give the impurity-independent part and the impurity-dependent part of the inverse relaxation time, respectively, for sample A.

sample A it is found that $\gamma \sim 1$. That is, the linewidth consists of a part that is independent of the carrier density and a part that is linearly dependent on the carrier density. The linearly dependent part reflects the contribution of the carrier–carrier interaction in the nonadiabatic region (Kawamura *et al.*, 1964), where the cyclotron radius l_c is much larger than the screening length q^{-1}. Numerical estimates show some problems, however. If the cyclotron radius is obtained for $B = 4.33$ T, corresponding to the FIR wavelength of 172 μm, it becomes 1.2×10^{-6} cm, and the Debye–Hückel approximation at 4.2 K yields the screening length of 1.1×10^{-6} cm even at the high electron density of 2×10^{15} cm^{-3}. Thus we find that the nonadiabatic condition cannot be guaranteed. It cannot be made readily clear whether the application of the Debye–Hückel theory is inappropriate or whether the intuitive classification of the carrier–carrier scattering theory between adiabatic and nonadiabatic cases is not well-grounded. Be that as it may, one can derive from the result for sample A that

$$b = 1.9 \times 10^{-4} \text{ cm}^3/\text{sec}. \tag{18}$$

Nearly the same value is obtained for sample C. The density-independent term may be compared with the prediction based on electron–neutral-donor scattering. The conventional Erginsoy formula (Erginsoy, 1950)

$$1/\tau_N = 20\hbar a_B N_D^0/m^* \tag{19}$$

may be applied for fitting. Inserting the values of a_B and m^* appropriate to GaAs, one finds

$$1/\tau_N = 3.1 \times 10^{-4} N_D^0/\text{sec}, \tag{20}$$

if N_D^0 is in cm^{-3}. Our experimental result, however, yields the relation

$$1/\tau_N = 3.6 \times 10^{-5} N_D^0/\text{sec}. \tag{21}$$

The discrepancy between theory and experiment is obvious, but one should note that the Erginsoy formula is actually far from being applicable to the present case. Use of the formula is based on the assumption that the cyclotron orbit is much larger than the Bohr orbit of the neutral donor. Under the present circumstance of $l_c \sim a_B$, an entirely new treatment should be looked for. It should be mentioned further that the scattering rate becomes somewhat larger at the lower magnetic field corresponding to the FIR wavelength of 220 μm, showing somewhat better agreement with the Erginsoy formula. If we go to an even lower magnetic field, such as that used for observing cyclotron resonance in Si or Ge, the applicability of the Erginsoy formula becomes almost perfect (Otsuka et al., 1966, 1968; Ohyama et al., 1970). It must be emphasized that the linewidth should contain the contributions of the nonparabolicity and of the k_z broadening, which are practically missing in the electron cyclotron resonance in Si or Ge. Nevertheless the observed linewidth is much smaller than that obtained by the straightforward application of the Erginsoy formula.

2. Zeeman Transition

The Zeeman transition also gives rise to varieties of linewidth, depending on the sample, intensity of photoexcitation, etc. Generally speaking, four possible contributions can be considered for giving the extra linewidth: lattice vibrations, the Stark effect caused by ionized centers (Ohyama, 1980), overlapping of the impurity wave functions, and internal strains. In photoexcitation where the lifetime of neutralized donors is extremely long, the Stark effect due to ionized centers can practically be neglected. The effect of lattice vibrations is temperature-dependent, and we may put it aside by fixing the temperature. Internal strains would perhaps be very small. Thus the most important contributing factor would be the overlapping of the impurity wave functions. If this is the case, the linewidth must be critically dependent on the concentration of neutral donors. The observed features are given in Fig. 23 for three samples, at a fixed delay time of 30 μsec. Similar data are available at various delay times after photoexcitation.

It has been confirmed experimentally that the magnetic-field dependence of the ls $\rightarrow$ 2p$_{+1}$ Zeeman transition energy, which we are dealing with now, is given by 1.7×10^{-3} eV/T. The spread in energy can readily be derived then from the linewidth measured in the magnetic field. One typically finds that $\Delta B = 0.13$ T for sample A and 0.42 T for sample C at the delay time of 30 μs. These correspond to $\Delta E = 2.2 \times 10^{-4}$ eV and $\Delta E = 7.2 \times 10^{-4}$ eV, respectively. Thus increasing the neutral-donor concentration by a factor of 3 leads to a considerable broadening of the Zeeman transition line. There exists a theoretical estimate (Baltensperger, 1953) predicting the onset of the broadening at the inter-neutral-donor distance r_s^* given by

$$r_s^* = 12a_B. \tag{22}$$

Inserting $a_B = 96$ Å for GaAs and putting

$$r_s^* = (4\pi/3N_D)^{-1/3}, \tag{23}$$

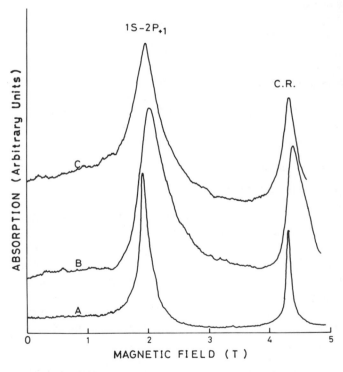

FIG. 23 Absorption features for three samples (A, B, C) at the fixed delay time 30 μsec after photoexcitation (172 μm, 4.2 K).

one finds the threshold concentration for broadening to be 2×10^{14} cm^{-3}. This is quite consistent with our experimental observation.

One more comment on the broadening of the Zeeman line is required. There is an appreciable broadening observed for a period of 30 μsec after photoexcitation, when a considerable number of conduction electrons still exist. The interaction between the conduction electrons and the donor electrons is evidently contributing to the linewidth. It is an interesting new feature in our observation, because this type of interaction has so far been observed only from the side of the conduction electrons, namely, from the transport study. In fact, in the preceding treatment of the cyclotron resonance linewidth (Section IV.F.1), we have made this type of argument. The same electron–neutral-donor collision phenomenon thus brings about not only the broadening in the cyclotron resonance of the conduction electron but also a similar broadening in the Zeeman transition line. Though this may not sound surprising, one has to note the lack of proper theoretical treatment of the problem.

G. FAR-INFRARED PHOTOCONDUCTIVITY

The final important characteristic of high-purity n-type GaAs is its photoresponse to far infrared, which has a technical importance in making a far-infrared detector. The response originates from the photoionization of donor electrons on receiving far-infrared radiation. Thus the sensitivity of the detector is proportional to the density of electrons at the donor states. In thermal equilibrium, the density at liquid-helium temperature is $N_D - N_A$. The ordinary use of the detector is based on this condition. We have seen, however, that the density of neutral donors can be increased by illumination with the band-gap light. This fact tells us that one can raise the sensitivity of the detector through an auxiliary application of intrinsic excitation. For demonstration, Fig. 24 shows a large difference in the far-infrared photoconductivity response with and without intrinsic excitation. The comparison is made for four different far-infrared wavelengths. The signals are taken at either 300 μsec or 500 μsec after the photopulse. A drastic improvement of the sensitivity through intrinsic excitation is obvious.

One should note that the overall sensitivity of a GaAs detector is far below that of a doped Ge detector. But the sharpness of the ls $\rightarrow$ 2p transition is frequently useful for spectroscopic analysis, such as for cyclotron emission (Gornik, 1978). Our experience of the improvement of the detector sensitivity through auxiliary illumination should be quite helpful for furthering the study of cyclotron emission.

Before closing this section, we should add a comment on the magneto-oscillations shown in Fig. 25. They are observed very clearly for wavelengths of 84 μm and 119 μm, along with the sharp peaks associated with transitions

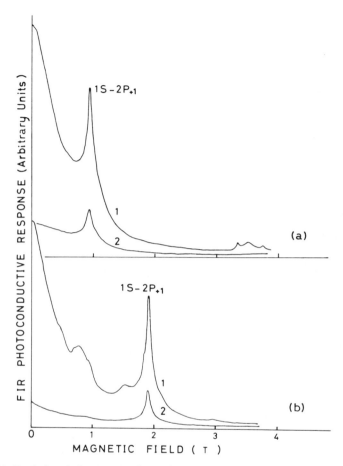

FIG. 24   Far-infrared photoconductive response with (1) and without (2) intrinsic photoexcitation; signals with photoexcitation are taken 300 μsec after the photopulse. (a) 220 μm, 4.2 K and (b) 172 μm, 4.2 K.

from the donor ground state to the excited states as indicated. The series of oscillations evidently has a quantum nature. It is believed to be the resonant series from the impurity state to the Landau levels. Such a clear observation as presented here will be a strong motivation for further detailed study.

V.   Exciton Dynamics in Si

A.   THE BACKGROUND

So far we have dealt with III – V compounds under photoexcitation. Now we discuss the more fundamental semiconductor, Si, also under photoexci-

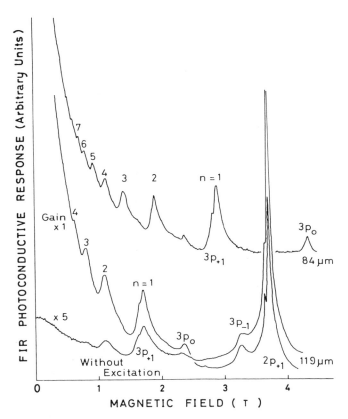

FIG. 25 Magnetic field dependence of the far-infrared photo-conductivity for wavelengths of 84 and 119 μm for sample A, 4.2 K, $t_d = 500$ μsec. The integer n is the Landau quantum number.

tation, to supplement article I which dealt with Ge. There are many common features between Si and Ge. Both are indirect-gap semiconductors, being of many-valley character in the conduction band and having nearly the same effective masses. Yet quantitatively one finds many differences. Minor differences can cause critical contrasts especially in low-temperature properties. One can see the best example of these contrasts in the behavior of the exciton system. The nature of the indirect gap makes the average lifetime of an exciton very long in both materials, of the order of a few microseconds at liquid-helium temperature. Such a long life is favorable, helped by the degeneracy in energy bands, for forming the electron–hole condensate, or the electron–hole drop (EHD). Indeed, Ge and Si are the only materials so far in which the formation of EHD has been convincingly shown. But the general characteristics of EHD in the two materials are very different. Above all, EHD is much more short-lived in Si than in Ge due to the higher carrier density. An obvious result of this is the smaller radius of the drop. Thus EHD in Si is characterized by a radius of ∼ 0.1 μm and a lifetime of ∼ 0.1 μsec. These values suggest that studying EHD is much harder in Si than in Ge. In fact, fewer physical quantities are confirmed for EHD in Si; for example, neither surface tension nor compressibility measurements have been reported. The sole favorable aspect would be the larger binding energy, or work function. In contrast to EHD in Ge, which does not exist above 6.5 K, EHD in Si is supposed to remain stable up to higher temperatures. The same is true also for excitons. Thus the kinetics of excitons and EHD in Si can be explored over a wider range of temperature than they can in Ge. In other words, more quantitative results can be expected. With this end in view, we have used the same experimental procedure as before for studying photoexcited Si.

B. SAMPLES

One undoped and five doped samples have been used for measurement. Their impurity characteristics are shown in Table III. The special quality common to these samples is the small degree of compensation, except for the undoped one. It has been confirmed through microwave cyclotron resonance in the manner described elsewhere (Otsuka *et al.*, 1968) that the lightest boron-doped sample ($N_B = 8.0 \times 10^{13}$ cm^{-13}) contains a donor impurity level considerably less than 2×10^{12} cm^{-3}. The more heavily doped samples contain their respective minority dopants at concentrations lower than the majority dopants by at least two orders of magnitude. These crystals were specially grown by Toshiba for the study of electron – neutral-impurity scattering by cyclotron resonance.

C. EXCITONIC ZEEMAN EFFECT

The free exciton in Si can be detected by far-infrared laser in terms of the Zeeman transition. The transition energy is tunable with the application of a

TABLE III

LIST OF THE Si SAMPLES

Dopant	Type	$\lvert N_D - N_A \rvert$ (/cm^3)
Undoped	p	$\lesssim 10^{12}$
Boron	p	8×10^{13}
	p	4.3×10^{14}
	p	2.0×10^{15}
Phosphorus	n	8.8×10^{13}
	n	8.0×10^{14}

magnetic field. Typical magnetoabsorption traces by the free exciton created in an undoped Si are shown in Fig. 26. The temperature is fixed at 4.2 K. Various laser wavelengths are used but the magnetic field is always applied along $\langle 111 \rangle$. For 96.5, 103, and 119 μm some resonance peaks emerge. For 84 μm we observe only a broad nonresonant response. We ascribe this to the photoionization of excitons. In other words, the binding energy of a free exciton is supposed to be less than 1.47×10^{-2} eV below 5 T. The absence of any absorption at the wavelength 146 μm means that the first excited state lies at an energy higher than 8.5×10^{-3} eV above the ground state. Because the absorption peaks are most numerous for 119 μm, we use this wavelength for anisotropic measurement. The absorption trace is taken for three typical crystallographic directions, namely, $\langle 111 \rangle$, $\langle 110 \rangle$, and $\langle 001 \rangle$. The result obtained at 1.7 K (Fig. 27) shows that the major absorption lines have a little dependence on the orientation of the applied magnetic field. These transitions are considered to reflect the nature of the excitonic Zeeman transitions. To see the global feature of the Zeeman transition we assume the transition energy $\hbar\omega$ to be of the form

$$\hbar\omega = a \pm bB + cB^2, \tag{24}$$

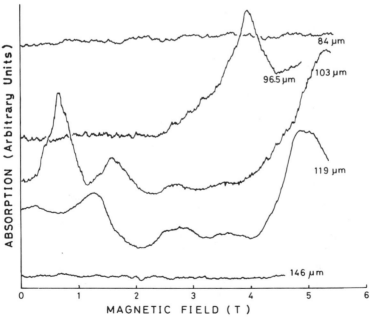

FIG. 26 Typical magnetoabsorption traces by the free exciton in indoped Si for various wavelengths, $B \parallel \langle 111 \rangle$, 4.2 K (Ohyama, 1981a).

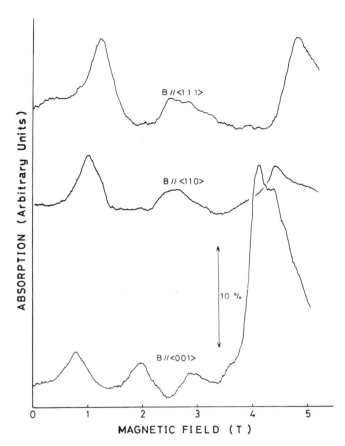

FIG. 27 Typical magnetoabsorption traces for a wavelength of 119 μm and for $B \parallel \langle 111 \rangle$, $B \parallel \langle 110 \rangle$, and $B \parallel \langle 001 \rangle$ at 1.7 K (Ohyama, 1981a).

where a, b, and c are the parameters to be determined in accordance with theory. Then the magnetic-field dependence of the transition energy for each peak is sought for a fixed orientation $B \parallel \langle 111 \rangle$ and wavelength of 119 μm. The observed transition energies, plotted in Fig. 28 against the magnetic field, are fitted by the calculation according to Eq. (24) with proper parameters. The experimental points include the results by Timusk *et al.* (1978) at zero field. Variation of the excitonic level with magnetic field is obvious, but a complete fit between experimental points and calculated curves is not there. Because in Si the spin–orbit splitting is only $\sim 4 \times 10^{-2}$ eV, all the six-fold valence-band states may contribute to the excitonic energy. Lipari and Altarelli (1976) predict that two $1s$-like ground states and

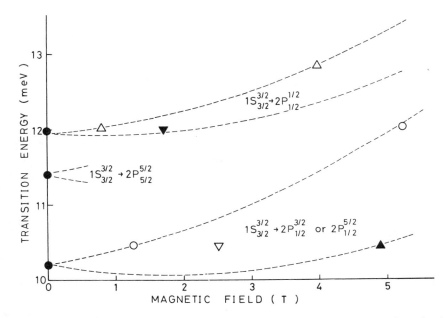

FIG. 28   Magnetic field dependence of the transition energy for each peak for a fixed magnetic field orientation $B \parallel \langle 111 \rangle$. The dashed lines represent our calculations using the parameters appearing in Eq. (24). Solid circles are after the data by Timusk *et al.* (Ohyama, 1981a).

at least six of the 2p-like states exist at zero magnetic field. Once the magnetic field is introduced, exact fitting becomes almost hopeless. With this difficulty in mind, we shall examine the feature of the excitonic Zeeman transition later.

The suggested dependence of transition energy on the magnetic field Eq. (24) is based on the form introduced by Button *et al.* (1969) for Ge,

$$E = D(J_z^2 - J^2/3) + \beta(g_e S - g_h J_z)B + cB^2, \qquad (25)$$

where J is the total angular momentum of the hole, S is the spin angular momentum of the electron, D is the crystalline field splitting constant, β is the Bohr magneton, and g_e and g_h are the g-factors for electrons and holes, respectively. In accordance with Knox (1963) we assume the effective g-factor for a given excitonic state to be of the form

$$g_{\text{eff}} = a_\mu g_\mu + a_e g_e + a_h g_h. \qquad (26)$$

The coefficients a_μ, a_e, and a_h depend on the symmetries of the electron, hole, and envelope functions. Writing $F = L + J$, where L is the orbital angular momentum, Lipari and Altarelli employ the (F, F_z) representation for expressing the wave function of the exciton. F_z is the projection of F in the direction of the magnetic field. In this representation one can rewrite the effective g-factor

$$g_{\text{eff}} = a_e g_e + a_F g_F. \tag{27}$$

The selection rules in the Faraday configuration require $\Delta S = 0$ and $\Delta F_z = \pm 1$. The double sign in Eq. (24) preceding the second term on the right-hand side comes from the two-way change of F_z. The term itself represents the difference in Zeeman energy of the hole for $1s$ and $2p$ excitonic states, respectively. The first term represents the zero-field splitting and the third a diamagnetic contribution. The latter becomes dominant above $B = 2$ T.

We try to obtain the values of the coefficients in Eq. (24) for the particular transition from the lowest excitonic state $1S_{\pm 3/2}^{3/2}$ (Timusk $et\ al.$, 1978) with the help of our experimental data given in Fig. 28, i.e., those obtained for the geometry $B\|\langle 111\rangle$. We find that $b = 0.062$ meV/T for $1S_{\pm 3/2}^{3/2} \rightarrow 2P_{\pm 1/2}^{1/2}$, 0.15 meV/T for $1S_{\pm 3/2}^{3/2} \rightarrow 2P_{\pm 1/2}^{3/2}$ or $1S_{\pm 3/2}^{3/2} \rightarrow 2P_{\pm 1/2}^{5/2}$, and $c = 0.041$ meV/T^2. The coefficient for the diamagnetic term is common to all the transitions. The values of a and b should be different from transition to transition. Those for other transitions are hard to obtain from our limited data. Generally the transition energy ΔE should be expressed as

$$\Delta E = a^F \pm b_{F_z}^F B + c_{F_z}^F B^2 \tag{24a}$$

according to the (F, F_z) representation. In principle an exact interpretation of the coefficients a^F, $b_{F_z}^F$, and $c_{F_z}^F$ could be made from the effective-mass approximation. But in that case inclusion of the other anisotropic terms in the Zeeman Hamiltonian would be required. Assumption of the simpler form of Eq. (24) or Eq. (24a) seems to be more convenient for explaining the qualitative nature of the anisotropy observed in our experiment.

Now let us make use of the excitonic Zeeman transition for obtaining information on the collision of excitons with phonons and impurities.

D. EXCITON COLLISION WITH PHONONS

As in the case of cyclotron absorption for free carriers the linewidth of the excitonic Zeeman absorption can be used for studying the transport problem. We will see that collisions of excitons with phonons and impurities make the Zeeman transition line broader. To make things more explicit, we will focus our attention on the result obtained using the FIR wavelength of 119 μm. First we shall take the undoped Si sample and apply the magnetic field along the $\langle 001\rangle$ direction. Of all the Zeeman transitions we select the line appearing at 4.5 T for linewidth analysis, because this line is most easily observed up to high temperatures. It can readily be confirmed that the line becomes broader as the temperature is raised and that the intensity of absorption increases with increasing temperature. An important characteristic, however, is that the linewidth is affected neither by the excitation intensity nor by the delay time after the photoexcitation pulse. This fact tells us that line broadening is not due to the exciton–exciton interaction. The most probable interpretation would then be the exciton–phonon collision for the undoped sample. Thus we shall make two fundamental assumptions: first, the Zeeman transition line in question gives rise to a Lorentzian shape; second, the excitonic energy makes a linear shift with magnetic field over the linewidth interval. These two assumptions will always be underlying our interpretation of experimental results.

In Fig. 29 we plot the half-width ΔB of the transition line as a function of temperature. Above 5 K, ΔB changes almost linearly with temperature, but it tends to become independent of temperature below 5 K. The linear temperature dependence is consistent with the theory by Toyozawa (1958) dealing with the exciton–phonon interaction. The deviation from linear temperature dependence at lower temperatures may be due either to the contribution of exciton–impurity scattering or to the onset of the spontaneous emission of phonons.

According to Toyozawa the scattering of $1s$ excitons by acoustical phonons can be expressed as

$$1/\tau_{\text{ex-ph}} = 8m_{\text{ex}}^2 k_B T (C_v - C_c)^2 / 9\pi\hbar^4 \rho u; \tag{28}$$

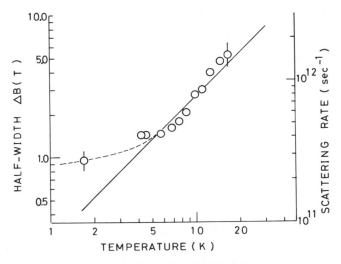

FIG. 29 Temperature dependence of the half-width ΔB of the excitonic absorption line (for undoped Si, 119 μm, $B \parallel \langle 001 \rangle$. Converted scattering rates are available on the right scale (Ohyama, 1981a).

where m_{ex} is the effective mass of an exciton, ρ is the density of the crystal, u is the velocity of sound, and C_v and C_c are the deformation potentials of the valence and conduction bands, respectively. The collision frequency given in Eq. (28) is proportional to temperature. This relation holds down to the characteristic temperature θ, which is given by

$$\theta = 2m_{ex}u^2/k_B. \tag{29}$$

For temperatures $T < \theta$, T in Eq. (28) should be replaced by θ. Putting $m_{ex} \sim 0.6\ m_0$ and $u = 5 \times 10^5$ cm/sec for Si, we get $\theta \simeq 2$ K.

We have information on the excitonic energy variation with magnetic field, so we can convert the observed ΔB to the energy width. For example, $\Delta B = 3.9$ T at 13 K should be understood as $\Delta E = 0.98$ meV. Making use of the relation $\hbar/\tau = \Delta E$, one finds that the energy-width corresponds to $1/\tau \simeq 1.5 \times 10^{12}$/sec. This energy-width compares fairly well with $\Delta E \simeq 0.5$ meV observed by Timusk *et al.* at 13 K. Their observation was made at zero magnetic field on the far-infrared absorption of the exciton system. The value shows a considerable deviation from the prediction by Toyozawa. Taking $\rho = 2.3$ g cm^{-3} and $C_v - C_c = 5$ eV, one finds that Eq. (28) gives $1/\tau_{ex-ph} = 7 \times 10^9$/sec at 13 K. This is far smaller than the experimental value. The reason for such a large discrepancy is not obvious. One possible explanation would be that we shall have to take into account the interband scatterings. Toyozawa considered only the intraband scattering of the $1s$ excitons. Perhaps the experimentally observed linewidth consists of two bandwidths, namely, those of the $1s$ and $2p$ excitonic states. If that is the case, the interband scattering of the $2p$ exciton to $1s$ states will make an overwhelming contribution to the observed linewidth, because this type of scattering will contain the contribution from additional modes of the lattice vibration. Another possibility is the contribution from the magneto-Stark effect due to the thermal motion of excitons in the presence of a magnetic field. This mechanism, however, would not account for the entire broadening.

E. Exciton Collision with Impurities

To see the effect of impurities the contribution from the exciton–phonon collision should be minimized, i.e., the temperature should be kept as low as possible. Again we employ the 119 μm wavelength. In Fig. 30 we give the traces of magnetoabsorption for various impurity concentrations, for both boron-doped and phosphorus-doped samples. The magnetic field is applied along $\langle 111 \rangle$, and the temperature is fixed at 1.7 K. The monitoring absorption line will be the one at 1.23 T. We look for the contribution of impurities to the linewidth by subtracting the linewidth of the undoped crystal for $B \| \langle 111 \rangle$. The result is shown in Fig. 31, for both boron and phosphorus

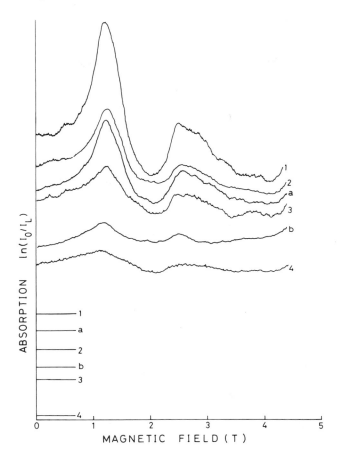

FIG. 30 Magnetoabsorption traces of the excitonic transition for samples with various impurity concentrations, for 119 μm, 1.7 K, $B \parallel \langle 111 \rangle$. The base line for each trace is also shown (Ohyama, 1981a). Curve 1, undoped $N_A - N_D \lesssim 10^{12}$ cm^{-3}; curve 2, 8.0×10^{13} boron cm^{-3}; curve 3, 4.3×10^{14} boron cm^{-3}; curve 4, 2.0×10^{15} boron cm^{-3}; curve a, 8.8×10^{13} phosphorus cm^{-3}; and curve b, 8.0×10^{14} phosphorus cm^{-3}.

impurities. One finds by writing the contribution of impurities to the half-width ΔB_I that

$$\Delta B_I = \alpha N, \qquad (30)$$

where α is the coefficient of proportionality and N is the concentration of impurities that are expected to have been neutralized after the band-gap light illumination. The contribution of the exciton–phonon collisions, which we write ΔB_{ex-ph}, can readily be obtained from the linewidth at

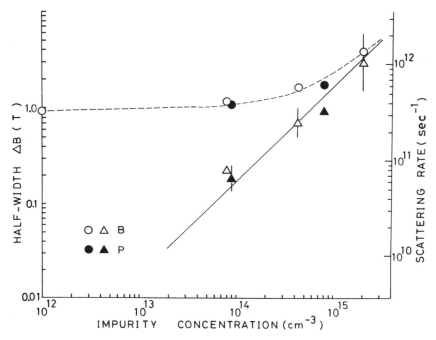

FIG. 31 Concentration dependence of the half-width ΔB of the excitonic absorption line for 119 μm, 1.7 K, $B \parallel \langle 111 \rangle$. Converted scattering rates are also shown on the right scale. Triangles represent the contribution of impurities to the linewidth, obtained by subtracting the linewidth of the undoped crystal (Ohyama, 1981a).

$N \rightarrow 0$. One can further write, as an approximation,

$$\alpha = v_{ex}\sigma_I, \tag{31}$$

where v_{ex} is the thermal velocity of the exciton and σ_I is the scattering cross section by the impurity. Putting $v_{ex} = (3k_B T/m_{ex})^{1/2}$ and inserting $T = 1.7$ K, one obtains $v_{ex} \cong 10^6$ cm/sec. The coefficient α obtained from our experimental result is 6.1×10^{-4} cm^3/sec. Thus one can derive for the scattering cross section $\sigma_I = 5.6 \times 10^{-10}$ cm^2 at 1.7 K.

We might as well look for some theoretical background for the exciton–impurity scattering. Because both an exciton and an impurity are quasi-hydrogen atoms, we may assume a dipole–dipole interaction between them; thus one can write

$$V(r) = -C/r^6 \tag{32}$$

for the interaction potential. In Eq. (32) C is the van der Waals constant and r is the distance between the exciton and an impurity. A calculation by

Elkomoss and Nikitine (1980) is available for the elastic collision between an exciton and a neutral impurity. Making use of the collision damping theory by Weisskopf (1932), they find that the cross section can be expressed as

$$\sigma_I = \pi (3\pi C / 8 v_{ex})^{2/5}. \tag{33}$$

The perturbation calculation for the van der Waals interaction yields

$$C = 24 E_{ex} a_{ex}^3 a_I^3 / [\hbar (1 + a_I / a_{ex})], \tag{34}$$

where E_{ex} is the binding energy of the exciton and a_I and a_{ex} are the effective Bohr radii for the neutral impurity and for the exciton, respectively. Introducing Eq. (34) into Eq. (33), we obtain a theoretical estimate for σ_I; i.e., $\sigma_I = 1.1 \times 10^{-12}$ cm^{-2}. We have substituted here $E_{ex} = 14.7$ meV, $a_{ex} = 41$ Å, $a_I = 13$ Å, and $v_{ex} = 10^6$ cm/sec. Again we are surprised to see such a large discrepancy, a factor of 500, between theory and experiment. The situation, however, would perhaps be the same for the exciton–phonon collision.

Inclusion of the $2p$ exciton for calculating the collision cross section would certainly improve the case. Another possible improvement would be consideration of the contribution of the collision between a free exciton and an exciton bound to a neutral impurity. At such a low temperature as 1.7 K a considerable number of excitons are expected to be bound at the impurity sites. The fraction will depend on the intensity of excitation. Our experience with the boron-doped sample ($N_B = 4.3 \times 10^{14}$ cm^{-3}) shows that nearly 20% of the boron impurities are capturing excitons at 1.5 K under illumination by the tungsten lamp (Otsuka, 1981). If this is the case, the expression in Eq. (33) for C and hence that for σ_I should be modified. The situation would be closer to the exciton–exciton collision, and a_I may be replaced with a_{ex} if all the impurities had captured the excitons. Even in such an extreme case, however, the enhancement of σ_I is still less than a factor of ten. The contribution of the $2p$ exciton will thus have to be taken into account to interpret the discrepancy.

Finally we shall comment on the apparent nondistinction between p-type and n-type materials for the exciton–neutral-impurity scattering. A large difference in cross section between donor and acceptor has been observed for electron scattering; in fact, the cross section is larger for electron–neutral-donor scattering by a factor of 40 than for electron–neutral-acceptor scattering (Otsuka *et al.*, 1968). The essential reason for such a large difference is that electron–neutral-donor scattering can be simulated with electron–hydrogen atom (e$^-$–H) scattering and electron–neutral-acceptor scattering with positron–hydrogen atom (e$^+$–H) scattering. Such a distinction should not be expected in exciton–neutral-impurity scattering. Because an exciton can be simulated with either a neutral donor or a neutral acceptor, whether a collision is with a neutral donor or a neutral acceptor should make little difference to the size of the cross section.

F. WORK FUNCTION OF THE ELECTRON–HOLE DROP

Though numerous data exist for the work function of the electron–hole drop (EHD) in Ge, relatively few are available for the work function in Si. A new contribution from the far-infrared experiment will be presented here.

We have seen in the foregoing description that the behavior of excitonic absorption is quite strongly affected by temperature, especially when the excitation intensity is strong. This is shown in Fig. 32 where the absorption data obtained for the FIR wavelength of 119 μm are presented. One notes first the amount of absorption at zero magnetic field, which gives the height of the plateau on which the absorption peak at 4.5 T is based. A similar sight has been observed, incidentally, at the same FIR wavelength for Ge (Ohyama and Otsuka, 1980). The behavior is characteristic of the exciton system that coexists with EHD. One has to keep in mind, however, that the

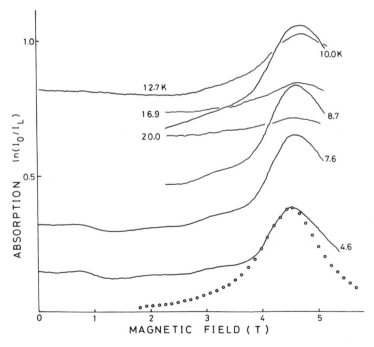

FIG. 32 Temperature dependence of magnetoabsorption of undoped Si at 119 μm for $B \parallel \langle 001 \rangle$. The curve shown by open circles is the absorption calculated according to the Lorentzian shape with $\Delta B = 1.4$ T (Ohyama, 1981a).

$1s \rightarrow 2p$ type transition of the exciton in Si is resonant with the energy of the 119 μm laser beam, or 10.4 meV, at $B = 0$. The height of the plateau, accordingly, should be understood as monitoring the density of excitons rather than the volume of the total EHD system. In Fig. 33 we plot the amount of absorption at $B = 0$ against the inverse temperature with the excitation level as a parameter. The figure clearly shows the existence of EHD. The diagram is bisected by a solid straight line called the *threshold line*. On the right side of the threshold line, the absorption decreases sharply with lowering temperature indicating that the excitons are adsorbed onto the surface of EHD. This region corresponds to the phase in which free excitons and EHD coexist. The left side of the threshold line corresponds to

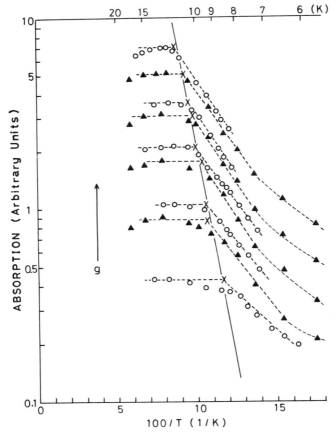

FIG. 33 Excitonic absorption intensity of undoped Si at 119 μm with $B = 0$ versus inverse temperature with the excitation level g as parameter. The solid threshold line yields the value of the work function $\phi = 7.6$ meV according to Eq. (39) (Ohyama, 1981a).

the phase at which no more EHD exists and only free excitons contribute to absorption. The slope of the threshold line gives the value of the work function for EHD. The standard kinetics on which this interpretation is based are as follows: denoting the radius of EHD by R, the steady flow of excitons adsorbed by EHD will be given by $\pi R^2 v_{ex} n_{ex}$, where n_{ex} is the density of free excitons. This quantity will be equated, under the steady-state assumption, to the loss of electron–hole pairs from EHD. This loss consists of two parts, internal recombination and evaporation from the surface. Thus one can write

$$\pi R^2 v_{ex} n_{ex} = 4\pi R^3 n_0 / 3\tau_0 + 4\pi R^2 a T^2 \exp(-\phi/k_B T), \tag{35}$$

where n_0 is the concentration of the electron–hole pairs within EHD, τ_0 is the internal recombination time, ϕ is the desired work function, and a is the coefficient appearing in the theory of thermionic emission. In the meantime, one has to expect the condition

$$g = n_{ex}/\tau_{ex} + 4\pi R^3 n_0 N_d / 3\tau_0 \tag{36}$$

for steady state. Here q is the excitation level, or the generation rate of electron–hole pairs, τ_{ex} is the lifetime of excitons, and N_d is the density of EHD. Combining the last two equations, we obtain

$$\begin{aligned} g - (4aT^2/v_{ex}\tau_{ex}) \exp(-\phi/k_B T) \\ = (4n_0/3v_{ex}\tau_0)(1/\tau_{ex} + N_d v_{ex} \pi R^2) R. \end{aligned} \tag{37}$$

In order to keep EHD, one has to hold

$$g - (4aT^2/v_{ex}\tau_{ex}) \exp(-\phi/k_B T) \geqq 0, \tag{38}$$

because only for this condition do we have $R \geqq 0$. The threshold excitation level can then be defined as

$$g_{th} = (4aT^2/v_{ex}\tau_{ex}) \exp(-\phi/k_B T). \tag{39}$$

The threshold line in Fig. 33 corresponds to this equation. We find from fitting procedures that $\phi = (7.6 \pm 0.5)$ MeV. A strict argument requires consideration of the temperature dependence of ϕ as well as the correction due to the surface energy. The latter, however, will be of the order of 0.1 meV for $R = 0.1$ μm. The temperature dependence also is usually ignored. Typical of experimental data for ϕ available elsewhere are those by Dite *et al.* (1977), who give $\phi = 7.9$ meV, and by Hammond *et al.* (1976), who give $\phi = 8.2$ meV. The former value was obtained by means of thermodynamic method like the present one. The latter value was obtained by emission spectroscopy which, in the case of Ge, always gives a larger value than the thermodynamic value.

G. FREE EXCITONS AND FREE CARRIERS

We have already seen in Fig. 32 that the density of excitons first increases with temperature up to a certain temperature and then starts to decrease. The turning point seems to occur near 15 K. The result of a more careful study using a wavelength of 119 μm is presented in Fig. 34. The decrease in excitons above 15 K is followed by the emission of free carriers as a result of the dissociation of excitons. The density of free carriers is also determined at 220 μm by watching the free-carrier absorption.

We are concerned here about the coexistence of free excitons and free

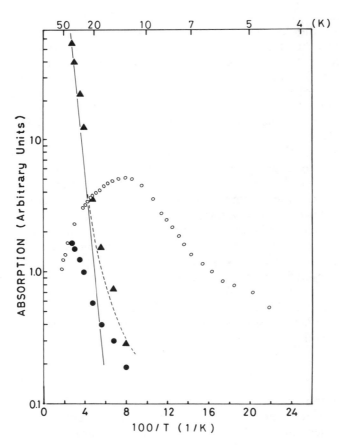

FIG. 34 Temperature dependence of the excitonic ($\bigcirc = 119$ μm, n_{ex}) and the free-carrier ($\bullet = 220$ μm, n_{e}) absorptions of undoped Si. Triangles show the value $K = n_{\text{e}}^2/n_{\text{ex}}$. Solid line [$T^{3/2} \exp(-E_{\text{ex}}/kT)$] yields $E_{\text{ex}} = 14.7$ meV (Ohyama, 1981a).

carriers above 15 K where no EHD exists. The chemical reaction inherent to this state can be written symbolically as

$$e + h \rightleftarrows \text{ex.} \tag{40}$$

One can then expect from the law of mass action that

$$n_e n_h / n_{ex} = K(T), \tag{41}$$

with

$$K(T) = (g_e g_h / g_{ex})(m_e m_h / 2\hbar^2 m_{ex})^{3/2}(k_B T)^{3/2} \times \exp(-E_{ex}/k_B T), \tag{42}$$

where n_e and n_h are the densities of electrons and holes, respectively; g_e, g_h, and g_{ex} are the degeneracy factors for the corresponding constituents as indicated by the subscripts; and E_{ex} is the binding energy of the exciton. In an intrinsic material one can put $n_e = n_h = n$. The solid straight line in Fig. 34 gives $K(T) = n^2/n_{ex}$ with $E_{ex} = 14.7$ meV. This fitting value for E_{ex} is quite in agreement with the binding energy obtained elsewhere by the optical method (Shaklee and Nahory, 1970). Thus one can see the existence of the free-carrier plasma that is in quasi-equilibrium with free excitons above 15 K.

So far all the absorption data have been taken during excitation. In other words, no delay time has been set in the measurement. The time-resolution method is applied here to explore the lifetime of excitons. In Fig. 35, the time variation of the excitonic absorption obtained at $B = 0$ and wavelength 119 μm is shown for three temperatures. The initial decays are more or less similar to each other despite the differences in temperature. This indicates the occurrence of the Auger process. The time constant for the measurement at 1.7 K is found clearly to be 3.6 μsec. At higher temperatures the suggested Auger recombination process seems to be followed by the slower decay process. This secondary process gives a longer time constant at higher temperatures, reaching 30 μsec above 20 K. Probably some unidentified impurity-associated recombination will be found to be responsible for the decay process in the second stage.

VI. Concluding Remarks

The combination of band–gap photoexcitation with far-infrared laser beam application leads to an unexpected development in semiconductor physics which supplements the luminescence experiment that served for many years as the primary approach to the property of a photoexcited carrier system in a semiconductor. One of the greatest merits of the new approach is its ability to obtain information on the carrier-relaxation mechanism. Momentum-relaxation and energy-relaxation times are obtainable

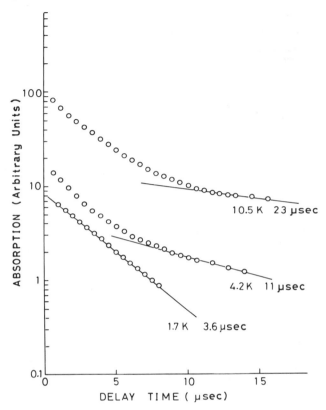

FIG. 35 Time variation of the excitonic absorption of undoped Si for 119 μm and $B = 0$ for three temperatures (Ohyama, 1981a).

by measuring the linewidth of the far-infrared resonance absorption and the time variation of the resonance intensity, respectively, for the photoexcited carrier system, in the form of either cyclotron resonance or the impurity Zeeman transition. The same idea extends to the Zeeman transition of the free-exciton system in Si. It should be emphasized that the excitonic Zeeman transition itself has been observed for the first time, by the technique described here. Joint observation of electrons and holes in InSb has brought us to a deeper insight into carrier kinetics. It has been found that even in the presence of opposite carriers cyclotron emission can take place. It could be a surprise to nonspecialists that electron resonance was observed more clearly in p-type material than in n-type material and hole resonance more clearly in n-type material than in p-type material because of the difference in electron scattering cross section by photoneutralized impurities. Owing to the slow relaxation of neutralized donors, one can devise, for GaAs in particular, a high-sensitivity FIR detector. The neutralization of impurities further enables us to determine the absolute densities of donors and acceptors. This is a promising substitute for the conventional Hall measurement, because it requires neither sample shaping nor electrode attachment. Elaborate analysis after carrying out measurements at different temperatures can also be avoided.

The band-gap illumination has been a simple and conventional technique, widely used with semiconductors in order to produce photo-carriers. Re-evaluation of this technique is nowadays stressed in association with the time-resolution technique in dealing with the nonthermal equilibrium problem of the photocarrier system. In conjunction with the far-infrared magnetooptical resonances, one meets various new relaxation processes of photocarriers, sometimes surprisingly slow, sometimes unclear in mechanism, and sometimes unstable in the form of electron–hole coexistence. In contrast to the fairly long history of photo-excitation technology and far-infrared laser spectroscopy, the approach proposed here, which combines these two methods on the time-resolution scale axis, has almost been ignored so far but now seems to have a wide-open future.

ACKNOWLEDGMENTS

The authors are indebted to H. Nakata for discussions. Assistance by K. Fujii in carrying out experiments in InSb has been very helpful. This work has been supported by a Grant-in-Aid for Scientific Research by the Ministry of Education, Science and Culture.

REFERENCES

Baltensperger, W. (1953). *Phil. Mag.* **44,** 1355–1363.
Blatt, F. J. (1957). *J. Phys. Chem. Solids* **1,** 262–269.
Brooks, H. (1951). *Phys. Rev.* **83,** 879.
Button, K. J., Lax, B., and Bradley, C. C. (1968). *Phys. Rev. Lett.* **21,** 350–352.
Button, K. J., Roth, L. M., Kleiner, W. H., Zwerdling, S., and Lax, B. (1969). *Phys. Rev. Lett.* **2,** 161–162.
Chamberlain, J. M., Simmonds, P. E., Stradling, R. A., and Bradley, C. C. (1971). *J. Phys. C 4,* L38–L42.
Conwell, E. M., and Weisskopf, V. F. (1950). *Phys. Rev.* **77,** 388–390.
Dite, A. F., Kulakovskii, V. D., and Timofeev, V. B. (1977a). *Zh. Eksp. Teor. Fiz.* **72,** 1156–1170 [*Sov. Phys.-JETP 45,* 604–612].
Elkomoss, S. G., and Nikitine, S. (1980). *J. Phys. Chem. Solids* **41,** 413–420.
Erginsoy, C. (1950). *Phys. Rev.* **79,** 1013–1014.
Fetterman, H. R., Larsen, D. M., Stillman, G. E., and Tannenwald, P. E. (1971). *Phys. Rev. Lett.* **26,** 975–978.
Fujii, K. (1982). Master's Thesis, Graduate School of Science, Osaka University, unpublished.
Gornik, E. (1972). *Phys. Rev. Lett.* **29,** 595–597.
Gornik, E. (1978). *In* "The Application of High Magnetic Fields in Semiconductor Physics," pp. 329–345. Clarendon Laboratory, Oxford.

Hammond, R. B., McGill, T. C., and Mayer, J. W. (1976). *Phys. Rev. B* **13**, 3566–3575.

Hasegawa, H., and Howard, R. E. (1961). *J. Phys. Chem. Solids.* **21**, 179–198.

Hensel, J. C., and Suzuki, K. (1974). *Phys. Rev. B* **9**, 4219–4257.

Katana, P. K., Tiron, S. D., and Cheban, A. G. (1970). *Fiz. Tek. Poluprovodn.* **4**, 260–263 [*Sov. Phys.-Semicond.* **4**, 210–213].

Kawamura, H., Saji, H., Fukai, M., Sekido, K., and Imai, I. (1964). *J. Phys. Soc. Jpn.* **19**, 288–296.

Knox, R. S. (1963). *Solid State Phys.* **5**, 83.

Kobayashi, K. L. I., Komatsubara, K. F., and Otsuka, E. (1973). *Phys. Rev. Lett.* **30**, 702–705.

Kobayashi, K. L. I., Komatsubara, K. F., Otsuka, E., Ohyama, T. and Matsuda, O. (1974). *Jpn. J. Appl. Phys.* **43**, Supplement, 301–304.

Kobayashi, K. L. I., and Otsuka, E. (1974). *J. Phys. Chem. Solids* **35**, 839–489.

Krieger, J. B. (1969). *Phys. Rev.* **178**, 1337–1339.

Larsen, D. M. (1968). *J. Phys. Chem. Solids* **29**, 271–280.

Lipari, N. O., and Altarelli, M. (1976). *Solid State Commun.* **18**, 951–954.

Matsuda, O., and Otsuka, E. (1979). *J. Phys. Chem. Solids* **40**, 809–829.

Nakata, H., and Otsuka, E. (1981). *Phys. Rev. B* **23**, 5428–5444.

Nakata, H., and Otsuka, E. (1982). *Appl. Phys. B* **27**, 207–209.

Ohyama, T. (1980). *Phys. Status. Solidi (B)* **98**, 373–378.

Ohyama, T. (1981a). *Phys. Rev. B* **23**, 5445–5453.

Ohyama, T. (1981b). *Jpn. J. Appl. Phys.* **20**, 2423–2424.

Ohyama, T., Murase, K., and Otsuka, E. (1970). *J. Phys. Soc. Jpn.* **29**, 912–924.

Ohyama, T., and Otsuka, E. (1980). *J. Phys. Soc. Jpn.* **48**, 1550–1558.

Otsuka, E. (1980). *In* "Infrared and Millimeter Waves," Vol. 3 (K. J. Button, ed.) Academic Press, New York.

Otsuka, E. (1981). *J. Phys. Soc. Jpn.* **50**, 2631–2642.

Otsuka, E., Fujii, K., and Kobayashi, K. L. I. (1973). *Jpn. J. Appl. Phys.* **12**, 1600–1605.

Otsuka, E., Murase, K., Iseki, J., and Ishida, S. (1964). *Phys. Rev. Lett.* **13**, 232–233.

Otsuka, E., Murase, K., and Iseki, J. (1966). *J. Phys. Soc. Jpn.* **21**, 1104–1111.

Otsuka, E., Ohyama, T., and Murase, K. (1968). *J. Phys. Soc. Jpn.* **25**, 729–739.

Pidgeon, C. R., and Brown, R. N. (1966). *Phys. Rev.* **146**, 575–583.

Ranvaud, R., Trebin, H.-R., Rossler, U., and Pollak, F. H. (1979). *Phys. Rev. B* **20**, 701–715.

Roth, L., Lax, B., and Zwerdling, S. (1959). *Phys. Rev.* **114**, 90–104.

Shaklee, K. L., and Nahory, R. E. (1970). *Phys. Rev. Lett.* **24**, 942–945.

Thomas, D. G., Hopfield, J. J., and Augustyniak, W. M. (1965). *Phys. Rev.* **140**, A202–A220.

Timusk, T., Navarro, H., Lipari, N. O., and Altarelli, M. (1978). *Solid State Commun.* **25**, 217–219.

Toyozawa, Y. (1958). *Prog. Theor. Phys.* **20**, 53–81.

Ulbrich, R. (1973). *Phys. Rev. B* **8**, 5719–5727.

Weisskopf, V. F. (1932). *Z. Phys.* **75**, 287–301.

CHAPTER 7

High-Temperature Infrared Reflectivity Spectroscopy by Scanning Interferometry

F. Gervais

Centre de Recherches sur la Physique
des Hautes Températures
Centre National de la Recherche Scientifique
Orléans, France

LIST OF SYMBOLS

$a(\mathbf{k}j)$	phonon annihilation operator	k	lattice force constant
$a^+(\mathbf{k}j)$	phonon creation operator	$\mathbf{k}$	wave vector
$A(\mathbf{k}j)$	phonon operator	k_{ij}	real coupling constant
α_v	volume expansion coefficient	k_B	Boltzmann constant
$\mathbf{D}$	displacement vector	K	extinction coefficient
δ	optical path difference	χ	dielectric susceptibility
$\delta(u)$	delta function	ℓ_i	abbreviation for $(\mathbf{k}_i j_i)$
$\delta\omega_j$	frequency shift due to pure-volume effect	m_k	mass of the kth ion
$\Delta\omega_j(\omega)$	anharmonic frequency shift function	M	dipolar moment component
		μ	reduced mass of ions
$\Delta\varepsilon_j$	oscillator strength	n	mean number of phonons
e_S^*	Szigeti effective charge	N	refractive index
e_T	transverse effective charge	ν	electronic signal frequency
$\mathbf{E}$	electric field of IR radiation	ω	IR radiation frequency
ε	complex dielectric function	$\omega(0j)$	harmonic phonon frequency
ε_v	dielectric constant of vacuum	$\overline{\omega}_j$	quasi-harmonic frequency
ε_0	static dielectric constant	Ω_j	phonon frequency
ε_∞	high-frequency dielectric constant	Ω_p	plasma frequency
		$\mathbf{p}$	instantaneous dipole moment
f	constant that has the dimension of a frequency	$\mathbf{P}$	polarization
		$P_j(\omega, T)$	(proper) phonon self-energy
g_j	mode Grüneisen parameter	R	reflectivity
$g(0j, \omega)$	free-phonon propagator	$\rho(\omega, T)$	one-phonon density of states
$G(0j, \omega)$	phonon propagator	t	time
γ_j	classical damping constant	T	temperature
$\Gamma_j(\omega)$	quantum damping function	T_C	phase transition temperature
$\hbar$	Planck's constant	v	mobile mirror velocity
H_0	harmonic Hamiltonian	V	volume of the unit cell
I	IR light emission	$V^{(n)}(\cdots)$	nth order anharmonic potential
j	phonon branch index		
J	interferometric signal	w	coupling constant
$\overline{J}$	interferometric signal returning to its source	$\mathbf{x}$	instantaneous displacement of ion
		Ze	effective charge of an ion

I.   Introduction

The subject of this chapter is best introduced by discussing certain merits and applications of a technique, scanning interferometry, and an experimental method, infrared reflectivity spectroscopy, in comparison with other spectroscopic techniques such as Raman and neutron scattering. Concerning the first point, the principles of this technique have been known for at least two decades, but it is only recently that the availability and the reduction of the cost of fast electronic minicomputers with large storage capacities made the development of scanning interferometers possible. In the first generation of infrared Fourier spectrometers, the interferogram is recorded by moving the mirror in steps. However, the mirror may be moved at *constant velocity*. Because it is possible to record a single interferogram in a few seconds or even in less than one second, depending on the resolution needed, *repetition* is used to improve the signal-to-noise ratio. Whereas earlier Fourier spectrometers usually covered the wave number range from 10 to a few hundred reciprocal centimeters, new commercial Fourier transform "rapid-scan"—here referred to as "scanning"—interferometers (Bruker, Digilab, Nicolet) are now able to cover the range $10-4000$ cm^{-1} or even more. This range allows one to record the complete infrared (IR) vibrational spectrum of any compound that exists in nature. Only three pages were devoted to this technique in the book Bell published in 1972, for example. This relatively novel technique has been little applied in solid-state physics up to now.

Infrared reflectivity spectroscopy, on the other hand, allowed quantitative studies of solids in practice when computers (again!) were available, that is, at the end of the 1950s. Spitzer and Kleinman (1961) were the first, to the best of the author's knowledge, to perform a Kramers–Kronig analysis of polarized IR reflection spectra in a multiband case, α-quartz, and they beautifully fitted the classical dispersion theory, known for several decades and associated with the names of Helmholtz and Kettler, to experimental spectra. In particular, these methods were early applied to ferroelectric (FE) materials at the beginning of the 1960s and gave information, especially in perovskite oxide crystals, in relation to the soft mode concept (see Barker, 1967). But soon (Spitzer *et al.*, 1962) difficulties appeared in the analysis of the spectra. A review of these studies is given in the chapter by Perry (1971) in the book of Möller and Rothschild. In the 1960s, IR reflectivity measurements commonly needed two spectrometers, a conventional dispersive one for medium and near infrared plus a Michelson interferometer for the far infrared. Such a practice is still usual, as is shown by data published recently. However, the development of laser technology and of high-flux neutron reactors led experimentalists away from this rather tedious method—because computational analysis of spectra is also necessary—and to use Raman scattering preferentially, because it yields readily usable information, and neutron scattering, to a lesser extent, because it yields more information (but it is less accurate and much more "heavy" to perform). It may be noted nevertheless that (i) a fit of a convenient model to Raman profiles (also neutron spectra) was often performed in the studies published during the last decade, especially when phenomena of mode-coupling and/or overdamped modes complicate the spectrum; (ii) in noncentrosymmetric crystals, where a comparison can be made because modes are infrared- and Raman-active, Fourier-transform scanning interferometers now provide spectral information in a time comparable to what is needed in a Raman experiment, or even shorter when the crystal is large enough. Both techniques, therefore, are now competitive again. In fact, some advantages of IR reflectivity spectroscopy will be discussed in this chapter, among them its convenience for high-temperature studies and obviously its power in the understanding of phenomena in which the study of excitations that are IR-active only constitutes the main conclusive information. This is, for example, the case for the soft FE mode in the paraelectric phase of FE crystals.

But first, let us briefly recall that vibrational motions of atoms around their equilibrium position are responsible for (thermo)dynamical properties of solids. In crystals, such coherent motion, called a fundamental vibration mode, a normal mode, or a phonon mode, has a characteristic frequency that depends only on lattice force constants and masses of the atoms. The number of normal modes depends solely on the structure of the crystal lattice and may be predicted by group theory. In polar crystals — the widest class of materials in nature — a number of vibrational modes are also polar. Such modes consist of the motion of positive ions against negative ions and couple directly with the electromagnetic field of IR radiation. In other words, polar crystals selectively absorb but also reflect IR radiation. Both effects are complementary. To illustrate these facts, Fig. 1 shows the IR reflectivity of a thin slab of a single crystal of lead phosphate (unpolarized light) as obtained with a scanning interferometer BRUKER IFS 113 C. In the regions where the absorption coefficient is elevated, above $\sim 10^3$ cm^{-1}, the crystal, even 55 μm thick, totally absorbs the radiation, and the absorption coefficient has been determined by a Kramers–Kronig (KK) analysis of reflectivity data (see, for example, Mitra, 1969). The main absorptions correspond to (i) external modes at the lowest frequencies, i.e., the vibrational motions of Pb ions, which are heavy, against PO$_4$ entities; (ii) internal modes that are strongly IR-active, i.e., motions that derive from the v_3 and v_4 modes (in Herzberg's notation) of the free PO$_4$ tetrahedron. The v_2 modes that are IR-inactive in the free tetrahedron because of symmetry considerations become slightly active in the lattice. They do not yield any observable reflection band and are observed by transmission measurements only. Detailed analysis of polarized spectra of Pb$_3$(PO$_4$)$_2$ were reported by Luspin et al. (1978). The reflection spectrum of Fig. 1 also shows that interference fringes arise from multiple internal reflections in the frequency regions in which the crystal is partially transparent. This is an experimental test of the small solid angle of the incident IR beam. Such a small angle favors accurate measurements in polarized light. Besides, the transmission spectrum reveals additional structure above 1200 cm^{-1}, seen in the absorption coefficient profile of Fig. 1, that cannot be assigned to fundamental modes. This observation of features related to combinations and overtones calls for preliminary discussion.

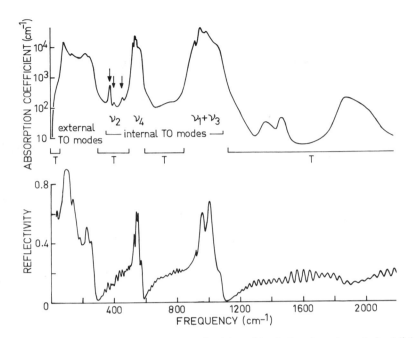

FIG. 1 IR absorption coefficient and reflectivity of lead phosphate (unpolarized light). Results obtained in the frequency ranges denoted by T were deduced from transmission measurements through a 55-μm-thick single crystal. Results in other regions are deduced from a KK analysis of the reflection spectrum shown below. Notice interference fringes at frequencies where the sample is semitransparent.

Vibrational modes are represented by delta functions within the *harmonic* approximation, and their frequency does not depend on temperature. This harmonic description, therefore, is unable to account for the actual thermophysical properties of solids. Vibrational modes are in reality *anharmonic*. The simplest way to account for anharmonicity is to consider that the modes are damped. Within the description of classical theory, a constant damping statistically accounts for all anharmonic coupling processes that relax the phonon mode under consideration. But nature is far from being so simple. If this description is often a good approximation in the immediate vicinity of the mode frequency, it is poor in other regions of the spectrum. This is because, actually, damping is a function of frequency and turns out to be the imaginary part of a complex function $P_j(\omega, T) = \Delta\omega_j(\omega, T) \pm i\Gamma_j(\omega, T)$ called the phonon self-energy, its real part being a frequency-shift function (Cowley, 1963; Maradudin and Fein, 1962; Wallis *et al.*, 1966). Perturbation treatments based on the Green's function formalism predict the temperature dependence of the phonon self-energy, and thus of the total energy of the modes, which assumes the approximate form

$\sim \hbar[\omega_j + \delta\omega_j(T) + \Delta\omega_j(\omega,\ T) \pm i\Gamma_j(\omega,\ T)$. The simple mechanical effect of change of lattice force constants due to lattice thermal expansion is also incorporated in the expression above via a frequency shift $\delta\omega_j(T)$.

After a simplified derivation of the interaction of IR light and matter, we will briefly review the concept of the phonon within the quantum approach of the phonon self-energy formalism. This preliminary seems to be necessary in order to state the bases on which approximate models, able to predict the experimental temperature dependence of the IR spectra of solids, can be retained. This is particularly important for the analysis of a class of phenomena in which temperature plays the dominant role: solid-state phase transitions. Generally, we will avoid excessive mathematics and closely follow the viewpoint of the experimentalist.

II. Interaction of Infrared Electromagnetic Field and Crystalline Solids

Because the subject is much too wide for detailed presentation here, we assume the reader is familiar with the fundamentals of lattice dynamics and the application of group theory to vibrational problems. In cubic diatomic crystals the motion of one type of ion, all in phase, against the other type, around their equilibrium position, is associated with a microscopic instantaneous dipole moment,

$$\mathbf{p} = e_T\mathbf{x} \tag{1}$$

where $\mathbf{x}$ represents the displacement of the cation relative to the anion in normal coordinates. The transverse effective charge e_T is so denoted because only transverse modes can interact with an electromagnetic field. The infrared dispersion (Born and Huang, 1954) describes the response of a crystalline lattice to an infrared (IR) electromagnetic radiation. Equation (1) is sufficient to describe the fundamentals of IR *absorption,* at least in the usual conditions, meaning we ignore measurements on very thin films to avoid the depolarizing effect of a slab (Berreman, 1963). Another equation is needed to describe IR *reflection,* which is related to the *dielectric function* ε (defined as the ratio of the displacement vector $\mathbf{D}$ to the electric field $\mathbf{E}$),

$$\text{div } \mathbf{D} = 0, \tag{2}$$

when there are no conduction electrons and where

$$\mathbf{D} = \varepsilon\varepsilon_v\mathbf{E}. \tag{3}$$

This is merely intended to review the main steps of the treatment of the interaction of IR radiation with matter, with emphasis on assumptions and approximations made in the derivation of analytical expressions that will be useful in the analysis of experimental results.

A. THE RESPONSE OF TRANSVERSE-OPTICAL (TO) MODES

In a cubic diatomic crystal the equation of motion of an undamped TO mode that interacts with the transverse electric field of electromagnetic radiation is simply that of an harmonic oscillator driven by the periodic field,

$$\ddot{x} + \Omega_{TO}^2 x = (e_T/\mu)E, \tag{4}$$

where μ is the reduced mass of the ions and $\Omega_{TO} = (k/\mu)^{1/2}$ is the TO resonance frequency of the oscillator.

We seek plane-wave solutions for each of the vector quantities

$$(\mathbf{E}, \mathbf{P}, \mathbf{x}) = (\mathbf{E}_0, \mathbf{P}_0, \mathbf{x}_0)e^{i(\omega t - \mathbf{k} \cdot \mathbf{r})}. \tag{5}$$

Taking the derivatives, we obtain

$$(\Omega_{TO}^2 - \omega^2)\mathbf{x} = (e_T/\mu)\mathbf{E}. \tag{6}$$

Utilizing the definition of the ionic susceptibility

$$\mathbf{P} = \chi_i \varepsilon_v \mathbf{E}, \tag{7}$$

where ε_v is the dielectric constant of vacuum, and Eqs. (1) and (6), we obtain the ionic susceptibility in the form

$$\chi_i(\omega) = \left(\frac{e_T^2}{\mu \varepsilon_v V}\right)\left(\frac{1}{\Omega_{TO}^2 - \omega^2}\right), \tag{8}$$

where V is the volume of the unit cell. The resonance term in the right side of Eq. (8) is characteristic of the response of the TO mode to the IR electromagnetic field.

The dielectric function is the response of all transverse excitations (not only ionic) to the macroscopic field. It is obtained by summing all independent susceptibilities and unity,

$$\varepsilon = 1 + \chi_{\text{electronic}} + \chi_{\text{ionic}} + \chi_{\text{plasma}} + \chi_{\text{relaxation}} + \cdots. \tag{9}$$

Equation (9) is known as the sum rule. If we confine ourselves to lattice effects and ignore plasmon and low-frequency relaxation effects in a first approximation, the dielectric function can be written

$$\varepsilon(\omega) = \varepsilon_\infty + \left(\frac{e_T^2}{\mu \varepsilon_v V}\right)\left(\frac{1}{\Omega_{TO}^2 - \omega^2}\right), \tag{10}$$

where $\varepsilon_\infty = 1 + \chi_{\text{el}}$ is the "high-frequency" dielectric constant from the viewpoint of the IR spectroscopist.

If a phenomenological damping is introduced into the equation of motion, Eq. (4), in the form

$$\ddot{x} + \gamma_{TO}\dot{x} + \Omega_{TO}^2 x = (e_T/\mu)E, \tag{11}$$

defining the oscillator strength as

$$\Delta\varepsilon = \left(\frac{1}{\Omega_{TO}^2}\right)\left(\frac{e_T^2}{\mu\varepsilon_v V}\right), \tag{12}$$

a definition that satisfies the sum rule when $\omega = 0$

$$\varepsilon(0) = \varepsilon_\infty + \Delta\varepsilon, \tag{13}$$

the classical dielectric function is obtained

$$\varepsilon(\omega) = \varepsilon_\infty + \Delta\varepsilon \frac{\Omega_{TO}^2}{\Omega_{TO}^2 - \omega^2 + i\gamma_{TO}\omega}. \tag{14}$$

B. LONGITUDINAL OPTICAL (LO) MODES

Equation (2), rewritten in the form $\varepsilon \, \text{div} \, E = 0$, has two solutions, $\text{div} \, E = 0$, a criterion that is satisfied by transverse waves, and $\varepsilon = 0$. For waves that propagate in a direction parallel to E (longitudinal waves), $\text{div} \, E \neq 0$ and thus $\varepsilon = 0$, necessarily. In other words, the longitudinal modes are the zeros of the dielectric function. The formulation of this concept yields

$$\varepsilon_\infty + \Delta\varepsilon \frac{\Omega_{TO}^2}{\Omega_{TO}^2 - \Omega_{LO}^2} = 0. \tag{15}$$

The TO–LO splitting of a polar vibrational mode is thus straightforwardly related to the oscillator strength

$$\frac{\Omega_{LO}^2 - \Omega_{TO}^2}{\Omega_{TO}^2} = \frac{\Delta\varepsilon}{\varepsilon_\infty}. \tag{16}$$

By combining Eqs. (13) and (16) we arrive at the important relation

$$\varepsilon_0/\varepsilon_\infty = \Omega_{LO}^2/\Omega_{TO}^2 \tag{17}$$

first derived by Lyddane et al. (LST) in 1941.

On summing over the ionic susceptibilities of independent oscillators that vibrate along a given polarization, we obtain the general formulation of the classical dispersion model

$$\varepsilon = \varepsilon_\infty + \sum_j \Delta\varepsilon_j \cdot \frac{\Omega_{jTO}^2}{\Omega_{jTO}^2 - \omega^2 + i\gamma_{jTO}\omega}. \tag{18}$$

Notice that a microscopic treatment of the TO–LO splitting can also be derived with emphasis on the Lorentz correction of the local electric field (see, for example, Slater, 1967; Barker, 1967).

C. A STRAIGHTFORWARD APPLICATION: MEASUREMENTS OF EFFECTIVE CHARGES

When one derives IR dispersion relations in a way that accounts for the Lorentz field mentioned above, Eq. (10) is rewritten as

$$\varepsilon(\omega) = \varepsilon_\infty + \left(\frac{e_s^{*2}}{\mu\varepsilon_v V}\right)\left(\frac{\varepsilon_\infty + 2}{3}\right)^2\left(\frac{1}{\Omega_{TO}^2 - \omega^2}\right). \tag{19}$$

It was first derived by Szigeti (1949) who introduced the effective charge e_s^* that bears his name.

If we set $\varepsilon = 0$, Eq. (19) becomes

$$\mu(\Omega_{LO}^2 - \Omega_{TO}^2) = \left(\frac{e_s^{*2}}{\varepsilon_v\varepsilon_\infty V}\right)\left(\frac{\varepsilon_\infty + 2}{3}\right)^2. \tag{20}$$

It is convenient to introduce (Gervais, 1976a, 1980; Hiraishi, 1973; Scott, 1971) a new effective charge Ze, related to the Szigeti effective charge by

$$Ze = e_s^*(\varepsilon_\infty + 2)/3\varepsilon_\infty^{1/2}, \tag{21}$$

so that Eq. (20) is rewritten in the very simple form

$$\Omega_{LO}^2 - \Omega_{TO}^2 = (Ze)^2/\mu\varepsilon_v V, \tag{22}$$

which clearly shows that an accurate measurement of the effective charge Ze can be deduced from the experimental knowledge of the TO–LO splitting.

Before we proceed any farther, notice that Lyddane et al. (1941) had defined two effective charges e^* and $e_n^* \equiv e_T$ the longitudinal and transverse effective charge, respectively, and are related to Ze by

$$e^* = \varepsilon_\infty^{-1/2}Ze \tag{23}$$

and

$$e_n^* = \varepsilon_\infty^{1/2}Ze. \tag{24}$$

The Szigeti effective charge does not differ from Ze by more than 10% when ε_∞ is lower than ~ 6 (the case in most crystals). Both charges are equal for $\varepsilon_\infty = 1$ and $\varepsilon_\infty = 4$. On dividing these charges by the nominal charge Z_0e, one obtains a value intermediate between 0 and 1 which can be compared with ionicities.

To generalize Eq. (22) to several modes, one has merely to sum over the dipolar moment components of the independent oscillators along a given polarization. The result is (Gervais and Arend 1983; Scott, 1971)

$$\sum_j (\Omega^2_{jLO} - \Omega^2_{jTO})_\alpha = \frac{1}{\varepsilon_v V} \sum_k \frac{(Ze)^2_{k\alpha}}{m_k}, \tag{25}$$

where α denotes a polarization and the k sum is over all atoms of mass m_k in the elementary volume V.† The electrical neutrality of the crystal provides another equation,

$$\sum_k (Ze)_{k\alpha} = 0. \tag{26}$$

The Szigeti effective charge has been determined in cubic diatomic crystals (see, for example, Mitra, 1969). More recently, tables of values for Ze have been reported for more complex binary crystals (Gervais, 1976a; Gervais and Baumard, 1979), based on available IR data. It is more important to emphasize that Eq. (25) allows one to measure the *temperature dependence* of the effective charge and therefore of the chemical bond, provided that the temperature dependence of all TO and LO frequencies has been obtained experimentally. This may be particularly important in the vicinity of structural phase transformations (Gervais, 1980).

Surprisingly, these concepts, which can be useful in a number of physical and chemical applications, have been little applied apart from cubic diatomic crystals. But interesting properties do not necessarily occur only in rocksalt-structured crystals; and the determination of the effective charge, which is straightforward with Eq. (25) in binary compounds of arbitrary complexity, and possible, with some approximations in ternary crystals, seems worthy to be carried out systematically.

III. Phonon Self-Energy, Theory and Practice

A. PHONON INTERACTIONS

Anharmonic couplings of phonon modes in crystals, also called phonon–phonon interactions, are mainly responsible for thermodynamic properties, thermal expansion, and heat-transport phenomena. Although the situation is not entirely clear (Migoni *et al.,* 1976), they are also believed (Cowley, 1965) to play an important role near phase transformations where small causes yield dramatic effects. Spectroscopic methods allow these effects to be analyzed, because phonon modes can be studied separately, contrary to

---

† Notice that the ω_j's are angular frequencies and should be expressed in radians per second in Eqs. (22) and (25). The practice of IR spectroscopists is that the ω's are usually referred to as *frequencies* and expressed in wave number units, namely reciprocal centimeters (cm^{-1}). Applications presented in this chapter comply with this use.

macroscopic measurements which involve statistics over the entire phonon spectrum.

Several treatises on lattice dynamics are available; those by Maradudin *et al.* (1971), Maradudin (1974), and Venkataraman *et al.* (1975) have been published recently. For our purpose it should be recalled that a free phonon is characterized by its energy $\hbar\omega(kj)$ and its wave vector **k**. The dispersion relation of the energy versus wave vector in the harmonic approximation gives rise to a *branch*, labelled by the index *j*, in the first Brillouin zone. After normal coordinate transformations, the harmonic Hamiltonian of the free phonon is

$$H_0 = \sum_{kj} \hbar\omega(kj)[a^+(kj)a(kj) + \tfrac{1}{2}], \qquad (27)$$

where $a^+(kj)$ and $a(kj)$ are the phonon creation and annihilation operators. The thermal average of the product of these operators is

$$\langle a^+(kj)a(k'j')\rangle = \Delta(\mathbf{k} - \mathbf{k}')\delta_{jj'}n(kj), \qquad (28)$$

$$\langle a(kj)a^+(k'j')\rangle = \Delta(\mathbf{k} - \mathbf{k}')\delta_{jj'}[n(kj) + 1], \qquad (29)$$

$$\langle a(kj)a(k'j')\rangle = \langle a^+(kj)a^+(k'j')\rangle = 0, \qquad (30)$$

where, because phonons are bosons,

$$n(kj) = \{\exp[\hbar\omega(kj)/k_BT] - 1\}^{-1} \qquad (31)$$

is the *mean number of phonons* in the vibrational mode (kj) at temperature T, a quantity that is most important in all of what follows. In Eq. (31) k_B is the Boltzmann constant, and $\Delta(\mathbf{k} - \mathbf{k}')$ and $\delta_{jj'}$ equal unity when $\mathbf{k} = \mathbf{k}'$ and $j = j'$, respectively, and vanish otherwise.

All anharmonic interactions involving n phonon modes are described by the nth order anharmonic Hamiltonian

$$H^{(n)} = \sum_{\mathbf{k}_1j_1}\sum_{\mathbf{k}_2j_2}\cdots\sum_{\mathbf{k}_nj_n} V^{(n)}(\mathbf{k}_1j_1, \mathbf{k}_2j_2,..., \mathbf{k}_nj_n)A(\mathbf{k}_1j_1)A(\mathbf{k}_2j_2)\cdots A(\mathbf{k}_nj_n) \qquad (32)$$

in the notation of Wallis *et al.* (1966), where it is convenient to define

$$A(\mathbf{k}_ij_i) = a^+(-\mathbf{k}_ij_i) + a(\mathbf{k}_ij_i), \qquad (33)$$

and where the $V^{(n)}(\cdots)$ anharmonic potentials are Fourier-transformed nth-order atomic force constants.

When a perturbation theory can be applied, i.e., when anharmonicity is weak enough, as mentioned in the introduction, all possible anharmonic interactions of a phonon mode $(\mathbf{k}_ij_i)$ with other phonons are described by a quantity called the (proper) *phonon self-energy*. In fact, we will be primarily concerned with phonons of approximately $\mathbf{k} = 0$ wave vector. This is because the wave vectors of infrared photons are very small in comparison with those of phonons, apart from those that belong to the "center" of the Brillouin zone and with which they interact. This is usually referred to as the long-wavelength approximation.

In quantum formulation the analog of the classical response function in Eq. (8), $\propto (\Omega_{TO}^2 - \omega^2)^{-1}$, is called the free-phonon propagator and at $\mathbf{k} \approx 0$ has the same form

$$g(0j, \omega) = [\omega(0j) + \omega]^{-1} + [\omega(0j) - \omega]^{-1} \tag{34}$$
$$= 2\omega(0j)[\omega^2(0j) - \omega^2]^{-1}.$$

It now seems necessary to deal with some physics of the many-body problem with the aid of the Green's function formalism. The anharmonic *phonon propagator* is the Fourier coefficient of the one-phonon thermodynamic Green's function. It may be evaluated by solving the Dyson equation (Cowley, 1963; Maradudin and Fein, 1962; Wallis *et al.,* 1966). Thus

$$G(0jj', \omega) = g(0j, \omega)\delta_{jj'} + g(0j, \omega) \sum_{j_1} P(0jj_1, \omega)G(0j_1j', \omega), \tag{35}$$

where P is the phonon self-energy

$$P(0j, \omega, T) = \Delta\omega(0j, \omega, T) \pm i\Gamma(0j, \omega, T), \tag{36}$$

which is approximated to be a matrix diagonal over the branch indices. The result is

$$G(0j, \omega) = \frac{2\omega(0j)}{\omega^2(0j) - \omega^2 \pm 2\omega(0j)P(0j, \omega, T)}. \tag{37}$$

Anticipating the dielectric function formulations of section IVA, we compare the classical response function of Eq. (18) and the phonon propagator of Eq. (37), which is nothing but another formulation of the response function; this allows us to write the following equivalence relations,

$$\Omega_j^2 \equiv \omega^2(0j) + 2\omega(0j)[\delta\omega_j(T) + \Delta\omega(0j, \Omega_j, T)] \tag{38}$$

and

$$[\gamma_j\omega \equiv 2\omega(0j)\Gamma(0j, \omega, T)]_{\omega=\Omega_j}, \tag{39}$$

where the frequency shift $\delta\omega_j(T)$ due to the mechanical effect of lattice thermal expansion has been incorporated. This shift is related to the volume expansion $\alpha_v(T)$ and the mode Grüneisen parameter g_j via (Jasperse *et al.,* 1966),

$$g_j = -\frac{d[\ln \omega(0j)]}{d[\ln V]} = -\frac{1}{\alpha_v(T)} \frac{1}{\omega(0j)} \frac{d\omega (0j)}{dT}, \tag{40}$$

and thus is approximately equal to

$$\delta\omega_j(T)/\omega(0j) = -\int_0^T g_j(T)\alpha_v(T)\,dT \tag{41}$$

when the shifts are small and simply equal to

$$\delta\omega_j(T)/\omega(0j) = -g_j\,\Delta V/V \tag{42}$$

if the mode Grüneisen parameter is assumed to be temperature independent. Equations (38) and (39) will provide analytical expressions for comparison with the experimental temperature dependence of mode frequencies Ω_j and damping γ_j as soon as the phonon self-energy has been evaluated.

B. DIAGRAMMATIC REPRESENTATION OF PHONON SELF-ENERGY CONTRIBUTIONS

The calculation of the phonon self-energy will not be developed here. The reader may refer to basic papers by Maradudin and Fein (1962) and Cowley (1963) in which the rules are indicated. Nevertheless, we will be concerned with a larger number of terms than the two contributions evaluated in the papers cited herein. We will restrict ourselves to the next higher order and consider the contributions in the vicinity of the fundamental frequency only. In the transparent regime of crystals at sufficiently high frequency, phonon self-energy should be summed over the higher orders to account for the observed results and, theoretically, even to infinite orders to account for the near-exponential frequency dependence of the absorption coefficient (Bendow, 1978; Billard *et al.*, 1976; Boyer *et al.*, 1975; Pohl and Meier, 1974). If we confine ourselves to terms that contribute to the phonon self-energy up to eighth order, the possible contributions are diagrammatically represented in Fig. 2 and classified according to their order, where η is an ordering operator. To understand the diagrams of Fig. 2 it should be kept in mind that (i) each line labeled l_i corresponds to a free phonon $(k_i j_i)$, (ii) an anharmonic potential $V^{(n)}$ is associated with each vertex or interaction of n lines, (iii) the conservation law for energy and wave vector is satisfied at each vertex, and (iv) one has to sum over all wave vectors, branch labels, and frequencies. Historically the lowest-order contribution (a) was not calculated first, because it derives from the quartic Hamiltonian $H^{(4)}$ whereas the subsequent one (b) involves the cubic Hamiltonian $H^{(3)}$, which was considered alone in the first derivations (Kascheev and Krivoglaz, 1961; Vinogradov, 1962). Kokkedee (1962) and Maradudin and Fein (1962) determined both $0(\eta^2)$ terms.

C. LOWEST-ORDER TERMS

Diagrams (a) and (b) of Fig. 2 correspond to the following contributions to

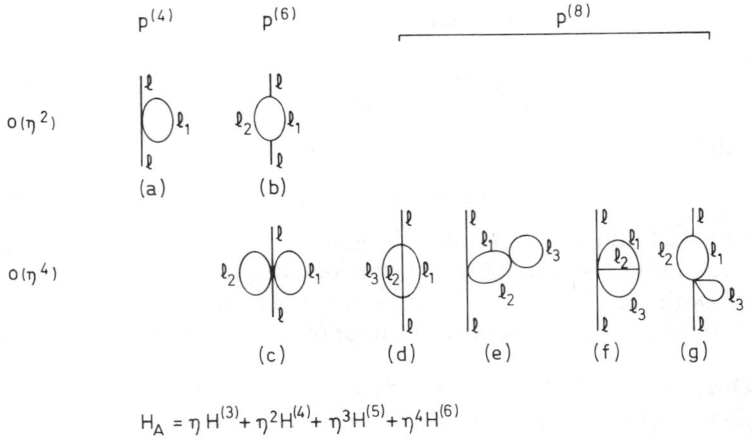

$$H_A = \eta\, H^{(3)} + \eta^2 H^{(4)} + \eta^3 H^{(5)} + \eta^4 H^{(6)}$$

FIG. 2 Diagrammatic representation of phonon–phonon interactions.

the phonon self-energy:

$$P_a^{(4)}(\mathbf{0}j,\, T) = \frac{24}{\hbar} \sum_{l_1} V^{(4)}(lll_1l_1)\left(n_1 + \frac{1}{2}\right), \qquad (43)$$

$$P_b^{(6)}(\mathbf{0}j,\, \omega,\, T) = -\frac{18}{\hbar^2} \sum_{l_1 l_2} |V^{(3)}(ll_1l_2)|^2 S(ll_1l_2), \qquad (44)$$

where

$$S(ll_1l_2) = (n_1 + n_2 + 1)\left[\frac{1}{(z + \omega_1 + \omega_2)_{\mathcal{P}}} - \frac{1}{(z - \omega_1 - \omega_2)_{\mathcal{P}}}\right]$$

$$+ 2(n_1 - n_2)\left[\frac{1}{(z - \omega_1 + \omega_2)_{\mathcal{P}}}\right]$$

and

$$\frac{1}{(z - \mu)_{\mathcal{P}}} = \mathcal{P}\,\frac{1}{\omega - \mu} \pm i\pi\delta(\omega - \mu). \qquad (45)$$

$\mathcal{P}$ denotes the principal part. In these simplified notations, l and l_i correspond to $(\mathbf{0}j)$ and $(\pm\mathbf{k}_i j_i)$ phonons, respectively, and n_i to $n(\mathbf{k}_i j_i)$, as given by Eq. (31). The first term of the right side of the expression for $S(ll_1l_2)$ corresponds to *additive* anharmonic processes and the second term to *difference* processes. At the order $O(\eta^2)$ the damping function reduces to

$$\Gamma_b^{(6)}(0j, \omega, T) = \frac{18}{\hbar^2} \sum_{l_1 l_2} |V^{(3)}(ll_1l_2)|^2[(n_1 + n_2 + 1)\delta(\omega - \omega_1 - \omega_2)$$

$$+ 2(n_1 - n_2)\delta(\omega + \omega_1 - \omega_2)]. \quad (46)$$

In the high-temperature limit ($\hbar\omega \ll k_BT$) the cubic damping linearly depends on temperature.

If the number of phonon branches is small (cubic diatomic crystals for example), the phonon self-energy $P_b^{(6)}$ can be a rapidly varying function of frequency. *Ab initio* calculations of such a term confirms this aspect in LiF (Eldridge and Howard, 1973; Ipatova *et al.*, 1967; Miskevich, 1963), KBr (Bruce, 1973; Cowley, 1963; Haque, 1975; Johnson and Bell, 1969); KI (Berg and Bell, 1971; Eldridge and Kembry, 1973), CsBr (Daubert *et al.*, 1973), CsI (Beairsto and Eldridge, 1973), lithium halides (Rastogi *et al.*, 1974), and other ionic crystals with rocksalt structure (Lowndes and Rastogi, 1976). An example of the dramatic frequency dependence of the cubic phonon self-energy, which yields a two-peak response for the TO mode in this case, has been reported by Krauzman *et al.* (1974) in CuCl and confirmed by Hennion *et al.* (1979). On the other hand, if one attaches some confidence to the accuracy of a KK analysis, $P_b^{(6)}$ can be deduced from a *single* IR reflection band with the aid of formulas derived in section IVA. Results obtained by Axe *et al.* (1965) on CdF$_2$ and PbF$_2$, Bilz (1966) on LiF, Zernik (1967) and Denham *et al.* (1970) on fluorite crystals, Le Toullec (1968) on ZnS, and Denham *et al.* (1973) on SrCl$_2$ indicate that the structure of $P_b^{(6)}$ experimentally is somewhat smoother than when it is calculated by anharmonic models. Cowley (1965), who performed calculations on the complicated crystal SrTiO$_3$, pointed out that the coarseness of the sampling for the summation of l_i's over the Brillouin zone in Eq. (44) may cause a spurious structure.

A quite general result may now be formulated. So long as the structure of the studied crystal involves a sufficient number of phonon branches, the one-phonon density of states usually displays less marked peaks. The two-phonon density of states, therefore, and *a fortiori* higher-order densities, and consequently the damping function, will tend to resemble a broad continuum. Within these conditions an approximation commonly made (Gervais and Piriou, 1974a, 1974b; Sakurai and Sato, 1971) consists of neglecting difference processes in Eq. (44) and assuming the frequencies of the phonons that interact with the phonon (0j) are dispersed on either side of the average frequency $\Omega_j/2$. An implicit renormalization of the frequencies ω_1 and ω_2 in Eq. (44) is to be noted to account for temperature effects (Semwal and

Sharma, 1974) when the damping function is simplified and rewritten in the form

$$[\Gamma_b^{(6)}(0j, \omega, T)]_{\omega=\Omega j} = \tfrac{1}{2}a_j\{[\exp(\hbar\Omega_j/2k_BT) - 1]^{-1} + \tfrac{1}{2}\}. \qquad (47)$$

Profiles calculated by Eq. (47) are plotted in Fig. 3 for several values of frequency, here assumed to be temperature independent. Quantum effects related to the term 1/2 in Eq. (31), which manifest themselves by a departure from linearity at low temperature, are visible in a wider temperature range as the frequency is higher. Equation (47) has been shown to fit quite satisfactorily the damping data obtained in a wide variety of compounds, even when $\Omega_j(T)$ significantly shifts with temperature as in the *FE* mode of rutile (Gervais and Piriou, 1974b). An example is shown in Fig. 4 for corundum in a wide temperature range. The adjustable parameter is a_j only. Such a procedure of simplification of the imaginary phonon self-energy gives a result that is nothing but the expression for the phonon decay evaluated by simple perturbation theory (Klemens, 1966). Equation (47) or similar expressions have been successfully used to fit the temperature dependence of phonon dampings in compounds with very different chemical bonds such as calcite (Park, 1967), diamond (Borer *et al.,* 1971; Klemens, 1975), silicon (Hart *et al.,* 1974), germanium (Safran and Lax, 1975), CdS (Bairamov and Kharshkhozhev, 1975), or other crystals (Mead and Wilkinson, 1977).

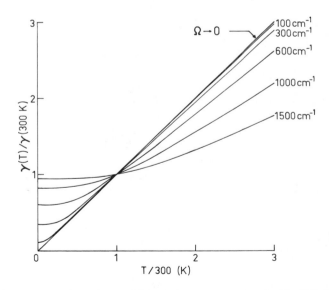

FIG. 3 Temperature dependence of damping constant as given by Eq. (47) for modes of different frequencies.

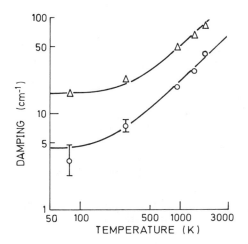

FIG. 4 Experimental temperature dependence of damping of TO (lower curve, 569 cm⁻¹) and LO (upper curve, 908 cm⁻¹) modes for the main mode in Al_2O_3 ($\mathbf{E} \perp \mathbf{c}$). Best fits (lines) of Eq. (47) to the data. (After Gervais and Piriou, 1974a.)

Certain authors among those cited herein used a slightly different procedure in that they identified the main combinations of phonons labelled 1 and 2 that relax the phonon ℓ, and directly used Eq. (44) on retaining the main phonon interaction processes only. This is the case of Pine and Tannenwald (1969) who studied phonon dampings in α-quartz in the temperature range 4–300 K. Phonon linewidth divergences observed near the $\alpha-\beta$ phase transition of quartz (Gervais and Piriou, 1975) will be discussed in Section VIE.

The contribution $P_a^{(4)}$ is real and frequency independent. Calculations in simple cases (Bruce, 1973; Cowley, 1963, 1965; Haque, 1975; Ipatova *et al.*, 1967; Lowndes and Rastogi, 1976; Rastogi *et al.*, 1974) indicate a positive sign for this frequency shift. Its temperature dependence is linear above the Debye temperature. Experimental and calculated data indicate that the cubic frequency shift $\Delta\omega_b^{(6)}(0j, \Omega_j, T)$ evaluated at the resonance frequency is negative. Thus both contributions balance each other (Lowndes, 1972; Lowndes and Rastogi, 1976; Portmus *et al.*, 1968; Samara and Peercy, 1973).

D. HIGHER-ORDER CONTRIBUTIONS

The calculation of the first $0(\eta^4)$ term (Fig. 2c) yields (Tripathi and Pathak, 1974)

$$P_c^{(6)}(0j, T) = \frac{360}{\hbar} \sum_{l_1 l_2} V^{(6)}(ll_1 l_2 ll_1 l_2) \left(n_1 + \frac{1}{2}\right)\left(n_2 + \frac{1}{2}\right). \qquad (48)$$

This is a contribution only to the frequency shift with a quadratic temperature dependence at high temperature (Gervais, 1973),† and is the sole contribution to the real phonon self-energy that is possibly positive at $\omega \approx \Omega_j$, among $0(\eta^4)$ terms considered here.

For 15 years the contribution represented in diagram (d) of Fig. 2 has been thought to explain the dependence of damping on temperature when the power-law exponent was intermediate between 1 and 2. Explicit results for the contributions of diagrams (d–g) of Fig. 2 are given by Tripathi and Pathak (1974); two among them had been calculated earlier by Wallis *et al.* (1966).‡ As discussed previously (Bairamov *et al.*, 1974; Gervais *et al.*, 1972, 1973; Sakurai and Sato, 1971) and in agreement with simple perturbation theory (Ashkin *et al.*, 1968; Chang and Mitra, 1972; La Combe and Irwin, 1970), the contribution $\Gamma_g^{(8)}(0j, \omega, T)$ may be simplified in a way similar to that for the cubic damping and rewritten in the form

$$[\Gamma_d^{(8)}(0j, \omega, T)]_{\omega=\Omega_j} = \frac{b_j}{2}\left\{\left[\frac{1}{\exp(\hbar\Omega_j/3k_BT) - 1} + \frac{1}{2}\right]^2 + \frac{1}{12}\right\}. \quad (49)$$

A combination of Eqs. (47) and (49) has been shown to describe satisfactorily the temperature dependence of damping over large temperature ranges. Results in alkali halides have been controverted by Mooij (1969) who argued that an implicit temperature dependence of certain parameters due to lattice thermal expansion was ignored. However, an analysis including *both* suitable corrections to account for lattice thermal expansion *and* the equivalence relation Eq. (60) indicates that quartic terms do contribute to limit the lifetimes of certain phonons in some ionic crystals (Bruce, 1973; Gervais *et al.*, 1973). *Ab initio* calculations confirm this assertion (Rastogi *et al.*, 1974; Ipatova *et al.*, 1967; Lowndes and Rastogi, 1976).

It is to be noted that the term $P_g^{(8)}$ also contributes to damping to the same order of magnitude as $P_d^{(8)}$. Both contributions are difficult to separate from the dependence of damping on temperature when it is found to involve a quadratic contribution at high temperature. Contributions labeled $P_e^{(8)}$ and $P_f^{(8)}$ are real and yield frequency-independent frequency shifts. Their temperature dependence is quadratic at high temperature as is that of every $0(\eta^4)$ term. Inspection of both terms (Tripathi and Pathak, 1974) and calculations performed by Rastogi *et al.* (1974) and Lowndes and Rastogi (1976) show that those contributions are negative.

---

† The factors $\delta_{s_1,0}$ and $\delta_{s_1+s_2,0}$ should be removed from Eqs. (8), (12), and (13) in this reference. As a result, the form of Eq. (14) is correct in the high-temperature approximation only, corresponding to the context of the paper.

‡ A factor $2n_3 + 1$ has been omitted in the expression corresponding to diagram (e), in both references.

Sparks and Sham (1972), and Pohl and Meier (1974), by extension of the perturbation theory outlined above, have obtained the form of the damping function due to n-phonon summation processes:

$$\Gamma^{(n)}(0j, \omega, T) \propto \sum_{l_1 \cdots l_n} |V^{(n+1)}(l, l_1, \ldots, l_n)|^2$$
$$\times \delta(\omega - \omega_1 - \ldots - \omega_n) \times \{(n_1 + 1) \cdots (n_n + 1)$$
$$- n_1 \cdots n_n\}. \tag{50}$$

This expression has been shown to be useful in the description of the residual absorption profiles in the transparent regime of crystals (Billard *et al.*, 1976; Pohl and Meier, 1974) and would account (at low orders) for the features experimentally observed above the fundamental mode region in the absorption coefficient shown as an example in Fig. 1.

IV. Dielectric Function Models

In principle the dielectric function can be obtained by a KK analysis of reflectivity data (see, for example, Mitra, 1969; Perry, 1971, for the case of FE perovskites). In practice, however, it appears that the errors involved in calculating the phase from the reflectivity lead to a dielectric function that is often less useful than when data are deduced from a fit of an appropriate dielectric function model to experimental data. So long as a model fits the data at any frequency within experimental error, one may have confidence in the model and the parameters that enter it. Besides, these parameters will be most useful for later analysis of their dependence on temperature.

According to Section II there are essentially two basic ways in which a dielectric function model can be derived: either by *summing* over independent (or coupled) oscillators or by *factorizing* over poles and zeros, using the property of the LO modes that are the zeros of the dielectric function according to Maxwell's equation, Eq. (2). These alternatives will be discussed separately in the following subsections.

A. THE QUANTUM DIELECTRIC FUNCTION AND THE THREE-PARAMETER MODEL

Cowley (1965) derived the expression of the linear dielectric susceptibility. This is the sum over all TO vibrational modes of products of dipolar moment tensor components and phonon propagators:

$$\chi_{\mu\nu}(\omega) = \chi_{\mu\nu}(\omega) \mp i\chi_{\mu\nu}''(\omega)$$
$$= \frac{1}{\hbar V} \sum_{0j} M_{\mu j} M_{\nu j} \left[\frac{2\omega(0j)}{\omega^2(0j) - \omega^2 \pm 2i\omega(0j)P(0j, \omega, T)} \right]; \tag{51}$$

when $j = 1$ the dielectric function reduces to

$$\varepsilon(\omega) = \varepsilon(\infty) + \frac{M^2}{\hbar V} \frac{2\omega_0}{\omega_0^2 + 2\omega_0[\Delta\omega(\omega) + \delta\omega] - \omega^2 \pm 2i\omega\Gamma(\omega)}, \quad (52)$$

where ω_0 is the harmonic TO mode frequency. If the functions $\varepsilon'(\omega)$ and $\varepsilon''(\omega)$ are known in a given spectral range, the real and imaginary phonon self-energies can be obtained in the form

$$\Delta\omega(\omega) = \frac{\omega^2 - \overline{\omega}_0^2}{2\omega_0} + \frac{M^2}{\hbar V} \frac{\varepsilon'(\omega) - \varepsilon(\infty)}{[\varepsilon'(\omega) - \varepsilon(\infty)]^2 + \varepsilon''^2(\omega)}, \quad (53)$$

$$\Gamma(\omega) = \frac{M^2}{\hbar V} \frac{\varepsilon''(\omega)}{[\varepsilon'(\omega) - \varepsilon(\infty)]^2 + \varepsilon''^2(\omega)}, \quad (54)$$

where $M^2/\hbar V = [\varepsilon'(0) - \varepsilon(\infty)]\Omega_0^2(0)/2\omega_0$ is evaluated in the limit $\omega = 0$, and $\overline{\omega}_0^2 = \omega_0^2 + 2\omega_0\delta\omega$ is the square of the quasi-harmonic frequency, renormalized to account for the pure-volume effect of thermal lattice expansion. See Eqs. (38)–(42).

A similar procedure has been used by Axe *et al.* (1965), Bilz (1966), Zernik (1967), Le Toullec (1968), and Denham *et al.* (1970, 1973) to evaluate either the damping function only or both real and imaginary parts of the frequency-dependent phonon self-energy. It is to be emphasized, however, that this treatment can be utilized in principle to analyze one-band spectra only and that the accuracy of the result depends on that of the KK inversion. This difficulty has been obviated by Rastogi *et al.* (1977) who directly obtained the real and imaginary dielectric functions by using *dispersive* Fourier-transform IR reflectivity spectroscopy.

In many-phonon-branch crystals the phonon self-energy is expected to be a slowly varying function of frequency. Besides, when the TO–LO splitting is weak (narrow IR reflection band), the phonon self-energy may be approximated by a constant in the spectral range where the mode contributes predominantly to reflectivity, i.e., the IR reflection band itself. The 3-parameter model thus consists of using the classical dielectric function

$$\varepsilon = \varepsilon_\infty + \sum_j \Delta\varepsilon_j \left(\frac{\Omega_{jTO}^2}{\Omega_{jTO}^2 - \omega^2 \pm i\gamma_{jTO}\omega} \right) \quad (55)$$

derived in Section IIA and adjusting three parameters per polar mode, Ω_{jTO}, γ_{jTO}, and $\Delta\varepsilon_j$, to fit experimental reflectivity. When the conditions outlined above are satisfied, the classical three-parameter model has been shown to fit simple and complicated experimental spectra in a wide variety of compounds, satisfactorily. When the agreement is good for all bands except one (most often the wide one), the procedure commonly used is to add a weak but very damped oscillator that may be considered as a multiphonon absorption. This is in fact a palliative used in order to overcome the difficulty related to the weakness of the classical dispersion theory which does not take the frequency-dependence of the phonon self-energy into account.

In the case of a single mode ($j = 1$) the resonance function of TO modes is simply the imaginary part of the dielectric function

$$\text{Im}(\varepsilon) = \Delta\varepsilon \frac{\omega\gamma_{\text{TO}}\Omega_{\text{TO}}^2}{(\Omega_{\text{TO}}^2 - \omega^2)^2 + \gamma_{\text{TO}}^2\omega^2}. \tag{56}$$

Because the LO mode is a zero of the dielectric function, its profile but not, properly speaking, its resonance (because LO modes are not excited by the IR radiation field) is visualized via

$$\text{Im}\left(\pm\frac{1}{\varepsilon}\right) = \frac{\Delta\varepsilon}{\varepsilon_0\varepsilon_\infty} \frac{\omega\gamma_{\text{TO}}\Omega_{\text{LO}}^2}{(\Omega_{\text{LO}}^2 - \omega^2)^2 + \gamma_{\text{TO}}^2\omega^2}. \tag{57}$$

A comparison of Eqs. (56) and (57) clearly shows that within the 3-parameter model description, the damping of the LO mode is implicity assumed to be equal to that of the TO mode. It is reasonable to assume approximately the same phonon decays for TO and LO modes when the TO–LO splitting is weak within the hypothesis of a broad continuum for the two-phonon density of states, but this assumption is no longer valid when the damping function is a rapidly varying function of frequency and/or, principally, when TO and LO energy levels are very distant. These considerations explain the failure of the three-parameter model when a reflection spectrum consists of wide bands (Spitzer et al., 1962).

Before we proceed any farther, notice that a comparison of Eqs. (51) and (55) yields

$$\Delta\varepsilon_j\Omega_j^2 \equiv 2M_{\mu j}M_{\nu j}\omega(0j)/\hbar V, \tag{58}$$

$$\Omega_j^2 \equiv \omega^2(0j) + 2\omega(0j)[\Delta\omega(0j, \Omega_j, T) + \delta\omega j(T)], \tag{59}$$

$$\gamma_j \equiv 2\omega(0j)[\Gamma(0j, \omega, T)/\omega]_{\omega=\Omega_j}. \tag{60}$$

B. THE FACTORIZED FORM OF THE DIELECTRIC FUNCTION

1. The 4-Parameter Model

Tests in realistic cases (Gervais and Piriou, 1974a, 1974b; Servoin et al., 1980a, 1980b) have shown that in the case of the wide reflection band just discussed, it is convenient to use the factorized form of the dielectric function

$$\varepsilon = \varepsilon_\infty \prod_j \frac{\Omega_{j\text{LO}}^2 - \omega^2 \pm i\gamma_{j\text{LO}}\omega}{\Omega_{j\text{TO}}^2 - \omega^2 \pm i\gamma_{j\text{TO}}\omega}, \tag{61}$$

following a procedure first employed by Berreman and Unterwald (1968). Equation (61) is an extension of the LST relation (Kurosawa, 1961) including damping (Barker, 1964; Chaves and Porto, 1973; Lowndes, 1970; Perry, 1971). We introduce $\gamma_{j\mathrm{LO}}$ as a fourth adjustable parameter per polar mode; it may be different from $\gamma_{j\mathrm{TO}}$ to account for different TO and LO phonon decays. The advantage of the four-parameter model compared with the 3-parameter model is shown in Fig. 5 for the case of strontium titanate. The "pole and zero" fit procedure also avoids the introduction of additional secondary oscillators in the vicinity of LO modes for corundum (Gervais and Piriou, 1974a) and quartz spectra (Gervais and Piriou, 1975), for example, as compared with previous analyses (Piriou, 1968; Spitzer and Kleiman, 1961).

When $j = 1$ a combination of Eqs. (55) and (61) yields, if $\omega \neq \Omega_{\mathrm{TO}}$ and if terms of order γ^2 are neglected (Gervais and Piriou, 1974a),

$$\frac{2\omega(0j)\Gamma(0j, \omega, T)}{\omega} = \gamma_{\mathrm{TO}} - (\gamma_{\mathrm{LO}} - \gamma_{\mathrm{TO}}) \frac{\Omega_{\mathrm{TO}}^2 - \omega^2}{\Omega_{\mathrm{LO}}^2 - \Omega_{\mathrm{TO}}^2}. \qquad (62)$$

Another way to obtain Eq. (62), even for several modes, is to use a development limited to the second order of the form $\gamma(\omega) = A + B\omega^2$, the constants A and B being obtained by equating γ_{TO} and γ_{LO} to $\gamma(\Omega_{\mathrm{TO}})$ and $\gamma(\Omega_{\mathrm{LO}})$ (Luspin $et\ al.$, 1980). Consequently, the factored form of the dielectric function implies an assumption about the form of $\Gamma(0j, \omega, T)$, which is supposed to increase monotonically between two values measured at the Ω_{TO} and Ω_{LO} frequencies. The model may qualify as a 4-parameter semi-quantum (FPSQ), because the frequency-dependence property of the phonon self-energy is roughly approximated and measured at only two

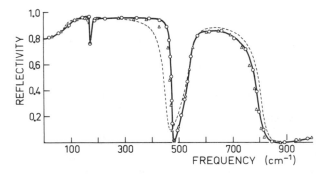

FIG. 5 Experimental and calculated reflectivity (three-parameter model) of F_{1u} modes of strontium titanate at 300 K (Δ experimental; --- best fit by Spitzer $et\ al.$, 1962) together with a fit on the basis of the four-parameter model Eq. (61) (solid line). Other symbols (O) represent data recently obtained with a scanning interferometer, for comparison.

points per polar mode whereas the temperature dependence may be analyzed at the TO and LO frequencies via Eqs. (59) and (60). Formally, γ_{LO} is the damping of the polar mode evaluated at the Ω_{LO} frequency. Actually, at the point of the Brillouin zone defined by $\omega = \Omega_{LO}$ and $\mathbf{k} = 0$, the conservation laws for the phonon wave vector and the energy of TO and LO phonon self-energies are the same, and both TO and LO modes have the same eigenvector at $\mathbf{k} = 0$; so both dampings are considred to be equal (see also Lowndes, 1970). Taking account of those considerations, one obtains another formulation of Eq. (61):

$$\varepsilon = \varepsilon_{\infty} \prod_{j} \left[\frac{\omega_{LO}^2(0j) - \omega^2 \pm 2\omega_{LO}(0j)P_{LO}(0j, \omega, T)}{\omega_{TO}^2(0j) - \omega^2 \pm 2\omega_{TO}(0j)P_{TO}(0j, \omega, T)} \right]. \tag{63}$$

A number of experimental justifications exists now to test this model. In particular, fair agreement is found between the Ω_{LO} and γ_{LO} parameters deduced from a fit of the model Eq. (61) to IR reflection spectra and Raman peaks and to linewidths in noncentrosymmetric crystals where polar modes are both infrared and Raman-active, such as α-quartz (Gervais and Piriou, 1975), $LiTaO_3$ (Gervais and Servoin, 1977), $BaTiO_3$ (Servoin et al., 1980a), or $SrTiO_3$ (Servoin et al., 1980b), as compared with hyper-Raman measurements in the latter case.

It should be emphasized that the four-parameter model Eq. (61), as opposed to the three-parameter model, accounts for asymmetric phonon line shapes as shown in Fig. 6. Although (i) the factored form of the dielectric

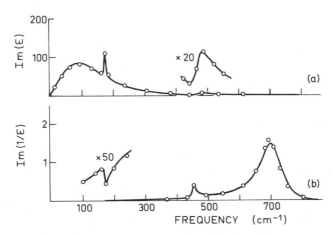

FIG. 6 Comparison between the imaginary parts of (a) the dielectric (TO modes) and (b) the inverse dielectric (LO modes) functions calculated (solid lines) with the factored form of Eq. (61) which yields best fit to reflectivity data and the results of a KK analysis. For $BaTiO_3$, F_{1u}-type modes, 1000 K. (After Luspin et al., 1980.)

function involves *only two adjustable parameters per observed TO or LO mode,* peak frequency Ω_j and linewidth γ_j, (ii) dampings are introduced as constant coefficients, and (iii) no coupling terms such as off-diagonal elements in the phonon response matrix (Barker and Hopfield, 1964; Scott, 1974) enter the model at this stage, the example of barium titanate reproduced in Fig. 6 clearly shows that even very asymmetric line shapes can be described quite correctly with this model, in agreement with KK analyses. This is a straightforward consequence of Eq. (62).

The model Eq. (61) has several other advantages from the viewpoint of the practitioner. Figures 5 and 6 have shown that the model correctly describes reflection spectra with wide bands because TO and LO dampings are allowed to be different, a situation that occurs in FE crystals, for instance. Anticipating section VA, because all TO and LO frequencies are located at the low- and high-frequency edges of IR reflection bands and those dampings correspond to the slopes at these edges, all parameters are already approximately adjusted for a first attempt at a fit. A small number of runs with a computer usually allows the best fit to be achieved by minor refinements of the parameters. Conversely, attempts to adjust any of the oscillator strengths within the sum-of-oscillators fitting procedure shifts the position of all LO modes so that many more runs are often necessary when IR reflection bands are not well separated. The model Eq. (61) also has the advantage of obviating any problem of renormalization of $M_{\mu j}M_{\nu j}$ terms due to volume thermal expansion or atom rearrangement when the temperature increases, because these factors do not enter Eq. (63).

Rather surprisingly, the four-parameter model has been used by very few experimentalists up to now (Clauws and Vennik, 1976; Lazarev *et al.,* 1975, for example) other than those cited throughout this section.

This model also has weaknesses in comparison with the actual dielectric function of the crystal. Because the damping function in Eq. (62) should be positive, the conditions $\gamma_{LO} \geq \gamma_{TO}$ and $\gamma_{TO}/\gamma_{LO} \geq (\Omega_{TO}/\Omega_{LO})^2$ should be satisfied, as first stated by Lowndes (1970). If this is not the case, the model is valid only in a limited range of frequency. At any rate, the actual imaginary phonon self-energy necessarily departs from the rough form of Eq. (62), at least in the transparent regimes in the far and near IR. In these regions $\gamma(\omega)$ is known to decrease rapidly for lack of phonons that can be anharmonically coupled to produce the desired relaxation effect. Such a rapid fall off is not detectable in reflection experiments, because it occurs in the two regions where the reflectivity is dominated by the real dielectric function in which dampings play a negligible role. As a result neither the product formula, Eq. (61), *nor* the sum formula, Eq. (55), are able to predict the frequency dependence of the absorption coefficient in both transparent regimes, which are dominated by the imaginary dielectric function and thus the imaginary phonon self-energy. This is the reason why both transmission and reflection measurements are complementary, just as mentioned in the discussion related to Fig. 1.

2. A Simplified Model of Phonon Self-Energy

As discussed previously there is a hierarchy of approximations that can be made to obtain the experimental level of the imaginary phonon self-energy. With the three-parameter model an average constant level is determined in the vicinity of the TO frequency. With the four-parameter model two points of the function are experimentally obtained in the vicinity of the frequencies Ω_{TO} and Ω_{LO}, and a monotonic frequency dependence is implicitly assumed between both values according to Eq. (62). This latter simple model is nevertheless able to account for asymmetric line shapes (Fig. 6). When both dielectric function models in their simplest formulation, described herein, do not fit experimental data, it most likely means that the frequency dependence of the damping function should be taken into account. The line profiles of both real and imaginary phonon self-energy can then be deduced from Eqs. (53) and (54) for a single-mode spectrum. In the case of a more complicated spectrum that is correctly fitted with the four-parameter model everywhere except in a region where the contribution of a neighbor main mode j clearly dominates the dielectric function, the following relations (Servoin and Gervais 1976, 1977),

$$\Delta\omega(0j, \omega) = \left[\omega^2 - \overline{\omega}^2(0j) + \frac{U(\omega)\varepsilon'(\omega) - V(\omega)\varepsilon''(\omega)}{\varepsilon'^2(\omega) + \varepsilon''^2(\omega)}\right]\bigg/ 2\omega(0j) \quad (64)$$

and

$$\Gamma(0j, \omega) = \frac{V(\omega)\varepsilon'(\omega) + U(\omega)\varepsilon''(\omega)}{2\omega(0j)[\varepsilon'^2(\omega) + \varepsilon''^2(\omega)]}, \quad (65)$$

can be useful to evaluate the phonon self-energy profile of the mode j. Here $\overline{\omega}(0j)$ is the quasi-harmonic frequency Eqs. (38)–(42), $\varepsilon(\omega)$ is obtained by a KK analysis, and

$$U(\omega) + iV(\omega) = (\Omega_{jLO}^2 - \omega^2 \pm i\gamma_{jLO}\omega) \prod_{k\neq j} \frac{\Omega_{kLO}^2 - \omega^2 \pm i\gamma_{kLo}\omega}{\Omega_{kTO}^2 - \omega^2 \pm i\gamma_{kTO}\omega} \quad (66)$$

is calculated from the best fit of Eq. (61) to experimental data, which is assumed to be correct in spectral regions where the modes labeled k dominate the spectrum.

The next step in stating an approximate dielectric function model is to consider a main two-phonon peak in the density of states. The contribution of this peak of frequency $\omega_{2ph} = \omega_2 \pm \omega_1$ to the phonon self-energy has the form $\propto (\omega - \omega_{2ph} - i\Gamma)^{-1}$, where a finite damping Γ has been conferred on the original delta functions in Eq. (45). The phonon self-energy now assumes the approximate form (Gervais and Servoin, 1977)

$$\Delta\omega(\omega) + \tfrac{1}{2}i\gamma(\omega) = \Delta\omega_0 + \tfrac{1}{2}i\gamma_0 + f^2/(\omega - \omega_{2ph} - i\Gamma), \qquad (67)$$

where f has the dimension of a frequency. Results of a treatment according to this procedure are reported in Fig. 7, with an intense two-phonon peak found near 100 cm^{-1} in the vicinity of the soft FE mode of lithium tantalate. Equation (66) in combination with the 3-parameter model has also been used by Bartsokas and Siapkas (1980) to fit reflection spectra of SbSI-type compounds in the vicinity of the soft mode frequency.

3. Contribution of Relaxational Effects to the Dielectric Function

If we consider an excitation, the frequency of which is much lower than its own damping and IR frequencies, a definition that applies to dielectric relaxation, its contribution to the dielectric function reduces to the form

$$\chi_{rel}(\omega) = \frac{\chi(0)}{1 \pm i\omega/\gamma_T}, \qquad (68)$$

which is a Debye equation. Such a term is added to ionic and electronic contributions according to the sum rule in Eq. (9).

Relaxational effects can also be described with the four-parameter model in a phenomenological way with the additional condition $\gamma_L > \gamma_T$ (Chaves and Porto, 1973). Equation (61) becomes

$$\varepsilon(\omega) = \varepsilon_\infty \frac{\omega \mp i\gamma_L}{\omega \mp i\gamma_T} \prod_j \frac{\Omega_{jLO}^2 - \omega^2 \pm i\gamma_{jLO}\omega}{\Omega_{jTO}^2 - \omega^2 \pm i\gamma_{jTO}\omega}. \qquad (69)$$

A comparison of Eqs. (68) and (69) yields

$$\chi(0) = \varepsilon_{lat}(\gamma_L/\gamma_T - 1), \qquad (70)$$

where $\varepsilon_{lat} = \varepsilon_\infty \Pi_j \Omega_{jLO}^2/\Omega_{jTO}^2$ is the contribution of high-frequency excitations (electrons + lattice) to the dielectric constant.

C. MODE COUPLING

1. Coupled TO Modes

The theory of the coupled-optical-phonon mode of IR dispersion was developed by Barker and Hopfield (1964) to overcome the difficulty that appeared in BaTiO$_3$, SrTiO$_3$, and KTaO$_3$ reflection spectra, namely, there is no satisfactory compromise when choosing the damping constant of a mode

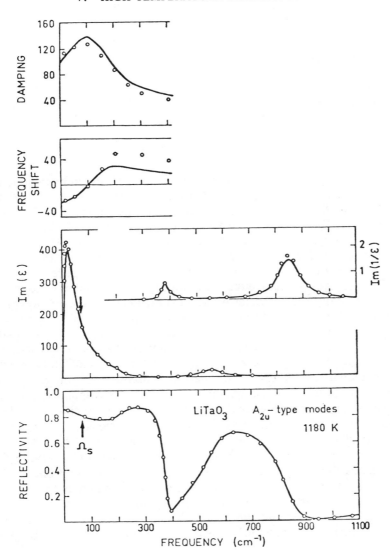

Fig. 7 An illustration of the spectral analysis developed in section IVB2. KK analysis of reflectivity data and experimental determination of the function $\Delta\omega(\omega)$ and $\Gamma(\omega)$ for the lowest-frequency intense soft mode via Eqs. (64)–(66). Solid lines are calculated on the basis of the model Eq. (61) with incorporation of Eq. (67) for the description of the soft mode (Servoin and Gervais, unpublished result).

that dominates the spectrum from a few up to several hundred reciprocal centimeter. After inspection of the various formulations of the coupled-mode formalism (Scott, 1974), it turns out that the approximation made to solve the Dyson's equation in Section IIIA is no longer retained, because the phonon self-energy matrix need not be diagonal over the branch indices; and in the case of two coupled modes, the phonon response function assumes the form

$$G(\omega) \propto \begin{bmatrix} \Omega_1^2 - \omega^2 \pm i\gamma_1\omega & k_{\bar{1}2}^2 \pm i\gamma_{12}\omega \\ k_{\bar{1}2}^2 \pm i\gamma_{12}\omega & \Omega_2^2 - \omega^2 \pm i\gamma_2\omega \end{bmatrix}^{-1}. \tag{71}$$

Formally, the terms written as constants in Eq. (71) are in reality phonon self-energy tensor components and are therefore temperature- and frequency-dependent. In practice, however (Alain and Piriou, 1977; Barker and Hopfield, 1964; Bozinis and Hurrel, 1976; Chaves et al., 1974; Katiyar et al., 1971; Lagakos and Cummins, 1974; Lowndes et al., 1974; Peercy, 1975; Scalabrin et al., 1977; Siapkas, 1978; Steigmeier et al., 1975; Zawadovski and Ruwalds 1970, and so on), a pure–real or pure–imaginary coupling constant is retained merely to minimize the number of adjustable parameters and avoid overdetermination of Eq. (71). It is the opinion of this reviewer that one should be careful with the application of the mode-coupling formalism (but not the formalism itself). It should be kept in mind that Eq. (71) derives from a perturbation treatment and that phonon self-energy terms, diagonal and off-diagonal, therefore, should be small with respect to harmonic energies. Otherwise, the harmonic frequencies found from the application of Eq. (71) could be not at all close to the resonances observed experimentally, when pure–real coupling parameters are choosen to be of the same order of magnitude as harmonic frequencies (see, for example, Chaves et al., 1974; Sanjurjo et al., 1980; Scalabrin et al., 1977). Apparent inconsistencies about mode coupling were also pointed out by Servoin et al. (1980a). It is more important to emphasize that when the vibrational modes, analyzed at room temperature in the three papers cited herein, "cross" each other at high temperature, the anticrossing region is found to be restricted to a few wave numbers, and the actual coupling parameter is consistently small, as will be detailed in the last section (Luspin et al. 1980; Servoin et al., 1980b).

Takahashi (1975) has shown the compatibility of a coupled system of many modes with damping and the factorized form of the dielectric function.

2. Coupled LO Phonon–Plasmon Modes

Due to the longitudinal character of electronic excitations and to the sensitivity of reflectivity to a zero of the dielectric function, this technique is quite useful for analyzing the coupling of free carriers and longitudinal lattice vibrations. The LO phonon–plasmon coupling, predicted by Yokota (1961), has been the subject of studies by Varga (1965) and Singwi and Tosi (1966) who performed calculations to predict the deformation of IR reflection bands in the presence of plasmons. This phenomenon has been observed experimentally on GaAs by both Raman scattering (Mooradian and Wright, 1966) and IR reflectivity (Olson and Lynch, 1969; Piriou, 1968). Analogous studies on $Pb_{1-x}Sn_xTe$ (Dionne and Woolley, 1972), HgTe (Grynberg *et al.*, 1974), CdTe (Perkowitz and Thorland, 1974), PbSe (Burkhard *et al.*, 1974), and so on, were performed later with the aid of the reflectivity technique discussed here. Spectra were analyzed with the 3-parameter model to which a Drude term

$$\chi_{\text{plasmon}} = -\varepsilon_\infty \frac{\Omega_p^2}{\omega(\omega \mp i\gamma_0)} \tag{72}$$

is added to describe the plasma contribution, where Ω_p is the plasmon frequency and γ_0^{-1} a lifetime. The dielectric function then becomes

$$\varepsilon = \varepsilon_\infty + \sum_j \Delta\varepsilon_j \frac{\Omega_{jTO}^2}{\Omega_{jTO}^2 - \omega^2 \pm i\gamma_{jTO}\omega} - \frac{\varepsilon_\infty \Omega_p^2}{\omega(\omega \mp i\gamma_0)}. \tag{73}$$

Kukharskii (1973) used the factorized form of the dielectric function, in which one of the TO frequencies is put to zero, to treat GaAs data in terms of coupled LO phonon–plasmon modes. This method was subsequently applied to GaAs and CdTe IR reflection spectra by Perkowitz and Thorland (1975), to PbTe by Perkowitz (1975), and to CdS and GaP by Giehler and Jahne (1976). But in the analyses cited above, the systems were decoupled by means of Eq. (73). Gervais and Baumard (1977) studied the LO phonon–plasmon coupling in nonstoichiometric rutile TiO_{2-x} (also Baumard and Gervais, 1977) and decoupled the system with the following model:

$$\varepsilon/\varepsilon_\infty = \prod_j \frac{\Omega_{jLO}^2 - \omega^2 \pm i\gamma_{jLO}\omega}{\Omega_{jTO}^2 - \omega^2 \pm i\gamma_{jTO}\omega} - \frac{\Omega_p^2 \pm i(\gamma_p - \gamma_0)\omega}{\omega(\omega \mp i\gamma_0)}. \tag{74}$$

Equation (74) is the 4-parameter analog of Eq. (73). The first term of the right side of Eq. (75) was shown to describe correctly the pure-phonon IR reflection of stoichiometric rutile (Gervais and Piriou, 1974a, 1974b), whereas the three-parameter model failed (Spitzer *et al.*, 1962). The second term reduces to the Drude term when $\gamma_0 = \gamma_p$. Both Eqs. (61) and (74) are strictly equivalent for an appropriate choice of parameters that correspond to coupled and uncoupled modes, respectively, and fit experimental data equally well. Best fits with different dielectric function models are compared

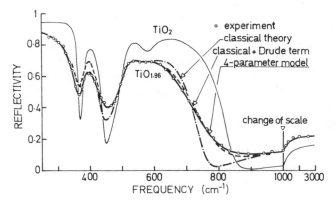

FIG. 8   A comparison of best fits to $TiO_{1.96}$ reflectivity data (O) with classical theory Eq. (55) (chain curve), classical theory plus Drude term Eq. (74) (broken curve), and the 4-parameter model Eq. (61) or equivalently Eq. (74) (heavy solid line). (After Gervais and Baumard, 1977.)

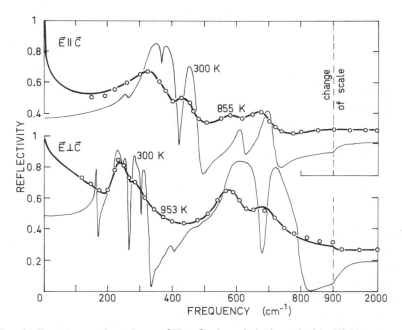

FIG. 9   Temperature dependence of IR reflection of niobium dioxide. High-temperature data (O) were fitted (solid lines) with Eq. (74).

in Fig. 8. Another example of fit with Eq. (74) is shown in Fig. 9. At room temperature, the IR reflectivity of niobium dioxide was fitted correctly with a lattice contribution alone (Gervais and Baumard, 1979). On heating towards the semiconducting–metal phase transition, which occurs at 810°C, one finds a plasma contribution that begins to mask the pure-lattice contributions. In this case Eq. (74) was used, but the fit was performed with the additional condition that the γ_{jLO} parameters of *weak* polar modes are constrained to be equal to the corresponding γ_{jTO} dampings, consistent with the discussions in Sections IVA and IVB. This procedure minimizes the number of adjustable parameters without loss of fit quality.

V. Reflectivity Spectroscopy with a Scanning Interferometer

A. REFLECTIVITY AND DIELECTRIC FUNCTION

The IR reflectivity is related to the dielectric function via the Fresnel formula

$$R = |(\tilde{N} - 1)/(\tilde{N} + 1)|^2, \tag{75}$$

where

$$\tilde{N} = N \mp iK = \sqrt{\varepsilon(\omega)} \tag{76}$$

is the refractive index and may assume complex values. Other formulations of reflectivity thus are

$$R = |(\sqrt{\varepsilon} - 1)/(\sqrt{\varepsilon} + 1)|^2, \tag{77}$$

or

$$R = [(N - 1)^2 + K^2]/[(N + 1)^2 + K^2]. \tag{78}$$

The dispersion of both functions $\varepsilon(\omega)$ and $R(\omega)$, illustrated in Fig. 10, depends on an increase of damping that is roughly proportional to the increase of temperature.

When a vibrational mode is soft, that is, when its frequency decreases substantially as a structural phase transition is approached from above or below or shifts much more than that of other modes, then the net effect of lowering frequency and increasing damping is that the ratio γ_s/Ω_s can become higher than $\sqrt{2}$. The mode is then called *overdamped*. In that case the frequency Ω_j no longer corresponds to the maximum of the response function of the classical oscillator. Figure 11 shows the effect of such a phenomenon displayed at constant Ω_{TO} frequency. It is seen that a low-frequency, highly damped or overdamped TO mode manifests itself by a nearly flat reflection level, and the dip that is characteristic of the TO

F. GERVAIS

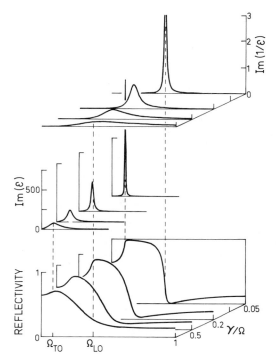

FIG. 10 Example of evolution of reflectivity and imaginary parts of the dielectric (TO mode) and inverse dielectric (LO mode) functions with increasing ratio γ/Ω.

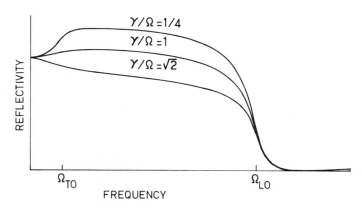

FIG. 11 Low-frequency reflection profiles for three values of γ_{TO}/Ω_{TO}.

resonance in the low-frequency edge of the IR band is no longer visible. It should be emphasized that, conversely, knowledge of the low-frequency flat reflection level R_o and of its increase or decrease with temperature with the down-shift or up-shift of the highly damped soft mode, respectively, allows an accurate determination of the soft mode frequency parameter via the generalized LST relation in the form (Gervais *et al.*, 1980)

$$\Omega_{\text{soft}} = \sqrt{\varepsilon_\infty} \frac{1 - \sqrt{R_0}}{1 + \sqrt{R_0}} \left(\prod_j \Omega_{j\text{LO}} \Big/ \prod_{j \neq \text{soft}} \Omega_{j\text{TO}} \right), \tag{79}$$

where all TO and LO modes on the right are assumed to be weakly damped and thus the Ω_j's are measured accurately. This is another advantage of IR reflectivity spectroscopy applied to soft-mode systems. A fit of the factored dielectric function to reflectivity is obviously more accurate than the straightforward method described here.

When two oscillators have very different oscillator strengths, a *dielectric singularity* can occur and seems worth mentioning (Gervais, 1977). Normally the effect of Coulomb forces is to shift the LO mode to a frequency *higher* than that of the TO mode (see Eq. 16). As pointed out by Scott and Porto (1967) there are situations that may qualify as unusual because a zero of the dielectric function, and therefore a LO mode, has its position at a slightly *lower* frequency than the corresponding pole. This situation is encountered when a very weak TO mode oscillates at a frequency intermediate between the TO and LO frequencies of another mode having a large oscillator strength. The application of Eq. (16) to the narrow spectral range, in which the weak mode "sees" the real dielectric function ε' due to the main mode, yields

$$\Omega_{w\text{TO}}^2 - \Omega_{w\text{LO}}^2 \approx \Delta\varepsilon_w \Omega_{w\text{TO}}^2 / \varepsilon'. \tag{80}$$

Because ε' is negative and possibly $|\varepsilon'| \ll \varepsilon_\infty$, the LO frequency of the weak mode is lower than the TO_w frequency. Besides, the strong mode can play the role of an "amplifier" for the weak mode, because the $\text{TO}_w - \text{LO}_w$ splitting can be larger than in the absence of main mode when $|\varepsilon'|$ is smaller than ε_∞. This effect can be shown more rigorously. The positions of two LO modes are obtained by solving the equation

$$\varepsilon_\infty + \Delta\varepsilon_1 \frac{\omega_{\text{TO1}}^2}{\omega_{\text{TO1}}^2 - \omega^2} + \Delta\varepsilon_2 \frac{\omega_{\text{TO2}}^2}{\omega_{\text{TO2}}^2 - \omega^2} = 0, \tag{81}$$

where damping is ignored for simplicity. The result is

$$\Omega_{\text{LO}\pm}^2 = \tfrac{1}{2}\{ \omega_{\text{LO1}}^2 + \omega_{\text{LO2}}^2 \pm [(\omega_{\text{LO1}}^2 - \omega_{\text{LO2}}^2) + 4(\Delta\varepsilon_1 \Delta\varepsilon_2 / \varepsilon_\infty^2)\omega_{\text{TO1}}^2 \omega_{\text{TO2}}^2]^{1/2} \}, \tag{82}$$

where $\omega_{LOi} = \omega_{TOi}(1 + \Delta\varepsilon_i/\varepsilon_\infty)^{1/2}$ are uncoupled LO mode frequencies. When $\Delta\varepsilon_2 \ll \varepsilon_1$, LO frequencies are given approximately by

$$\Omega_{LO+}^2 \approx \omega_{LO1}^2 + \frac{\Delta\varepsilon_2}{\varepsilon_\infty} \omega_{TO2}^2 \frac{\omega_{LO1}^2 - \omega_{TO1}^2}{\omega_{LO1}^2 - \omega_{TO2}^2}, \tag{83}$$

$$\Omega_{LO-}^2 \approx \omega_{TO2}^2 - \frac{\Delta\varepsilon_2}{\varepsilon_\infty} \omega_{TO2}^2 \frac{\omega_{TO2}^2 - \omega_{TO1}^2}{\omega_{LO1}^2 - \omega_{TO2}^2}. \tag{84}$$

The second term on the right-hand side of Eqs. (83) and (84) is positive. When $\Delta\varepsilon_2$ increases (starting from zero) but remains small, the LO+ frequency of the former mode is shifted up while the LO − frequency of mode 2 shifts down on starting from ω_{TO2}. A pattern of the coupling of a strong mode 1 and a weak mode 2 is shown in Fig. 12. Profiles are calculated with Eq. (81) combined with (Kukharskii, 1976)

$$1/\varepsilon_\infty - S_{LO1} \frac{\omega_{LO1}^2}{\omega_{LO1}^2 - \omega^2} - S_{LO2} \frac{\omega_{LO2}^2}{\omega_{LO2}^2 - \omega^2} = 0, \tag{85}$$

the solutions of which are coupled TO modes.

B.   SAMPLE SURFACE TREATMENTS

A practical problem that arises when IR reflectivity experiments are planned is that one wants to deduce bulk properties from the study of a

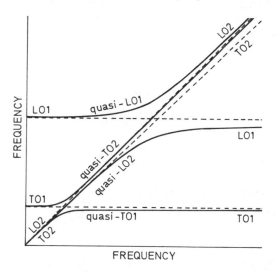

FIG. 12   Pattern of coupling of a strong mode 1 and a weak mode 2. Coupled mode (solid line) frequencies are plotted versus uncoupled mode-2 frequencies (dashed line) (Le Faou and Gervais, unpublished result).

"surface." The optical quality of this surface is thus of great importance. "Surface" means a penetration depth of IR radiation between 0.1 and 100 μm, in the spectral regions that correspond to lattice reflection bands. These conditions are less severe than those that are imposed when backscattering Raman experiments are performed on absorbing crystals (Evans and Ushioda, 1974). Crystal surfaces usually need optical polishing, but keep in mind that the higher the wavelength of the IR radiation the less important are the remaining defects of the surface. The IR reflection spectra of freshly polished surfaces, however, are not identical to the bulk properties in the vicinity of TO frequencies where the penetration depth is small (Barker, 1963; Piriou, 1964). Annealing samples in a furnace for several hours at about 75% of the melting point is highly desirable to regenerate the surface altered by polishing (Jasperse *et al.,* 1966). Figure 13 shows the increase of reflection maxima, related to phonon damping decrease, and the improvement of the resolution of certain narrow IR bands of a crystal of gadolinium molybdate after annealing up to 1100 K (performed without manipulation of the sample, which was kept in the measurement position between records

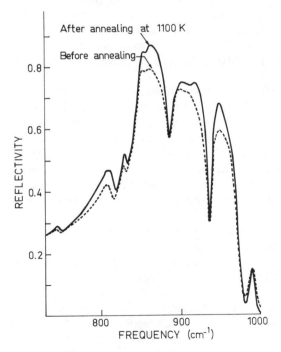

FIG. 13   Example of the effect of annealing on a polished single crystal surface of $Gd_2(MoO_4)_3$ (Kaczmarek and Gervais, unpublished result).

of the spectra). Otherwise, freshly cleaved surfaces usually yield good reflectivities.

IR absorption measurements give additional data, but experiments performed even on very thin slabs cannot replace reflectivity data, as mentioned in the discussion related to Fig. 1. This is also because only the structure of TO modes is then obtained, and besides, high-temperature studies appear difficult to perform (Chang and Mitra, 1972).

C. FOURIER TRANSFORM SCANNING INTERFEROMETRY

It is essential for the understanding of scanning interferometric techniques to keep in mind that the Fourier transform of a peak (infinitely narrow peak: delta function) is a sinusoid and, conversely, that the Fourier transform of a sinusoid is a peak. When the movable mirror of an interferometer shifts at constant velocity, then owing to the sequence of constructive and destructive interference, any optical energy at a given wavelength that enters the interferometer will be seen by the detector as a sinusoid of given period and amplitude. A spectrum can be decomposed into adjacent infinitely narrow peaks of variable amplitudes. Such a spectrum will be seen by the detector as the superposition of corresponding sinusoids with different periods and amplitudes. This superposition is called an *interferogram.* When one takes the Fourier transform of the interferogram — feasible since a computer is available to record the interferometric data rapidly — one reconstructs the spectrum.

The advantages of Fourier-transform spectroscopy over conventional spectroscopy are related to the amount of optical energy that reaches the detector. When the desired resolution requires a width δx at the output slit of the monochromator in a dispersive system, the energy available is limited to $\delta x/x$, where x is the length of the spectrum spatially dispersed by the prism or grating; the higher the desired resolution the lower this ratio. In contrast to this, about 50% of the energy of the source reaches the detector after passing through an interferometer. This is known as the multiplex or Felgett advantage. Because of its narrow shape, the entrance slit of a monochromator usually images only a small part of the surface of a sample of arbitrary shape, whereas in inferometers the spot is circular. This is known as the throughput or Jacquinot advantage. Other main advantages of Fourier spectrometers are (i) there is nothing that is equivalent to stray light, (ii) the wave number accuracy can be extremely good (typically 0.01 cm^{-1}), because the computer calculates the frequency by comparison with that of helium–neon laser light, and (iii) very high resolution can be obtained. If one wants to distinguish two extremely close wavelengths, the length of displacement of the movable mirror of the interferometer should be large enough so that both sinusoids will be distinct. The resolution is therefore proportional to this displacement length. Resolutions of 0.1 cm^{-1} are currently available. We refer the reader to some basic books, for example, Bell (1972), for detailed information about Fourier spectroscopy, in particular the mathematical treatments of the interferogram, such as phase correction, apodisation, zero filling, and so on.

All IR reflection and transmission measurements of crystals performed at the author's institution since 1978 have been made with the aid of a BRUKER IFS 113C spectrometer. The choice of this commercial instrument was partly dictated by the ease of its adaptation to high-temperature measurements, as will be discussed in the next subsection. The optical beam arrangement developed by Professor Genzel and his collaborators, detailed reflection device, and block diagram are shown in Figs. 14–16, respectively. The movable mirror shifts on an air bearing, and the system is purged with dry nitrogen. Interferograms obtained without samples but typical of the six beam splitters are shown in Fig. 17, together with their Fourier transform. Changes of sources, apertures, optical filters, beam splitters, samples or reference mirrors, and detectors are digitally controlled so that a complete experiment can be programmed into the computer and performed in the absence of the operator. Long computer-controlled experiments performed at night are useful when samples are very small and thus give extremely low spectral signals; then the interferogram, which is not necessarily visible amid the electronic noise, must be recorded repetitively for a long time before results can be averaged to get the desired signal-to-noise ratio.

An application of the principles of scanning interferometry, as opposed to early interferometric techniques that worked with a chopper, is electronic

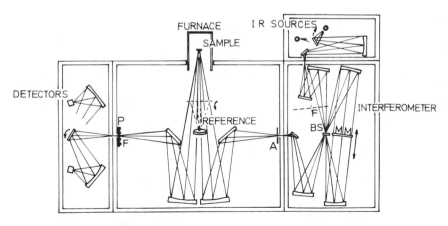

FIG. 14 Geometry of beams in the BRUKER IFS 113 scanning interferometer equipped for high-temperature reflection measurements. (Reproduced by permission.)

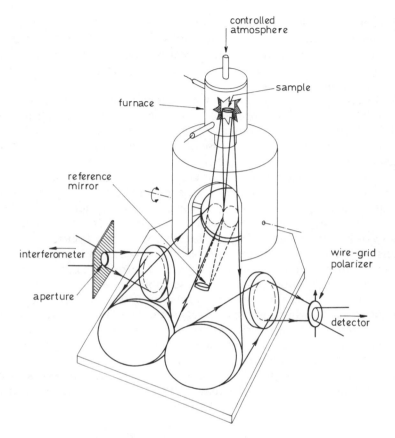

FIG. 15   Details of IR reflection device.

filtering of the spectrum. It follows from the discussion at the beginning of
this section that for a given wavelength a linear relation exists between the
electronic frequency of the detected signal and the wave number of the IR
light:

$$v \propto v\omega, \tag{86}$$

where v is the movable mirror velocity. High-pass and low-pass electronic
filters allow one to select a spectral range. An example is shown in Fig. 17.
This is useful, for example, in eliminating unwanted low-frequency mechanical vibrations that may be recorded by pyroelectric TGS detectors, or
for improving the resolution in a given spectral range when the computer
storage capacity is limited. Fringe frequencies in the interferogram are

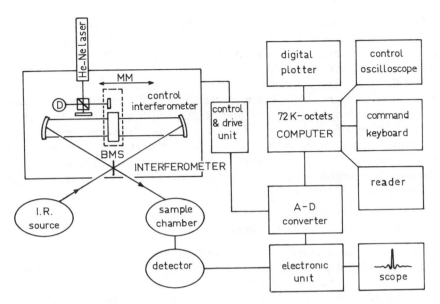

FIG. 16 Block diagram of the scanning interferometer BRUKER IFS 113 C.

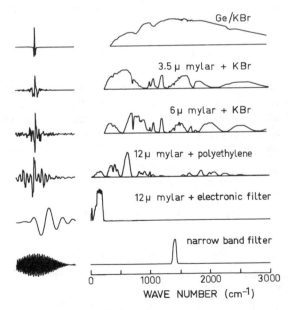

FIG. 17 Examples of interferograms and corresponding spectra.

intermediate between 10 and 10^4 Hz, depending on the adjustable velocity v. Such frequencies are easy to handle electronically. In addition, scanning interferometers are twice as efficient as spectrometers with choppers simply because there are no occulted phases.

But the main characteristic of the interferometer, shown in Fig. 14, is the focusing of the light on the beam splitter. This arrangement has no disadvantage in the opinion of this author (a practitioner) and has several advantages, among which are (i) reducing the size of the beam splitters; (ii) quadrupling the optical path difference with respect to the length of displacement of the movable mirrors, coupled back-to-back; and mainly (iii) allowing the elimination of the ac component of the thermal emission of a heated sample, discussed later.

D. ADAPTATION OF A SCANNING INTERFEROMETER FOR HIGH-TEMPERATURE EXPERIMENTS

In reflection measurements with a conventional or Fourier spectrometer that works with a chopper, the IR flux is chopped ahead of the sample to make the thermal radiation emitted by the crystal appear as a constant flux and thus stay undetected. There is no chopper in a scanning system; the heated sample plays the role of an IR source and emits a signal $I_{th}(\omega)$, and the dc component stays undetected. About 50% of the radiation emitted towards the interferometer comes back to the crystal after modulation and gives rise to an interferogram:

$$\bar{J}_{th}(\delta) = \int_{-\infty}^{\infty} I_{th}(\omega)(1 - e^{2i\pi\delta\omega}) \, d\omega, \tag{87}$$

where δ is the optical path difference (expressed in cm); $J_{th}(\delta)$ is an ac signal, because $\delta = 4 \, vt$ is a function of time t. This interferogram can be observed by putting a mask in front of the IR source as shown in Fig. 19. This signal exists as soon as the sample is above room temperature and *also when it is cooled.* Otherwise the IR source alone gives rise to an interferogram:

$$J(\delta) = \int_{-\infty}^{\infty} I(\omega)(1 + e^{2i\pi\delta\omega}) \, d\omega. \tag{88}$$

The signal that reaches the detector after reflection from the sample is therefore

$$S(\delta) = R(\omega)[J(\delta) + \bar{J}_{th}(\delta)]. \tag{89}$$

It follows from Eqs. (87) to (89), within a constant term,

$$S(\delta) = \int_{-\infty}^{\infty} R(\omega)[I(\omega) - \bar{I}_{th}(\omega)]e^{2i\pi\delta\omega} \, d\omega. \tag{90}$$

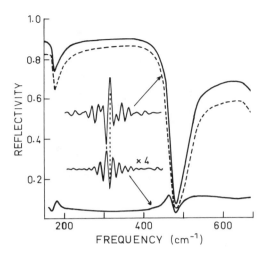

FIG. 18 An example of a spectrum related to thermal emission $\bar{J}_{th}(\delta)$, Eq. (87), of a single crystal sample (SrTiO$_3$, 700°C) when the IR source is masked (lower part), compared with the signal corresponding to Eq. (90) (dotted line). Notice that corresponding interferograms are opposed peak to peak. The solid line is the spectrum when the thermal emission of the sample is eliminated and is consistently found equal to the sum of both contributions below.

After Fourier transform (FT) the spectral intensity is thus obtained lower than in the absence of I_{th} as shown in Fig. 18. According to Eqs. (87) and (88), both interferograms are opposed peak to peak (Fig. 18). The spectrum of the heated sample alone — proportional to $R(\omega)[1 - R(\omega)]$ — is found positive in Fig. 18 because the phase-correction procedure included in the FT program acts so as to avoid negative parts in the spectrum; consequently it has changed the sign of this signal which was in antiphase. In principle this contribution could be subtracted (actually, added, because its sign is already changed) from the spectrum obtained after FT of $S(\delta)$ (Gervais and Servoin, 1978). However, when $I_{th}(\omega)$ is large enough, $I(\omega) - \bar{I}_{th}(\omega)$ consists of negative parts; then the phase corrections will make the final spectrum wrong after subtraction of the thermal-emission contribution, as verified experimentally. This major difficulty can be overcome with interferometers like the BRUKER system in which the optical beams are focused on the beam splitter and positioned between the beam splitter and the sample. As suggested by Gast (1980), an optical arrangement such as shown in Fig. 19, with an aperture that lies at the intermediate focus, allows the source signal to reach the sample but the returning sample emission signal $\bar{I}_{th}$ is masked.

All this remains true in transmission measurements. Besides, in a transmission device the optical arrangement of Fig. 19 avoids the unwanted effect of the part of the beam that is reflected from the crystal, even at room

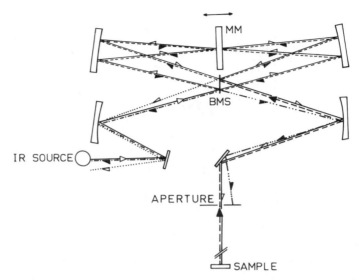

FIG. 19 Optical arrangement that can be achieved to eliminate unwanted sample thermal emission when focusing on the beam splitter is available. (Gast, 1980, reproduced by permission.)

temperature, as with the thermal emission discussed herein (Billard *et al.,* 1980).

E. HIGH-TEMPERATURE REFLECTIVITY SPECTROSCOPY AND TEMPERATURE CONTROL

The deformation of an IR band with increasing ratio γ/Ω, as shown in Fig. 10, is compared with the profiles of TO and LO modes as they would be obtained in a scattering experiment. The broadening of the imaginary part of the dielectric and inverse dielectric functions is found to occur more rapidly than that of the reflectivity. In the calculation shown in Fig. 10, which assumes a mode that is polar enough, the signal intensity measured in the IR experiment is lowered by a factor of less than 2 while the corresponding scattered intensities would be attenuated by a factor as large as 20. It is also worth pointing out that damping corresponds to the difference between the top of the IR reflection band and unity, as soon as the band is wide enough. The ratio $(1 - R_{max})/R_{max}$ has been shown to be approximately linear with damping (Hass, 1960). When damping is very low, reflectivity tends toward unity in the frequency range roughly from Ω_{TO} to Ω_{LO}, and this constitutes a very unfavorable case for obtaining damping accurately. But the accuracy of the damping parameter *increases with temperature* following the increase of the difference between reflectivity maximum and unity. As a result, reflectivity may be studied experimentally up to high temperature in better conditions than for scattering response signals.

It is clear from these considerations that one should not be satisfied with approximate experiments. If losses of reflected radiation occur due to imperfect sample surface or incorrect geometry of IR beams, the dielectric function deduced from the data will be erroneous and certain mode parameters may be meaningless.

Two techniques were used to heat the crystalline samples, either a small electric furnace or the radiation of a CO_2 laser. Laser heating is convenient for reaching elevated temperatures (1000–2000 K) with samples that are not too thick and when accurate temperature control is not needed. Any rapid variation of the sample temperature, however, should be carefully avoided because it would give rise to an ac signal. The temperature of the sample surface cannot be determined without measurement of the thermal emission of the surface itself when the crystal is heated by a CO_2 laser beam. Even when the sample is in a furnace, this method of temperature measurement is preferable (Servoin and Gervais, 1976). A removable chopper and a narrow-band filter put between the heated sample and the detector allow the measurement of the thermal radiation $[1 - R(\omega, T)]I(\omega, T)$ emitted by the crystal at a wavelength at which it emits like a blackbody, i.e., where the reflectivity is a minimum according to the Kirchhoff law. This radiation can be compared to that, denoted $[1 - R_{st}(\omega, T_{st})]I(\omega, T_{st})$, of a standard crystal heated to a known temperature T_{st} with the aid of the formula

$$\alpha = \frac{[1 - R(\omega, T)]I(\omega, T) - I(\omega, T_0)}{[1 - R_{st}(\omega, T_{st})]I(\omega, T_{st}) - I(\omega, T_0)} , \tag{91}$$

where α is the ratio of both detected signals and $I(\omega, T_0)$ is the blackbody radiation at room temperature T_0. The temperature of the crystal can then be deduced from the relation

$$\alpha = \frac{[1 - R(\omega, T)][\exp(\hbar\omega/k_B T) - 1]^{-1} - [\exp(\hbar\omega/k_B T_0) - 1]^{-1}}{[1 - R_{st}(\omega, T_{st})][\exp(\hbar\omega/k_B T_{st}) - 1]^{-1} - [\exp(\hbar\omega/k_B T_0) - 1]^{-1}}.$$

$$\tag{92}$$

Quartz heated to the temperature of the $\alpha-\beta$ phase transition has been used as calibration standard.

VI. Examples of Applications

The examples we will discuss to illustrate our subject will be restricted essentially to studies performed with IR reflectivity spectroscopy with a scanning interferometer. We will discuss barium titanate in some detail because it illustrates the merits of this technique well in an important case. Several studies performed with other IR techniques as a function of temperature have been reviewed by Perry (1971) and recently by Sherman and Wilkinson (1980).

A. STRUCTURAL PHASE TRANSITIONS

Over the two last decades numerous spectroscopic studies have been devoted to structural phase transitions. According to the concept developed independently by Cochran (1960) and Anderson (1960), an unstable, or soft, optical phonon mode is most often associated with the structural distortions when the crystal is ordered in both phases. The soft mode can be IR-active. This is the case in FE crystals in both FE and PE (paraelectric) phases, as will be seen in Sections VIB and C. But even in the case of quartz, for example, where the soft mode is Raman-active only in the α phase and becomes inactive in the β phase, Gervais and Piriou (1975) have shown that *all* polar modes exhibit unusual temperature dependencies.

To the best of the author's knowledge, the $\alpha-\beta$ phase transformation of lead phosphate $Pb_3(PO_4)_2$ seems to be the first that has been investigated by means of scanning interferometry (Luspin *et al.*, 1979). Although this material is ferroelastic at room temperature whereas quartz is piezoelectric and although structures and structural distortions are different in these materials, some similar behavior has been observed via IR reflectivity spectroscopy. In crystals that do not undergo structural transformations, mode frequencies are commonly found to shift linearly with temperature above the Debye temperature. Contrary to this behavior, most TO and LO modes exhibit a "critical" frequency shift, more or less marked, in the vicinity of the $\alpha-\beta$ phase transition of each material. The TO–LO splittings are related to the TO oscillator strengths via (Merten and Lamprecht, 1970)

$$\Delta\varepsilon_j = \varepsilon_\infty \Omega_{jTO}^{-2} \prod_k (\Omega_{kLO}^2 - \Omega_{jTO}^2) \Big/ \prod_{k \neq j} (\Omega_{kTO}^2 - \Omega_{jTO}^2), \qquad (93)$$

which is a generalization of Eq. (16) to several modes and derives from Eqs. (55) and (61), neglecting damping terms. The oscillator strengths, hence the polar characters of the modes, are expected to exhibit critical effects also. This is just what is observed. While certain modes consistently lose their polar character following a critical law, as shown in the examples of Fig. 20, the oscillator strengths of strong modes also increase critically. This is correlated with the fact that the structure of the high-temperature phase is generally of a higher symmetry than the low-temperature phase. Consequently the number of IR-active modes is smaller above the phase transition than below. In lead phosphate, A_u- and B_u-type modes degenerate into E_u-type modes on being heated to just above the phase transition. Besides,

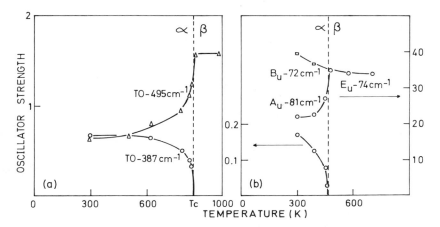

FIG. 20 Examples of temperature dependence of TO phonon mode strengths near structural phase transitions. More complete data were reported by (a) Gervais and Piriou (1975) for quartz and (b) Luspin *et al.* (1979) for lead phosphate.

phonon linewidth divergences are observed in both materials, contrary to the linear temperature dependence that is commonly observed above the Debye temperature in crystals that do not undergo structural phase transitions.

The range from liquid-helium to room temperature is obviously complementary to the medium and high temperatures chiefly discussed here. The theoretical background outlined in this chapter is, of course, applicable to low temperatures. High-temperature measurements simply require specific devices. Examples of similar IR reflection studies of phase transitions that occur below room temperatures were reported by Siapkas (1978) on thiourea, Bauer *et al.* (1977) on PbTe, and Strobel and Geick (1979) on $KMnF_3$.

B. OXIDIC PEROVSKITES, $BaTiO_3$, $SrTiO_3$, $KNbO_3$

Barium titanate has perhaps been one of the most experimentally and theoretically studied materials in solid-state physics since the discovery of its FE properties in the 1940s. In spite of an enormous literature concerning this compound, long considered a prototype FE material, owing to its simple structure, Scott (1974) mentioned in his review article that it has remained more enigmatic than other oxidic perovskite crystals. This opinion is generally shared by other reviewers (Lines and Glass, 1977). The first important spectroscopic studies of $BaTiO_3$ and $SrTiO_3$ were IR reflectivity measurements (Barker and Tinkham, 1962; Spitzer *et al.*, 1962). These experiments gave information that tended to confirm the soft-mode description of displacive ferroelectrics. Neutron-scattering measurements (Cowley, 1964; Harada *et al.*, 1971) further confirmed this interpretation. Some 23 reference papers concerning Raman investigations of $BaTiO_3$ or their analysis are quoted in a recent paper by Burns and Dacol (1978). Barium titanate has the cubic perovskite structure in the PE phase above ~ 400 K and thus three polar F_{1u}-type vibrational modes. Below the Curie temperature T_c it transforms successively to three FE phases (Jona and Shirane, 1962, Lines and Glass, 1977): first to tetragonal, then to orthorhombic at ~ 280 K, and finally to a rhombohedral phase below ~ 180 K. On cooling at the PE–FE phase transition, the F_{1u} modes split into three A_1 modes observed for a polarization along the FE axis and three E modes that vibrate in the plane perpendicular to this axis. An additional "silent" E mode that derives from a nonpolar F_{2u} mode of the PE phase is also observed in the tetragonal phase. As summarized by Burns and Dacol (1978), Raman investigations of $BaTiO_3$ have followed "a rather tortuous path." This is partly due to the fact that A_1 modes had not yet been observed by IR reflectivity so that the Raman assignments were not obvious. In noncentrosymmetric crystals, indeed, the reflection of laser light on the back face of the crystal can excite phonons that propagate in different directions and may give rise to mixed TO–LO modes. What seemed to be commonly accepted in 1978 was that (i) the frequency of the soft overdamped E–TO mode increases on heating up to the FE–PE phase transition, a result that is opposite to what is expected near a Curie temperature; (ii) it is not the lowest A_1–TO mode that contributes most to the dielectric constant but the next TO mode which shifts little with temperature; and (iii) it was still controverted (Verble *et al.*, 1978) that all normal modes abruptly disappear at T_c consistent with group-theory prediction but broad features persist above T_c.

Sanjurjo *et al.* (1979) and Servoin *et al.* (1980a) independently reported the IR reflection spectrum of A_1 modes at room temperature. Data analyses (Servoin *et al.*, 1980a) confirmed the most recent Raman assignments. Luspin *et al.* (1980) studied the temperature dependence of F_{1u} modes in the temperature range 400–1350 K with the type of scanning interferometer described in Sections VC–E. The silent E mode was also observed by infrared for the first time (Servoin *et al.* 1980a). Some of their results are reproduced in Fig. 21. A soft-mode behavior (Fig. 22) with a Curie–Weiss law,

$$\Omega_s^2 = a(T - T_c), \tag{94}$$

which holds over 800°C, was unambiguously observed. As discussed in Section VA, the up-shift of the overdamped soft-mode frequency shown in Fig. 21 and detailed in Figs. 22 and 23, manifests itself by a decrease of the

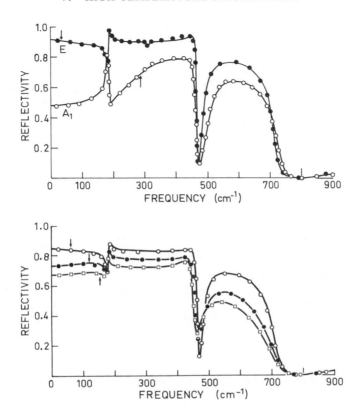

FIG. 21 IR reflection spectra of barium titanate and best fit (solid lines) to the data with the four-parameters model Eq. (61), (a) in the tetragonal FE phase, 300 K, parallel (A_1) and perpendicular (E) to the FE axis, and (b) in the cubic PE phase, F_{1u}-type modes (○ 400 K, ● 850 K, □ 1150 K). Arrows indicate the soft-mode frequency. (After Servoin *et al.*, 1980a; Luspin *et al.*, 1980.)

low-frequency reflection level on heating. The same kind of experiment performed in strontium titanate (Servoin *et al.*, 1980b) yields a similar behavior, as shown in Fig. 22.

Both low-frequency modes clearly are little coupled in $BaTiO_3$ because the anticrossing region is restricted to only a few reciprocal centimeters. No feature characteristic of interference effects between the two modes, which are merely superimposed, is observed consistently in Fig. 23. Both modes have been decoupled in $SrTiO_3$ (Servoin *et al.*, 1980b) with the aid of the formula (Scott, 1974)

$$\Omega_{\pm} = \tfrac{1}{2}(\omega_1 + \omega_2) \pm \tfrac{1}{2}[(\omega_1 - \omega_2)^2 + 4w^2]^{1/2}, \tag{95}$$

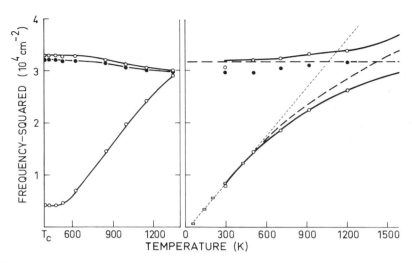

FIG. 22 Temperature dependence of the soft TO mode frequency squared together with TO (O) and LO (●) frequencies of the next polar mode in the cubic phase of barium and strontium titanate (see Luspin *et al.,* 1980; Servoin *et al.,* 1980b). For $SrTiO_3$, square symbols are the neutron data of Cowley (1964), the dotted line is the extrapolation of low-temperature data according a Curie–Weiss law Eq. (94), and dashed lines are decoupled frequencies according to Eq. (95).

where Ω_+ and Ω_- represent the experimental coupled-mode frequencies shown in Fig. 22. Again, the coupling parameter $w = 10\ cm^{-1}$ is found to be small.

The decoupled soft-mode frequency in $SrTiO_3$ clearly departs from the simple Curie–Weiss law, Eq. (94), at high temperature, and instead obeys a modified Curie–Weiss law consistent with the early interpretation of the temperature dependence of the static dielectric constant (Rupprecht and Bell, 1964). This result seems to support the theory based on the strong anisotropy of the oxygen polarizability, as proposed by Migoni *et al.* (1976) and Bilz *et al.* (1980), (see also Balkanski *et al.,* 1980 for SbSI), rather than a phonon self-energy mechanism alone (Cowley, 1965, 1980). After the works by Silverman and Joseph (1963) and Cowley (1965), quartic anharmonicity $P_a^{(4)}(0j,\ T)$, Eq. (43), was thought to play the dominant role in the stabilization of an FE mode in the PE phase of FE crystals. Otherwise all TO and LO modes except the soft mode behave "normally" or nearly so in both $SrTiO_3$ and $BaTiO_3$, i.e., their dependence on temperature is compatible with the expressions for the phonon self-energy presented in Section III (Luspin *et al.,* 1980; Servoin *et al.,* 1980b).

In order to explain the "anomalies" found in $BaTiO_3$, it seems worth-

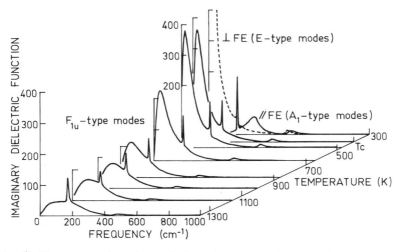

FIG. 23 Temperature dependence of TO mode resonances in barium titanate as obtained from a KK analysis of reflectivity data. (Adapted from Luspin *et al.,* 1980; Servoin *et al.,* 1980a.)

while first to recall the basic concepts of displacive and order–disorder mechanisms. The main points are schematically represented in Fig. 24 (second-order phase transition). In the *displacive* picture the ions that are responsible for the spontaneous polarization in the FE phase occupy the bottom of a double-well potential, all ions being on the same side in the relevant domain. This potential becomes flat at T_c and becomes a single-well potential on heating in the PE phase. As a result, the mode that consists mainly of the vibration of this type of ion against surrounding oxygens along the FE axis is soft in the FE phase and becomes completely unstable at T_c. Its frequency shifts down and possibly falls to zero. Above T_c the ion now occupies the centrosymmetric position, the soft mode stabilizes on heating in the PE phase, and its frequency increases according to Eq. (94). In the *order–disorder* picture, there is a double-well potential above and below T_c, and no vibrational mode displays any dramatic temperature dependence. But relaxational effects are observed that are associated with the jumps of thermally activated ions from one well to another. The PE phase is disordered, because statistically the ions are distributed equally on both sites. The FE phase becomes partially ordered on cooling because one site is preferentially occupied. Both mechanisms are indistinguishable from macroscopic measurements.

Results deduced from scanning interferometric measurements up to high temperature in BaTiO$_3$ (Luspin *et al.,* 1980) and the isomorphous com-

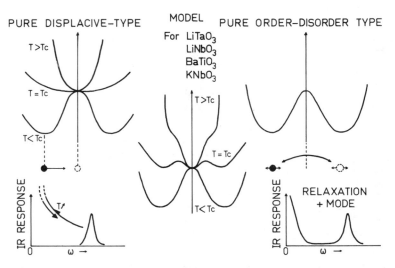

FIG. 24   Schematic representation of the temperature dependence of the spatial potential for the mobile ion along the FE axis and spectroscopic characterization of FE mechanisms (see text).

pound $KNbO_3$ (Fontana *et al.,* 1981, Quittet *et al.,* 1981) are schematically summarized in Fig. 25, together with some complementary data obtained by other techniques. On cooling from the PE phase, the A_1 component of the F_{1u} triply degenerate soft mode is abruptly stiffened at a frequency of ~ 280 cm^{-1} at T_c, whereas the E component continues to soften down to the next transition. At that transition a B_1-type component is suddenly stiffened, and the last B_2 component softens. Abrupt changes of frequency are consistent with the first-order character of all phase transitions. There is no more soft mode in the lowest-temperature phase. However, while the soft mode roughly explains the static dielectric constant behavior according to the LST relation in the orthorhombic phases, this becomes less and less true on approaching T_c from below along the FE axis and from above, in both materials, as shown in Fig. 25. Such large discrepancies of static and "lattice" dielectric constants near T_c, observed early, added fresh fuel to the controversy about the mechanism that drives the FE – PE phase transitions, which was believed by certain authors to be of the order–disorder type rather than the displacive type (see, for example, the review paper by Bozinis *et al.,* 1976) and which has its origin in the observation by Comes *et al.* (1968) of anomalous strong diffuse x-ray scattering in $BaTiO_3$. Comes and Shirane (1972) later demonstrated in $KTaO_3$ that such scattering is related to the large amplitude of vibration of underdamped soft phonons whose

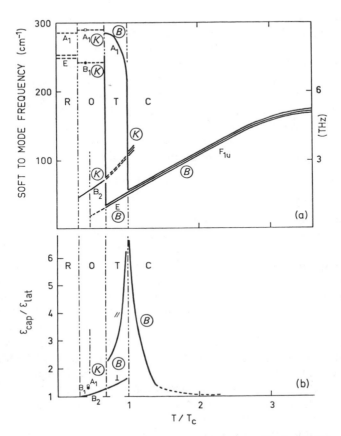

FIG. 25   Schematic representation of (a) the temperature dependence of the soft-mode components in BaTiO₃ (B) and KNbO₃ (K), together with (b) the ratio of the static dielectric constant over that determined via the LST relation.

dispersion is flat with respect to wave vector. Results reproduced in Figs. 22 and 25 have recently shown that at sufficiently high temperature the Ti (or Nb) ion is stabilized at the center of the unit cell, vibrating in a single-main-well potential, and that this vibration accounts for the static dielectric constant. *The PE phase, therefore, tends to become ordered at high temperature.* On cooling down to T_c the ion appears to hesitate about which site to occupy and has a very anharmonic motion in a potential that is not flat but consists of shallow secondary minima (Fig. 24) at T_c. Below T_c the soft mode is frozen into static displacements along the FE axis, and the ion motion is still more unstable in the plane perpendicular to the FE axis.

Fontana *et al.* (1979) concluded that the phase transitions are of the displacive type in KNbO$_3$. Quittet *et al.* (1981) in KNbO$_3$ independently and Luspin *et al.* (1980) in BaTiO$_3$ arrived at the same conclusion but moderated their statement in that this is not the soft mode alone that triggers the structural transformation but rather a minimization of the free energy in which relaxational motions associated with secondary minima of the spatial potential play a dominant part near T_c. The picture proposed by Luspin *et al.* (1981) thus is one of a displacive mechanism complicated by a diffuse step of dynamical disorder near T_c that would, in fact, intrinsically coexist in the case of any displacive *first-order* phase transition.

C. THE FE MECHANISM IN LiTaO$_3$ AND LiNbO$_3$: DISPLACIVE OR ORDER–DISORDER?

The same kind of controversy arose in LiNbO$_3$ and LiTaO$_3$ when Abrahams *et al.* (1973) published neutron-diffraction measurements and concluded that Li ions at T_c ($\sim$ 890 K in LiTaO$_3$) occupy two equivalent noncentrosymmetric sites with equal probabilities, the PE–FE phase transition being associated with the ordering of Li atoms. Early Raman experiments, performed by Johnston and Kaminow (1968) who reported on a soft mode in both materials and decided on a displacive mechanism, were then controverted by Penna *et al.* (1976a, 1976b, 1977) for LiTaO$_3$ and Chowdhury *et al.* (1978) for LiNbO$_3$ who claimed there is no evidence of a soft mode but rather relaxation phenomena related to disorder. IR reflectivity experiments in LiTaO$_3$ and LiNbO$_3$ were performed as a function of temperature (300–1250 K) for the first time by Servoin and Gervais (1979) by means of scanning interferometry and clearly showed that the low-frequency reflection level increases up to T_c and decreases above T_c for the polarization along the FE axis (A_1-type modes in the FE phase, A_{2u}-type above). A straightforward application of Eq. (79) then gives information on the temperature dependence of the soft mode. This dependence is shown more precisely in Fig. 26, as deduced from a KK analysis. Notice that any down-shift of a soft-mode frequency is accompanied by an increase of its oscillator strength according to $\Delta\varepsilon_s\Omega_s^2 \approx$ constant. This relation is derived from Eq. (93) where $\Omega_{sTO}^2 \ll \Omega_{sLO}^2$ and where all terms but $\Delta\varepsilon_s$ and Ω_s are weakly temperature dependent. Fits of an appropriate dielectric function model to reflectivity data at several temperatures, as shown at 1180 K for LiTaO$_3$ (Fig. 7), indicate that (i) the soft-mode frequency follows a law (T_c-T)$^{1/2}$, as might be expected (Johnston and Kaminow, 1968), over the range 550–850 K in LiTaO$_3$ and 750–1250 K in LiNbO$_3$ and (ii) the mode becomes overdamped only in the range T_c- $T \le 70°$C for LiTaO$_3$ and above 1200 K in LiNbO$_3$ ($T_c \approx 1480$ K). It was also found that within experimental error the soft mode alone is responsible for the static dielectric constant

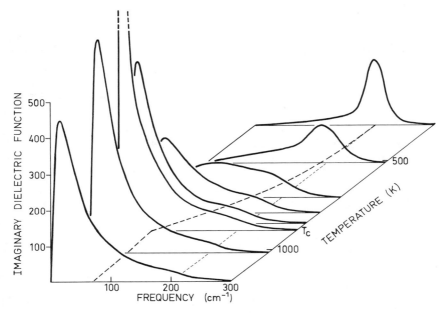

FIG. 26 Temperature dependence of the soft-mode response in lithium tantalate, $\mathbf{E} \parallel$ FE axis. The evolution of the TO frequency that allows the best fit to reflectivity data according to the procedure described in Section IV B2 and the example of Fig. 7 is also shown as a dotted line and compared with the room temperature value (Servoin and Gervais, unpublished result.)

behavior, as given by the LST relation, in both FE and PE phases of $LiTaO_3$, except in the immediate vicinity of T_c. Within a few tens of degrees on either side of T_c an additional relaxation term should be added to fit the reflection spectrum correctly. It is thus concluded (Servoin and Gervais, 1981) that the PE phase is ordered above ~ 1000 K, the phase transition mechanism is essentially displacive, and the spatial potential of Li ions is not flat but rather exhibits at least three shallow secondary minima in the immediate vicinity of T_c, just as is found in $BaTiO_3$ and $KNbO_3$ and as shown in Fig. 24. Again a diffuse evolution order–disorder–order complicates the displacive mechanism, although the disordered stage is restricted to a few tens of degrees in this case and seems to account for all observed results. This conclusion is also in agreement with the statistical theory of Lines (1969, 1970, 1972) and does not go against the experimental data of Abrahams et al. (1973). Results in $BaTiO_3$, $KNbO_3$, and $LiTaO_3$ support the displacive-order-disorder crossover concept discussed by Müller et al. (1982).

D.   OXIDES WITH RUTILE STRUCTURE

The IR spectra of materials with rutile structure consist of one single A_{2u} mode for the electric field polarized along the c axis and three E_u modes perpendicular to this axis. The responses of TO modes deduced from dispersion parameters that allowed the best fit to the reflectivity data of the oxides SnO_2 (Hargreave, 1971; Summit, 1968), GeO_2 (Roessler and Albers, 1972), and TiO_2 (Gervais and Piriou, 1974b) are reproduced in Fig. 27. Notice that the main modes of TiO_2 are more intense and have a lower frequency than those observed in the other oxides. This is due to the partially unstable character of those modes in relation to the incipient FE properties of the rutile TiO_2 (Samara and Peercy, 1973).

The structure of *niobium dioxide* NbO_2 is tetragonal at room temperature. In fact, this is a distorted-rutile structure. The analysis of the IR

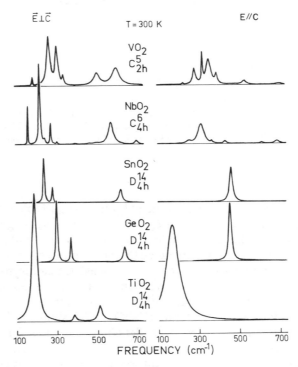

FIG. 27   A comparison of frequencies, dampings, and intensities of TO phonon modes in oxide crystals with rutile or distorted-rutile structure as calculated from dispersion parameters reported by Barker *et al.* (1966) for VO_2, Gervais and Baumard (1979) for NbO_2, Hargreave (1971) for SnO_2, Roessler and Albers (1972) and Summit (1968) for GeO_2, and Gervais and Piriou (1974b) for TiO_2.

reflectivity of NbO_2 single crystals yields the phonon spectra reproduced in Fig. 27 (Gervais and Baumard, 1979). A comparison with results found in other crystals indicates that the distortion of the rutile structure is small, because the spectra display the characteristic main modes plus additional but weak or very weak modes. Niobium dioxide has a number of features in common with vanadium dioxide VO_2. The semiconducting–metal phase transition of VO_2 (Goodenough, 1971), which occurs at 67°C and is accompanied by a transformation from monoclinic to rutile structure, has been extensively studied. IR reflection measurements performed by Barker et al. (1966) yield the room-temperature phonon spectra, shown in Fig. 7, related to the direction of the c axis of the rutile structure. Many intense modes are observed in both polarizations, and the spectra are very different from those of rutile. Although the results of Fig. 27 were obtained only ∼ 40° below the phase transformation, IR data thus indicate an important distortion with respect to the rutile structure consistent with crystallographic data that principally showed pairing of V–V atoms. By comparison, the spectra of NbO_2, which also undergoes a semiconducting–metal phase transition at 810°C, indicate a distortion effect that is much less pronounced even 800°C below the transition, although the pairing of Nb–Nb atoms is also known to occur (Goodenough, 1971). A progressive appearance of electronic excitations in IR spectra is found at intermediate temperatures, as shown in Fig. 9.

E. PHONON LINEWIDTH DIVERGENCE NEAR STRUCTURAL PHASE TRANSITIONS

As shown in section III the dependence of damping on temperature is linear above the Debye temperature, in other words, above the low-temperature region where quantum effects take place. Contrary to this common behavior, the experimental damping divergence of soft modes at the approach of structural phase transitions from below has been reported several times. This will not be discussed here because the phonon self-energy theory in its simplest formulation, is not in principle, appropriate for soft modes, particularly when they are overdamped. But there are a number of materials in which all phonon linewidths, deduced either from scattering experiments or analyses of IR reflection spectra, recorded as a function of temperature, exhibit a "diverging" behavior, more or less marked, in the vicinity of phase transitions (Fig. 28). This is the case in quartz (Gervais and Piriou, 1975), $LiIO_3$ (Peyrard et al., 1975), KDP (Laulicht, 1978), thiourea (Delahaigne, 1977), lead phosphate (Luspin et al., 1979), lithium tantalate and niobate (Servoin and Gervais, unpublished), and several ammonium compounds (for example, Iqbal and Christoe, 1976; Peyrard and Perret, 1979; Wang and Wright, 1973). Phonon linewidth divergence likely exists in other materials, and it seems to be a general phenomenon that is observed irrespective of the

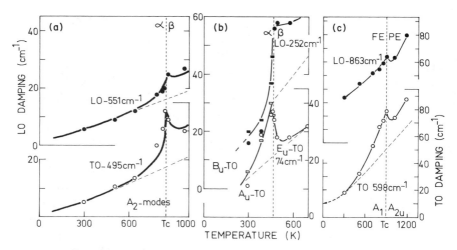

FIG. 28 Examples of phonon linewidth divergences of nonsoft modes near structural transitions. (a) Quartz, heavy line is a fit of Eq. (96) to IR dampings. (b) $Pb_3(PO_4)_2$. (c) $LiTaO_3$. More complete data were reported by Gervais and Piriou (1975), Luspin *et al.* (1979).

crystalline structures, types of phase transition, or chemical bonds. We do not assert there is a single reason for such abnormal temperature dependence, we would merely like to recall that the damping function, Eq. (46), renormalized to account for the actual temperature dependence of the one-phonon density of states $\rho(\Omega, T)$ in the form

$$
\begin{aligned}
\Gamma(0j, \Omega_j, T)/\Gamma(0j, \omega_j, 0) = \sum_l \{ & \rho(\Omega_j - \Omega_l)\rho(\Omega_l)[n(\Omega_l) \\
& + n(\Omega_j - \Omega_l) + 1] \\
& + 2\rho(\Omega_j + \Omega_l)\rho(\Omega_l) \\
& - n(\Omega_j - \Omega_l)]\}
\end{aligned}
\tag{96}
$$

yields a profile that fits the temperature dependence of any experimental damping of TO and LO modes in quartz without any adjustable parameter on the right side of Eq. (96) (Gervais, 1976b, Gervais *et al.*, 1975). This is due to the weight of terms $n(\Omega_s) \approx k_B T/\hbar\Omega_s$ which do diverge when Ω_s shifts down on heating in the low-temperature phase and up in the high-temperature phase, even partially. On the other hand, disorder, which exists in other crystals mentioned above, has been invoked, and pseudospin–phonon interactions have been shown to account for the anomalous broadening of phonon lines (Laulicht, 1978; Matsushita, 1976; Peyrard and Remoissenet, 1979).

ACKNOWLEDGMENTS

I am indebted to Dr. Anne-Marie Anthony, Director of the Centre de Recherches sur la Physique des Hautes Températures at that time, and to Prof. F. Cabannes and Prof. J. J. Fripiat who encouraged and aided us in our efforts to acquire the equipment and develop the applications of IR scanning interferometry. Dr. B. Piriou introduced me to conventional IR spectroscopy. Particular mention must be made of Dr. J. L. Servoin for his many contributions to the work from which matter in this chapter was drawn. My thanks also go to Prof. M. Lambert and to Drs. A. M. Quittet and Y. Luspin for contributing greatly to our knowledge of FE mechanisms, and to Mrs. A. Blin for her technical assistance.

REFERENCES

Abrahams, S. C., Buehler, E., Hamilton, W. C., and Laplaca, S. J. (1973). *J. Phys. Chem. Solids,* **34**, 521.

Alain, P., and Piriou, B., (1977). *J. Physique* **12**, C7, 389–394.

Anderson, P. W. (1960). *In* "Fizika dielektrikov" (G. I. Skanavi, ed.), p. 920, Acad. Nank, Moscow.

Ashkin, M., Parker, J. H., and Feldman, D. W. (1968). *Solid State Commun.* **6**, 343.

Axe, J. D., Gaglianello, J. W., and Scardefield, J. E. (1965). *Phys. Rev. A* **139**, 1211–1215.

Bairamov, B. Kh., Kitaev, Yu. E., Negodinko, V. K., and Kharshkhozhev, Z. M. (1974). *Fiz. Tverd. Tela.* **16**, 2036–2043.

Bairamov, B. Kh., and Kharshkhozhev, Z. M. (1975). *Fiz. Tverd. Tela.* **17**, 1358–1362.

Balkanski, M., Teng, M. K., Massot, M., and Bilz, H. (1980). *Ferroelectrics* **26**, 737–740.

Barker, A. S. (1963). *Phys. Rev. 132,* 1474–1481.

Barker, A. S. (1964). *Phys. Rev. A* **136**, 1290–1295.

Barker, A. S. (1967). *In* "Ferroelectricity" (E. F. Weller, ed.), pp. 213–250. Elsevier, Amsterdam.

Barker, A. S., and Hopfield, S. S. (1964). *Phys. Rev. A* **135**, 1732–1737.

Barker, A. S., and Tinkham, M. (1962). *Phys. Rev.* **125**, 1527–1530.

Barker, A. S., Verleur, H. W., and Guggenheim, H. J. (1966). *Phys. Rev. Lett.* **17**, 1286–1289.

Bartsokas, A., and Siapkas, D. (1980). *Ferroelectrics* **25**, 561–564.

Bauer, G. *et al.* (1977). *In* "Lattice Dynamics" (M. Balkanski, ed.), pp. 669–672. Flammarion, Paris.

Baumard, J. F., and Gervais, F. (1977). *Phys. Rev. B* **15**, 2316–2327.

Beairsto, J. A. B., and Eldridge, J. E. (1973). *Can. J. Phys.* **51**, 2550–63.

Bell, R. J. (1972). "Introductory Fourier Transform Spectroscopy." Academic Press, New York.

Bendow, B. (1978). *In* "Solid State Physics" (H. Ehrenreich, F. Seitz and, D. Turnbull, eds.), Vol. 33, p. 249, Academic Press, New York.

Berg, J. I., and Bell, E. D. (1971). *Phys. Rev. B* **4**, 3572–3580.

Berreman, D. W. (1963). *Phys. Rev.* **130**, 2193.

Berreman, D. W., and Unterwald, F. C. (1968). *Phys. Rev.* **174**, 791–799.

Billard, D., Gervais, F., and Piriou, B. (1976). *Phys. Status Solidi B* **75**, 117–126.

Billard, D., Gervais, F., and Piriou, B. (1980). *Int. J. Infrared Millim. Waves* **1**, 641–647.

Bilz, H. (1966). *In* "Phonons in Perfect Lattices and in Lattices with Point Imperfections" (H. Stevenson, ed.), p. 208–234. Oliver and Boyd, Edinburgh.

Bilz, H. *et al.* (1980). *Ferroelectrics* **25**, 339–342.

Borer, W. J., Mitra, S. S., and Namjoshi, K. V. (1971). *Solid State Commun.* **9**, 1377–1381.

Born, M., and Huang, K. (1954). "Dynamical Theory of Crystal Lattices." Oxford, London.

Boyer, L. L., Harrington, J. A., Hass, M., and Rosenstock, H. B. (1975). *In* "Optical Properties of Highly Transparent Solids" (S. S. Mitra and B. Bendow, eds.), pp. 59–69. Plenum, New York.

Bozinis, D. G., and Hurrell, J. P. (1976). *Phys. Rev. B* **13**, 3109.

Bozinis, D. G., Penna, A. F., and Porto, S. P. S. (1976). *In* "Proceedings 5th International Conference on Raman Spectroscopy," p. 569–570, Verlag, Freiburg.

Bruce, A. D. (1973). *J. Phys. C* **6**, 174–188.

Burkhard, H., Geick, R., Kastner, P., and Unkelbach, K. H. (1974). *Phys. Status Solidi B* **63**, 89–96.

Burns, G., and Dacol, F. M. (1978). *Phys. Rev. B* **18**, 5750.

Chang, I. F., and Mitra, S. S. (1972). *Phys. Rev. B* **5**, 4090–4101.

Chaves, A. S., and Porto, S. P. S. (1973). *Solid State Commun.* **13**, 865–868.

Chaves, A. S. Katiyar, R. S., and Porto, S. P. S. (1974). *Phys. Rev. B* **10** 3522–3533.

Chowdhury, M. R., Peckam, G. E., and Saunderson, D. H. (1978). *J. Phys. C.* **11**, 1671.

Clauws, P., and Vennik, J. (1976). *Phys. Status Solidi B* **76**, 707–713.

Cochran, W. (1960). *Adv. Phys.* **9**, 387.

Comes, R., and Shirane, G. (1972). *Phys. Rev. B* **5**, 1886.

Comes, R., Lambert, M., and Guinier, A. (1968). *C. R. Acad. Sci. B* **266**, 959–962.

Comes, R., Lambert, M., and Guinier, A. (1968). *Solid State Commun.* **6**, 715–719.

Cowley, R. A. (1963). *Adv. Phys.* **12**, 421–480.

Cowley, R. A. (1964). *Phys. Rev. A* **134**, 981–997.

Cowley, R. A. (1965). *Philos. Mag.* **11**, 673–706.

Cowley, R. A. (1980). *Adv. Phys.* **29**, 1–110.

Daubert, J., Jex, M., and Mullner, M. (1973). *Phys. Status Solidi B* **57**, 477–483.

Delahaigne, A. (1977). PhD. thesis, University of Reims.

Denham, P., Field, G. R., Morse, P. L. R., and Wilkinson, G. R. (1970). *Proc. R. Soc. Ser. A* **317**, 55–77.

Denham, P., Morse, P. L. R., and Wilkilson, G. R. (1973). *J. Phys. C* **6**, 2066–2075.

Dionne, G., and Woolley, J. C. (1972). *Phys. Rev. B* **6**, 3898–3913.

Eldridge, J. E., and Howard, R. (1973). *Phys. Rev. B* **7**, 4652–4665.

Eldridge, J. E., and Kembry, K. A. (1973). *Phys. Rev. B* **8**, 746–755.

Evans, D. J., and Ushioda, S. (1974). *Phys. Rev. B* **9**, 1638–1645.

Fontana, M. D., Dolling, G., Kugel, G. E., and Carabatos, C. (1979). *Phys. Rev. B* **20**, 3850–3858.

Gast, J. (1980). *Bruker Report, FT-IR Application Note 9.*

Gervais, F. (1973). *Solid State Commun.* **13**, 1211–1214.

Gervais, F. (1976a). *Solid State Commun.* **18**, 191–198.

Gervais, F. (1976b). *Ferroelectrics* **13**, 555–557.

Gervais, F. (1977). *Opt. Commun.* **22**, 116–118.

Gervais, F. (1980). *Phys. Status Solidi B* **100**, 337–381.

Gervais, F., and Arend, H. (1983). *Z. Phys. B* **50**, 17–22.

Gervais, F., and Baumard, J. F. (1977). *Solid State Commun.* **21**, 861–865.

Gervais, F., and Baumard, J. F. (1979). *J. Phys. C* **12**, 1977–1983.

Gervais, F., and Piriou, B. (1974a). *J. Phys. C* **7**, 2374–2386.

Gervais, F., and Piriou, B. (1974b). *Phys. Rev. B.* **10**, 1642–1654.

Gervais, F., and Piriou, B. (1975). *Phys. Rev. B* **11**, 3944–3950.

Gervais, F., and Servoin, J. L. (1977). *Phys. Rev. B* **15**, 4532–4536.

Gervais, F., and Servoin, J. L. (1978). *Infrared Phys.* **18**, 883–886.

Gervais, F., Piriou, B., and Cabannes, F. (1972). *Phys. Status Solidi B* **51**, 701–712.

Gervais, F., Piriou, B., and Cabannes, F. (1973). *J. Phys. Chem. Solids* **34**, 1785–1796.
Gervais, F., Piriou, B., and Billard, D. (1975). *Solid State Commun.* **17**, 861–865.
Gervais, F., Luspin, Y., Servoin, J. L., and Quittet, A. M. (1980). *Ferroelectrics* **24**, 285–288.
Giehler, M., and Jahne, E. (1976). *Phys. Status Solidi B* **73**, 503–516.
Goodenough, J. B. (1971). *In* "Progress in Solid State Chemistry" (H. Reiss, ed.), Vol. 5, pp. 145–399. Pergamon, Oxford.
Grynberg, M., Le Toullec, R., and Balkanski, M. (1974). *Phys. Rev. B* **9**, 517–526.
Haque, M. S. (1975). *Phys. Rev. B* **12**, 1501–1515.
Harada, J., Axe, J. D., and Shirane, G. (1971). *Phys. Rev. B* **4**, 155.
Hargreave, M. M. (1971). *J. Phys. C* **4**, 174–180.
Hart, T. R., Aggarwal, R. L., and Lax, B. (1974). *Phys. Rev.* **1**, 638–642.
Hass, M. (1960). *Phys. Rev.* **117**, 1497–1499.
Hennion, B. *et al.* (1979). *J. Phys. C* **12**, 1609–1624.
Hiraishi, J. (1973). *Bull. Chem. Soc. Japan* **46**, 1334–1338.
Ipatova, I. P., Maradudin, A. A., and Wallis, R. F. (1967). *Phys. Rev.* **155**, 882–895.
Iqbal, Z., and Christoe, C. W. (1976). *Solid State Commun.* **18**, 269–273.
Jasperse, J. R., Kahan, A., Plendl, J. L., and Mitra, S. S. (1966). *Phys. Rev.* **146**, 526–542.
Johnston, K. W., and Bell, E. E. (1969). *Phys. Rev.* **187**, 1044–1052.
Johnston, K. W., and Kaminow, I. P. (1968). *Phys. Rev.* **168**, 1045.
Jona, F., and Shirane, G. (1982). "Ferroelectrics." MacMillan, New York.
Kascheev, V. N., and Krivoglaz, M. A. (1961). *Fiz. Tverd. Tela.* **3**, 1528.
Katiyar, R. S., Ryan, J. F., and Scott, J. F. (1971). *Phys. Rev. B* **4**, 2635.
Klemens, P. G. (1966). *Phys. Rev.* **148**, 845.
Klemens, P. G. (1975). *Phys. Rev. B* **11**, 3206–3207.
Kokkedee, J. J. J. (1962). *Physica* **28**, 374–408.
Krauzman, M., Pick, R. M., Poulet, M., Hamel, G., and Prevot, B. (1974). *Phys. Rev. Lett.* **33**, 528–530.
Kukharskii, A. A. (1973). *Solid State Commun.* **13**, 1761–1765.
Kukharskii, A. A. (1976). *Opt. Spectrosk.* **41**, 499–501.
Kurosawa, T. (1961). *J. Phys. Soc. Japan* **16**, 1298.
La Combe, J. L., and Irwin, J. C. (1970). *Solid State Commun.* **13**, 1761–1765.
Lagakos, N., and Cummins, H. Z. (1974). *Phys. Rev. B* **10**, 1063–1069.
Laulicht, I. (1978). *J. Phys. Chem. Solids* **39**, 901–906.
Lazarev, A. N., Mirgorodskii, A. P., and Ignatiev, I. S. (1975). "Kolebatelnie Spectri Slojnic Okislov." Nauka, Leningrad.
Le Toullec, R. (1968). Thesis, Paris.
Lines, M. E. (1969). *Phys. Rev.* **117**, 797–819.
Lines, M. E. (1970). *Phys. Rev. B* **2**, 690–698.
Lines, M. E. (1972). *Phys. Rev. B* **5**, 3690.
Lines, M. E., and Glass, A. M. (1977). "Principles and Applications of Ferroelectrics and Related Materials." Oxford Univ. Press, London and New York.
Lowndes, R. P. (1970). *Phys. Rev. B* **1**, 2754–2763.
Lowndes, R. P. (1972). *Phys. Rev. B* **6**, 1490–1498.
Lowndes, R. P., and Rastogi, A. (1976). *Phys. Rev. B* **14**, 3598–3620.
Lowndes, R. P., Tornberg, N. E., and Leung, R. C. (1974). *Phys. Rev. B* **10**, 911–931.
Luspin, Y., Servoin, J. L., and Gervais, F. (1978). *Solid State Commun.* **27**, 1101–1104.
Luspin, Y., Servoin, J. L., and Gervais, F. (1979). *J. Phys. Chem. Solids* **40**, 661–668.
Luspin, Y., Servoin, J. L., and Gervais, F. (1980). *J. Phys. C* **13**, 3761–3773.
Luspin, Y., Servoin, J. L., Gervais, F., and Quittet, A. M. (1981). *In* "Symmetries and broken symmetries in condensed matter physics (N. Boccara, ed.), IDSET, Paris.

Lyddane, L. H., Sachs, R. G., and Teller, E. (1941). *Phys. Rev. 59,* 673.

Maradudin, A. A. (1974). *In* "Dynamical Properties of Solids" (G. K. Horton and A. A. Maradudin, eds.), Vol. 1, pp. 1–82. North-Holland Publ., Amsterdam.

Maradudin, A. A., and Fein, A. E. (1962). *Phys. Rev.* **128,** 2589–2608.

Maradudin, A. A., Montroll, E. W., Weiss, G. H., and Ipatova, I. P. (1971). "Theory of Lattice Dynamics in the Harmonic Approximation." Academic Press, New York.

Matsushita, M. (1976). *J. Chem. Phys.* **65,** 23–28.

Mead, D. G., and Wilkinson, G. R. (1977). *J. Raman Spectrosc.* **6,** 123–129.

Merten, L., and Lamprecht, G. (1970). *Phys. Status Solidi B* **39,** 573.

Migoni, R., Bilz, H., and Bauerle, D. (1976). *Phys. Rev. Lett.* **37,** 1155–1158.

Miskevich, V. V. (1963). *Sov. Phys. Solid State* **4,** 2224–2232.

Mitra, S. S. (1969). *In* "Optical Properties of Solids" (S. Nudelman and S. S. Mitra, eds.), p. 333–453. Plenum, New York.

Mooij, J. E. (1969). *Phys. Lett. A* **29,** 111–112.

Mooradian, A., and Wright, G. B. (1966). *Phys. Rev. Lett.* **16,** 999–1001.

Müller, K. A., Luspin, Y., Servoin, J. L., and Gervais, F. (1982). *J. Physique-Lettres* **43,** L 537–42.

Olson, C. G., and Lynch, D. W. (1969). *Phys. Rev.* **177,** 1231–1234.

Park, K. (1967). *Phys. Lett. A* **25,** 490–491.

Peercy, P. S. (1975). *Phys. Rev. B* **12,** 2725–2740.

Penna, A. F., Chaves, A. S., and Porto, S. P. S. (1976a). *Solid State Commun.* **19,** 491.

Penna, A. F., Porto, S. P. S., and Chaves, A. S. (1976b). *In* "Light Scattering in Solids" (M. Balkanski, R. C. C. Leite and, S. P. S. Porto, eds.), p. 890. Flammarion, Paris.

Penna, A. F., Porto, S. P. S., and Wiener-Avnear, E. (1977). *Solid State Commun.* **23,** 377.

Perkowitz, S. (1975). *Phys. Rev. B* **12,** 3210–3214.

Perkowitz, S., and Thorland, R. H. (1974). *Phys. Rev. B* **9,** 545–550.

Perkowitz, S., and Thorland, R. H. (1975). *Solid State Commun.* **16,** 1093–1096.

Perry, C. (1971). *In* "Far Infrared Spectroscopy" (K. D. Möller and W. G. Rothschild, eds.), pp. 557–591. Wiley, New York.

Peyrard, M., Garaudet, J., and Remoissenet, M. (1975). *Solid State Commun.* **16,** 227–231.

Peyrard, M., and Perret, R. (1979). *Phys. Status Solidi B* **52,** 521–528.

Peyrard, M., and Remoissenet, M. (1979). *J. Chem. Phys.* **71,** 2732–2733.

Pine, A. S., and Tannewald, P. E. (1969). *Phys. Rev.* **178,** 1424.

Piriou, B. (1964). *C. R. Acad. Sci.* **260,** 841–844.

Piriou, B. (1968). PhD. thesis, University of Paris.

Portmus, C., Ferraro, J. R., and Mitra, S. S. (1968). *Phys. Rev.* **174,** 983–987.

Pohl, D. W., and Meier, P. F. (1974). *Phys. Rev. Lett.* **32,** 58–61.

Quittet, A. M., Servoin, J. L., and Gervais, F. (1981). *J. Physique* **42,** 493–498.

Rastogi, A., Hawrenek, J. P., and Lowndes, R. P. (1974). *Phys. Rev. B* **9,** 1938–1950.

Rastogi, A., Pai, K. F., Parker, T. J., and Lowndes, R. P. (1977). *In* "Lattice Dynamics" (M. Balkanski, ed.), pp. 142–143. Flammarion, Paris.

Roessler, D. M., and Albers, W. A. (1972). *J. Phys. Chem. Solids* **33,** 293–296.

Rupprecht, G., and Bell, R. O. (1964). *Phys. Rev. A* **135,** 748–752.

Safran, S., and Lax, B. (1975). *J. Phys. Chem. Solids* **36,** 753.

Sakurai, T., and Sato, T. (1971). *Phys. Rev. B* **4,** 583–591.

Samara, G. A., and Peercy, P. S. (1973). *Phys. Rev. B* **7,** 1131–1148.

Sanjurjo, J. A., Porto, S. P. S., and Silberman, E. (1979). *Solid State Commun.* **30,** 55–57.

Sanjurjo, J. A., Katiyar, R. S., and Porto, S. P. S. (1980). *Phys. Rev. B* **22,** 2396.

Scalabrin, S., Chaves, A. S., Shim, D. S., and Porto, S. P. S. (1977). *Phys. Status Solidi B* **79,** 731–742.

Scott, J. F. (1971). *Phys. Rev. B* **4**, 1360–1366.

Scott, J. F. (1974). *Rev. Mod. Phys.* **46**, 83–128.

Scott, J. F., and Porto, S. P. S. (1967). *Phys. Rev.* **161**, 903.

Semwal, B. S., and Sharma, P. K. (1974). *Prog. Theor. Phys.* **51**, 639–655.

Servoin, J. L., and Gervais, F. (1976). *High Temp. High Pressures* **8**, 557–563.

Servoin, J. L., and Gervais, F. (1977). *Appl. Optics* **16**, 2952–6.

Servoin, J. L., and Gervais, F. (1979). *Solid State Commun. 31,* 387–391.

Servoin, J. L., and Gervais, F. (1981). *In* "Symmetrics and broken symmetries" (N. Boccara, ed.). IDSET, Paris.

Servoin, J. L., Gervais, F., Quittet, A. M., and Luspin, Y. (1980a). *Phys. Rev. B* **21**, 2038–2041.

Servoin, J. L., Luspin, Y., and Gervais, F. (1980b). *Phys. Rev. B* **22**, 5501.

Sherman, W. F., and Wilkinson, G. R. (1980). *In* "Advances in Infrared and Raman Spectroscopy" (R. J. J. Clark and R. E. Hester, eds.), pp. 158–336. Heyden, London.

Siapkas, D. I. (1978). *In* "Lattice Dynamics" (M. Balkanski, ed), pp. 692–695. Flammarion, Paris.

Silverman, B. D., and Joseph, R. J. (1963). *Phys. Rev.* **129**, 2062–2068.

Singwi, K. S., and Tosi, M. P. (1966). *Phys. Rev.* **147**, 658–662.

Slater, J. C. (1967). *In* "Ferroelectricity" (E. F. Weller, ed.), pp. 1–8. Elsevier, Amsterdam.

Sparks, M., and Sham, L. J. (1972). *Solid State Commun.* **11**, 1451–1456.

Spitzer, W. G., and Kleinman, D. A. (1961). *Phys. Rev.* **121**, 1324–1335.

Spitzer, W. G., Miller, R. C., Kleinman, D. A., and Howarth, L. E. (1962). *Phys. Rev.* **126**, 1710–1721.

Steigmeier, E. F., Auderset, H., and Harbeke, G. (1975). *Phys. St. Solidi B* **70**, 705.

Strobel, K., and Geick, R. (1979). *J. Phys. C* **12**, 3855–3870.

Summit, R. J. (1968). *J. Appl. Phys.* **39**, 3762.

Szigeti, B. (1949). *Trans. Faraday Soc.* **45**, 452.

Takahashi, H. (1975). *Phys. Rev. B* **11**, 1636–1639.

Tripathi, R. S., and Pathak, K. N. (1974). *Nuovo Cimento B* **21**, 289–302.

Varga, B. B. (1965). *Phys. Rev. A* **137**, 1896–1902.

Venkataraman, G., Feldkamp, L. A., and Sahni, V. C. (1975). "Dynamics of Perfect Crystals." MIT Press, Cambridge.

Verble, J. L., Gallego-LLuesma, E., and Porto, S. P. S. (1978). *J. Raman Spectrosc.* **7**, 7–9.

Vinogradov, V. S. (1962). *Sov. Phys. Solid State* **4**, 519–522.

Wallis, R. F., Ipatova, I. P., and Maradudin, A. A. (1966). *Fiz. Tved. Tela.* **8**, 1064.

Wang, C. H., and Wright, R. B. (1973). *J. Chem. Phys.* **58**, 1411.

Yokota, J. (1961). *J. Phys. Soc. Japan* **16**, 2075.

Zawadovski, A., and Ruwalds, J. (1970). *Phys. Rev. Lett.* **29**, 1111.

Zernik, W. (1967). *Rev. Mod. Phys.* **39**, 432–439.

CHAPTER 8

Millimeter and Submillimeter Waves Interacting with Giant Atoms (Rydberg States)

P. Goy

Laboratoire de Physique de l'Ecole Normale Supérieure
Paris, France

Very excited states of one-electron atoms are simple systems with unusual characteristics: "giant" size and extreme sensitivity to millimeter and submillimeter waves. We present here a very simple theoretical approach together with some experiments in the alkalis sodium and cesium. High-resolution spectroscopy has been performed in the range 50–500 GHz. Blackbody detection is also observed at millimeter wavelengths. Finally, the Rydberg maser is a new source, or quantum amplifier, observed in the range 70–1500 GHz.

I. Introduction

Electron transitions between atomic levels usually involve optical photons. They also involve millimeter and submillimeter waves, especially if the
atoms are prepared in very excited states.

If the energy E is measured from the ionization limit, the hydrogen energy
spectrum $E(n)$ obeys the very simple formula

$$E(n) = -R_H/n^2,   \tag{1}$$

where n is the principal quantum number and R_H the Rydberg constant (in
frequency units $R_H \simeq 3288$ THz). The distance between successive levels
$\Delta E = |E(n) - E(n + 1)|$ varies as

$$\Delta E \propto n^{-3}   \tag{2}$$

and can be as small as desired, if n is large enough. One can see in Fig. 1 that
the millimeter domain concerns transitions between neighboring states for
$n \simeq 30$. Such excited atoms, very close to ionization, are called Rydberg
states.

A few years ago there began an intense study of the interaction of atoms
with millimeter waves (Fabre *et al.*, 1978). This research was made possible
by the availability of tunable lasers that permit the excitation of a well-
defined n level from the ground state.

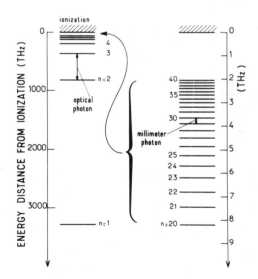

FIG. 1 Energy diagram of the hydrogen atom. Note the dramatic change in scale between
the ranges $1 \leq n \leq 7$ (left) and $20 \leq n \leq 40$ (right).

Because the diameter of the electron orbit varies as n^2, atoms with $n \simeq 30$ are about 1000 Å in diameter and can be called "giant atoms." Such atoms have a size very different from the usual atomic size (Fig. 2). These exotic species are very fascinating because of their simplicity. Two main features make them particularly valuable to study. First, their lifetime is very long, and as a consequence their natural width is very narrow, making possible a high-resolution spectroscopy (with potential applications to metrology). Second, they are very strongly coupled to the resonant millimeter waves, the transition probabilities being proportional to the square of the linear dimension of these giant atoms (making possible the application to sensitive detection and amplification).

In Section II we present the main characteristics of these excited states and explain how they couple to radiation.

In Section III we describe experimental aspects: preparation and detection of the Rydberg states, and the microwave setup.

In Section IV we present the spectroscopy of two alkali Rydberg states (sodium and cesium).

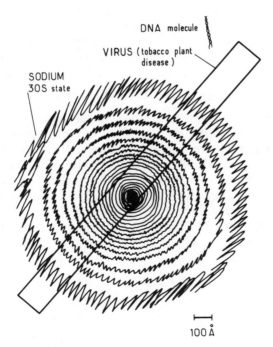

DNA molecule

VIRUS (tobacco plant disease)

SODIUM 30S state

⊢——⊣
100 Å

FIG. 2   Representation of the wave function extension of a sodium atom in the excited state 30S, together with biological objects at the same scale. (From Harocher, 1978.)

In Section V we present different kinds of experiments in which the interaction of Rydberg states with blackbody radiation has been studied.

In Section VI we present a new type of coherent source in the millimeter–submillimeter domain: the Rydberg maser–laser. Due to their extreme sensitivity to resonant radiation, Rydberg atoms can behave as self-oscillating systems with the number of atoms in the inverted population as small as 100 (this number could be further reduced). We present the study of such systems detected by monitoring the atomic population transfers and by looking at the emitted microwaves. In particular, a Rydberg maser triggered by a small impinging microwave can be considered as a coherent millimeter-wave amplifier that could be used to reduce the noise of coherent detectors close to the quantum limit.

II. One-Electron Rydberg Atoms

Let us consider hydrogen-like atoms (one valence electron). Although the study of the very high angular momentum states $l \simeq n$ seems very promising, we restrict ourselves to the case of small angular momentum $l \ll n$, because most of the experiments already performed deal only with this last case. (In fact large l states are difficult to prepare and to detect and are extremely sensitive, in particular to stray electric fields.)

Table I presents different physical properties of Rydberg atoms with $n \sim 30$ compared with the corresponding properties of ground-state atoms.

A. Rydberg Atoms Are Long-Lived Excited States

In the case of large n quantum numbers the classical approximation of the electron orbit becomes fairly realistic. Much of the behavior of very excited states can be understood from simple considerations about the corresponding classical trajectories. Consider an elliptical orbit (Fig. 3) representing the "path" of the electron in the state $|n,l\rangle$ of hydrogen. The major axis has a length of $2n^2 a_0$ where a_0 is the Bohr radius $a_0 = 0.53$ Å, and the minor axis is $2l\,(l+1)a_0$, related to the angular momentum l.

According to Kepler's third law the electron period is proportional to the $\frac{3}{2}$ power of the major axis, so the electron rotation frequency v varies as $(2n^2 a_0)^{-3} \propto n^{-3}$. Thus

$$v \simeq 2R_{\mathrm{H}} n^{-3}. \tag{3}$$

Equation (3) is naturally the classical limit of Eqs. (1) and (2).

The radiated power of the Rydberg electron is proportional to the square of its acceleration. In a very elliptical orbit ($l \ll n$) the acceleration is a maximum at perihelion, practically independent of the major axis (i.e., of n). Thus the electron radiates most when it is close to the nucleus, and the

TABLE I

PHYSICAL PROPERTIES OF RYDBERG ATOMS WITH PRINCIPAL QUANTUM NUMBER
AROUND 30, COMPARED WITH THE CORRESPONDING PROPERTIES OF
GROUND-STATE ATOMS[a]

Property	n dependence	Order of magnitude	
		$n \simeq 30$	$n \simeq 30$
Energy from ionization	n^{-2}	10 eV	10 meV
Frequency ν of the transition between successive levels	n^{-3}	500 THz	100 GHz
Corresponding wavelength, $\lambda = c/\nu$	n^3	6000 Å	3 mm
Fine structure	n^{-3}	1 THz	250 MHz
Orbital diameter	n^2	1 Å	1000 Å
Electric dipole between two excited states, $n' \simeq n$	n^2	1 Å × e⁻	1000 Å × e⁻
Electric dipole between an excited state and a low-lying level, $n' \simeq 1$	n^{-3}	1 Å × e⁻	10^{-4} Å × e⁻
Lifetime	n^3	1 nsec	30 μsec
One-photon transition probability	n^4	—	—
Stark effect	n^7	0.3 kHz/(kV/cm)²	10 MHz/(V/cm)²
Ionization electric field	n^{-4}	10^8 V/cm	10^2 V/cm

[a] From Fabre (1982).

radiative decay rate $\Gamma_{n,l}$ (for $l \ll n$) is proportional to the rate of the electron passage at perihelion $\Gamma_{n,l} \propto n^{-3}$. This explains the observed variation of the lifetime

$$\tau = An^3. \tag{4}$$

This law gives a very long lifetime (Gounand, 1979) for these excited states ($\tau = 32$ μsec for $30S$ in sodium).

From the perspective of very high resolution spectroscopy, it could be interesting to try to measure the resonances of states with very large quan-

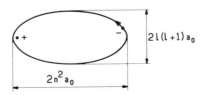

FIG. 3   The planetary model of the electron orbit in hydrogen. In highly excited states ($n \simeq 30$) with low angular momentum ($\ell \ll n$) the eccentricity is very high. In the representation of a $30S$ state the small axis has been enlarged 100 times.

tum number. But in fact, any transition between successive levels with $\Delta n = 1$ will give the same intrinsic quality factor $2\pi\nu\tau$ because $\nu \propto n^{-3}$, Eq. (3), and $\tau \propto n^3$, Eq. (4). Numerically,

$$2\pi\nu_{n\rightarrow n+1}\tau_{nl} \simeq 10^7 - 10^8. \qquad (5)$$

The Rydberg constant R_H is presently known to the relative accuracy $\pm 3 \times 10^{-9}$. Careful measurements of millimeter resonance between excited states of hydrogen should permit us to determine the Rydberg constant from Eq. (1),

$$R_H = \nu_{n\rightarrow n'}(n^{-2} - n'^{-2})^{-1}. \qquad (6)$$

In this way R_H can be determined directly in frequency units (Hz) instead wavelength units (cm^{-1}).

A comparison with deuterium will permit the precise determination of the neutron/proton mass ratio, because the Rydberg constant for a given atom X is

$$R_X = R_\infty(1 + m_{e^-}/m_{X^+})^{-1}, \qquad (7)$$

where m_{e^-} is the electron mass and m_{X^+} the mass of the ionized atom around which the excited electron is revolving.

B. ALKALI ATOMS ARE VERY HYDROGEN-LIKE

Alkalis have a single valence electron. All properties given for hydrogen are roughly valid for alkalis, the main difference being that the energy levels are not l-degenerate in alkalis.

The energy spectrum of alkalis, very similar to that of hydrogen, is given by the formula

$$E(n, l) = -R_X[n - \varepsilon_l(n)]^{-2} = -R_X n^{*-2} \qquad (8)$$

where the term $\varepsilon_l(n)$ is the quantum defect and $n^* = n - \varepsilon_l(n)$ the effective quantum number. For a given alkali X the quantum defect strongly depends upon l and only very slightly upon n (see Section IV). In sodium, for instance, $\varepsilon_s \simeq 1.348$, $\varepsilon_p \simeq 0.855$, $\varepsilon_D \simeq 0.015$, and $\varepsilon_F \simeq 0.002$. For a given l, ε_l depends upon the alkali and increases with the atomic number. For instance, $\varepsilon_s \simeq 1.348$ and 4.05, $\varepsilon_p \simeq 0.855$ and 3.57 in sodium and cesium, respectively.

To some extent the origin and properties of quantum defects can be understood from the planetary model (Fig. 4). They can also be explained fairly well within the frame of a WKB approximation. Both models indicate that the small l states have larger quantum defects and naturally lead to

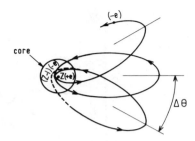

FIG. 4 Electron orbits in alkalis precess because the attractive charge cannot be reduced to a point as in hydrogen.

developing the quantum defect ε_l in powers of the binding energy,

$$\varepsilon_l(n) = \varepsilon_l(\infty) + \alpha_l E(n) + \beta_l E^2(n) + \cdots. \tag{9}$$

According to Eq. (8) this is equivalent to developing the quantum defect $\varepsilon_l(n)$ in even powers of the inverse of the effective quantum number n^*,

$$\varepsilon_l(n) = \varepsilon_l(\infty) + a_l n^{*-2} + b_l n^{*-4} + \cdots. \tag{10}$$

In our range of interest ($n > 20$) the high-power terms n^{*-4}, n^{*-6}, . . . are negligible (in sodium, for instance, $a_s \simeq 0.06$, $b_s \simeq 0.007$, $c_s \simeq 0.006$), and the variation of quantum defects needs high-resolution spectroscopy to become evident (in sodium (Goy, 1982) $\varepsilon_s(20) = 1.348146$ and $\varepsilon_s(45) = 1.348001$, for instance).

C. COUPLING TO RADIATION†

1. Single-Photon Transitions

The dipolar matrix element between two levels nl and $n'l'$ involved in a resonant transition at the frequency $\nu_{nl \to n'l'}$ is given by

$$R_{nl}^{n'l'} = \int \psi_{nl}^*(r) r \psi_{n'l'}(r) r^2 \, dr. \tag{11}$$

In order to evaluate this integral we must consider two very different cases.

(a) An excited state ($n \gg 1$) coupled to a low-lying state ($n' \simeq 1$). The two wave functions ψ_{nl} and $\psi_{n'l'}$ (Fig. 5a, b) have a very weak overlapping, and the dipolar matrix element is very small and decreases with n,

$$R_{nl}^{n'l'} \propto n^{*-3/2}. \tag{12}$$

(b) An excited state ($n \gg 1$) coupled to a nearby state ($n' \simeq n$). The two wave functions (Fig. 5c and d) have a very strong overlapping, extending over the full r range, so that $R_{nl}^{n' \simeq n, l'}$ is proportional to n^{*2}. However, the wave functions can constructively or destructively interfere, their phase at the origin being related to their effective quantum number n^*. For that reason,

† See Fabre (1982).

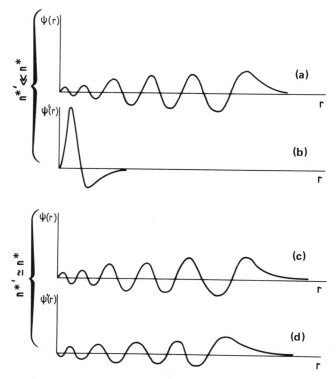

FIG. 5 Comparison between wave functions; (a) excited state, (b) ground state, (c) and (d) excited states.

one has

$$R_{nl}^{n'l'} \simeq Dn^{*2}a_0 G(\Delta n^*), \tag{13}$$

where $D = 0.707$ for a $nS \rightarrow n'P_{1/2}$ transition and $D = 1$ for a $nS \rightarrow nP_{3/2}$ transition, and where the function $G(\Delta n^*)$, given in Fig. 6, is a damped oscillatory term that depends upon the quantum-defect difference $\Delta n^* = |n^* - n'^*|$.

The transition probability per time unit $W_{nl \rightarrow n'l'}$ induced by resonant radiation having the spectral density $d\phi/dv$ is given by the following equation (two times greater in the case of an $nS \rightarrow nP_{3/2}$ transition):

$$W_{nS \rightarrow n'P_{1/2}} = 2.03 \times 10^{11} \frac{\max(l, l')}{2l + 1} n_c^{*4} G^2(\Delta n^*) \frac{d\phi}{dv}, \tag{14}$$

where n_c^* is the "average" effective quantum number $n_c^* = 2n^*n'^*/(n^* + n'^*)$, $W_{nl \rightarrow n'l'}$ is in sec^{-1}, and $d\phi/dv$ is in W/m^{2}Hz.

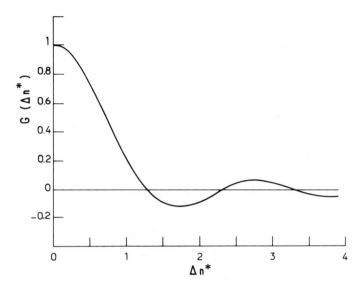

FIG. 6   Variation of the coefficient $G(\Delta n^*)$ as a function of Δn^*.

For example, the transition $30S \rightarrow 29P_{3/2}$ in sodium ($v \simeq 146$ GHz, $hv \simeq 0.6$ meV $\simeq 10^{-19}$ J) gives $\Delta n^* = 30 - 1.348 - (29 - 0.855) \simeq 0.5$; so $G(\Delta n^*) \simeq 0.72$ (Fig. 6), $n_c^* \simeq 28.4$, and $W_{30S \rightarrow 29P_{3/2}} \simeq 1.3 \times 10^{17} \, d\phi/dv$. For an effective bandwidth $dv \simeq 1$ MHz we compute a very small saturating flux $\phi = 7 \times 10^{-12}$ W/cm^2.

Another way of looking at this extremely small flux is to express it in terms of the number of photons impinging on the system per unit of surface, of the order of $\lambda^2 (\simeq 4$ mm^2), and for the lifetime $\tau \simeq 30$ μsec. The total energy of photons crossing the surface λ^2 during the time τ is $d\phi \times \tau \times \lambda^2 = 5 \times 10^{-18}$ J. If each photon has the energy $hv = 10^{-19}$ J, this total energy corresponds to only about 100 photons.

2. Two-Photon Transitions

The two-photon transition probability for radiation of frequency v given by Eq. (3) is

$$W_{nl \rightarrow n'l'} = 16\pi^2\alpha^2 \left| \sum_{n''l''} \frac{\langle nl|r|n''l'' \rangle \langle n''l''|r|n'l' \rangle}{E_{nl} - E_{n''l''} - hv} \right|^2 \frac{\phi^2}{\Gamma}, \qquad (15)$$

where α is the fine structure constant, ϕ the incident flux, and Γ the width of the observed transition, assuming the source is spectrally narrower than Γ. The summation is performed over all possible intermediate levels nl.

Alkalis have in fact quantum-defect differences $\varepsilon_s - \varepsilon_p$ very close to $\frac{1}{2}$

($\simeq 0.49$ for $\varepsilon_s - \varepsilon_{p_{1/2}}$ and $\varepsilon_s - \varepsilon_{p_{3/2}}$ in Na; $\varepsilon_s - \varepsilon_{p_{1/2}} \simeq 0.46$ and $\varepsilon_s - \varepsilon_{p_{3/2}} \simeq 0.49$ in Cs). Under these conditions the intermediate energy level $(n + q)P$ for an $nS \rightarrow (n + 2q + 1)S$ transition is very close to the middle of the interval $nS - (n + 2q + 1)S$. For this reason such two-photon transitions are very easy to induce. The frequency difference

$$\delta v = \left| \tfrac{1}{2} v_{nS \rightarrow (n+1)S} - v_{nS \rightarrow nP} \right| \tag{16}$$

can be as small as 984 MHz with the relay level $32P_{1/2}$ and 1162 MHz with the relay level $32P_{3/2}$ for the transition $32S \rightarrow 33S = 2 \times 108.876$ GHz in sodium. Both levels, $32P_{1/2}$ and $32P_{3/2}$, are effective intermediate levels. In cesium, on the contrary, only the $nP_{3/2}$ level can be considered to be an effective intermediate level. For instance, the transition $38S \rightarrow 39S$ in cesium (2×80.4963 GHz) gives $\delta v = 168$ MHz for the $38P_{3/2}$ level but $\delta v = 5073$ MHz for the $38P_{1/2}$ level.

Equation (15) gives the numerical expression

$$\phi = An_c^{*-4}\delta v^*, \tag{17}$$

where δv^* is the effective frequency difference

$$\delta v^{*-1} = \sum_{n''l''} \delta v^{-1}(n''l''). \tag{18}$$

For $nS \rightarrow (n + 1)S$ transitions, $\delta v^* \simeq \delta v^{-1}(P_{1/2}) + \delta v^{-1}(P_{3/2})$ in sodium and $\delta v^{*-1} \simeq \delta v^{-1}(P_{3/2})$ in cesium. With typical values $\Gamma = 100$ kHz and an interaction time 5 μsec one obtains $A \simeq 10^{-4}$ W/cm^2 MHz. This corresponds, for instance, with the $36S \rightarrow 37S$ transition of sodium (2×75.76912 GHz) to a very small predicted saturating flux $\phi \simeq 2 \times 10^{-8}$ W/cm^2.

III. Experimental Aspects

A. EXCITATION AND DETECTION OF RYDBERG ATOMS†

The experiment is performed on an alkali atomic beam (Na or Cs) effusing from an oven ($\simeq 350°$C) and propagating in a vacuum chamber with a background pressure of $\simeq 10^{-6}$ torr (Fig. 7).

From the ground state ($3S$ or $6S$) the atoms are excited by two pulsed-dye lasers to an nS or nD state (called i, initial state n,l, in Fig. 7) by a stepwise process via the intermediate $3P_{3/2}$ (Na) or $6P_{3/2}$ (Cs) level. These two dye lasers (5896 Å and $\simeq 4110$ Å for Na, 8521 Å and $\simeq 5100$ Å for Cs) are excited by the same pulsed pump laser (N$_2$ or Nd-YAG laser) at a repetition rate of 5–10 pps. The first dye laser saturates the transition $3S \rightarrow 3P_{3/2}$ (or

† See Fabre et al. (1978).

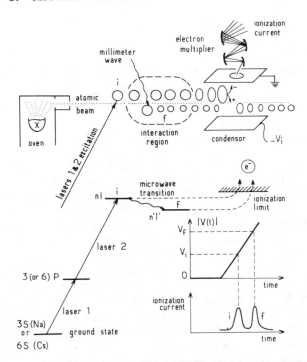

FIG. 7 Schematic diagram of the preparation and detection of the Rydberg states.

$6S \rightarrow 6P_{3/2}$) with less than 1 kW/mm². The second allows us to excite a maximum of $\simeq 10^6$ atoms per pulse.

After pulsed-laser excitation the Rydberg states are coupled in the interaction region (see Fig. 7) with a millimeter wave connecting the initial state i to the final state f. The interaction lasts for a variable length of time between 1 and 100 μsec.

After this interaction the atoms pass between the plates of a condensor. A time-varying electric field ramp is applied to the atoms. The field ionization can be roughly expressed (Fabre, 1982) as:

$$F_c(\text{V/cm}) \simeq 3.2 \times 10^8 n^{*-4}. \tag{19}$$

Therefore each Rydberg level is ionized for a characteristic value of the ionizing field, i.e., at characteristic time. For instance, in Fig. 7 the initial state i is ionized at the voltage V_i that is smaller than V_f, the voltage necessary to ionize the final state f, which is more strongly bound. The electrons produced are collected and amplified in an electron multiplier. This well-known field ionization technique (Fabre *et al.*, 1978) is selective and very sensitive: a single Rydberg atom gives a macroscopic current.

B. THE MICROWAVE SETUP

1. *The Carcinotron Sources*

As microwave sources we have used the Carcinotrons described in Table II, in which the indicated microwave power is the maximum obtained for each carcinotron; the average available power in the band is about 50% of this value (see Fig. 8).

2. *Carcinotron Frequency Measurement*

The frequency F_c of a Carcinotron (Fig. 8) is determined by the line voltage applied to it (Fig. 9). In order to count F_c and to stabilize it we used as a reference an X-band klystron (X-13) locked to a quartz oscillator (MOS-5 microwave system), as shown in Fig. 10. The frequency of the klystron F_k is measured directly on a counter, the clock reference of which comes from a long-wave receiver for a standard frequency with an accuracy better than 10^{-9} (Adret 4101 A). A wide-band Schottky diode harmonic mixer (type HM, Siemel or Micro Now) generates the pth harmonic of the klystron ($5 \le p \le 54$, see Fig. 11) and mixes it with the Carcinotron radiation. The beat note signal at the intermediate frequency $F_{if} = |F_c - pF_k|$ is amplified in the interval 0–250 MHz (low-noise amplifier Trontech W 110F) and is counted and monitored on a spectrum analyzer (HP 14IT, 8552B, 8553B).

Above the cutoff of the waveguide the same harmonic mixer is used for all frequencies (Goy, 1982); for instance, above 80 GHz we use the one with an F-band (90–140 GHz) waveguide (Siemel, Micro Now HM–F). Naturally the conversion loss increases with increasing harmonic mixing number p. The conversion loss expresses the ratio in dB of the millimeter input power

TABLE II

THOMSON–C.S.F. CARCINOTRONS USED IN OUR EXPERIMENTS

TYPE CO XX, where $XX = \lambda$ (in 0.1 mm units)	Frequency band (GHz)	Maximum power
CO 60	51–61	130 mW
CO 40	75–77	190 mW
CO 38	77–82	4 W
CO 33	88–98	200 mW
CO 30	91–100	200 mW
CO 25	107–125	1 W
CO 21	125–151	1 W
CO 12	226–274	600 mW
CO 11	265–279	1 W
CO 10	277–313	100 mW
CO 06	417–485	60 mW

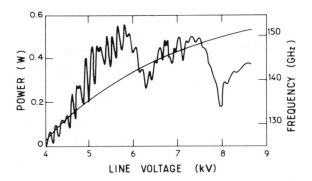

FIG. 8 Typical output of a CO 21 Thomson–C.S.F. Carcinotron.

at F_c to the low-frequency output power at F_{if} when the harmonic mixer is optimally biased and fed with a reasonable L.O. centimeter power at F_k (≤ 100 mW). The measured conversion loss goes from 25 dB with $p = 7$ at $F_c = 80$ GHz up to 85 dB with $p = 38$ at $F_c = 460$ GHz. In the interval 100–500 GHz the conversion loss has approximately the value (in dB) of $45 + 0.125[F_c(\text{GHz}) - 100]$. Nevertheless the signal-to-noise ratio is still

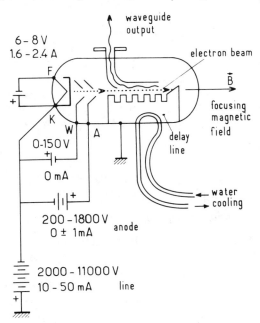

FIG. 9 Carcinotron power supply (Siemel Co.). The line voltage determines the frequency (see Fig. 8), and the anode voltage determines the microwave power.

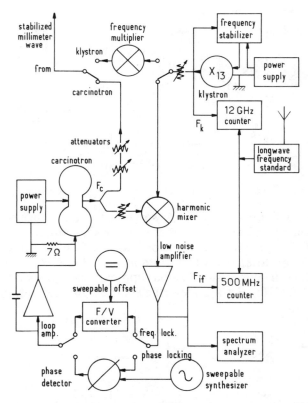

FIG. 10 General schematic diagram of the stabilized microwave source, which can be either the Carcinotron (phase- or frequency-locked on the stabilized klystron) or a harmonic of the stabilized klystron.

reasonable at the upper frequency (> 20 dB in a 100-kHz bandwidth) even for the very high harmonic mixing number $p = 54$ (Fig. 11).

3. *Carcinotron Frequency Stabilization*

When passively powered by a well-stabilized high-voltage power supply (manufactured by Siemel Co.) ($V = 2 - 11$ kV, $\Delta V < \pm 0.1$ V), the linewidth of the millimeter-wave source is typically of the order of 0.5 MHz for $F_c \sim$ 460 GHz, due mainly to the 50 Hz residual component, around 0.1 V peak-to-peak. The typical frequency drift is several MHz in a few minutes.

Because the klystron frequency F_k is stabilized, the beat frequency F_{if} varies in the same way as the millimeter frequency F_c and can be used for frequency locking or phase locking of the Carcinotron. In the frequency-

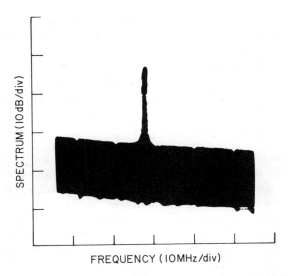

FIG. 11 Example of the performance of the Siemel or Micro Now HM–F harmonic mixer (Goy, 1982). Spectrum analysis of the beat around 60 MHz with 100-kHz bandwidth. The Carcinotron frequency is 460 GHz. The klystron frequency is 8.45 GHz. The harmonic mixing order p is thus particularly high: $p = 54$.

locking position (Fig. 10) a frequency/voltage converter gives a voltage directly proportional to the frequency F_{if}. This voltage is compared with a sweepable offset voltage, and their difference is sent to an amplifier that applies the correction voltage between the ground of the Carcinotron and the ground of the power supply. This feedback, applied on the "cold" side of the high voltage, is easy to operate, but it implies in particular the need to insulate the waveguide of the Carcinotron from the grounded waveguide of the experiment. The capacitance between the Carcinotron and the ground is rather large, due mainly to the water cooling of the tube. Nevertheless the frequency cutoff of the correction voltage is greater than 100 kHz when using the 7-Ω charge. In fact the amplifier works as an integrator (large gain for the dc correction) and generally has a cutoff around 100 kHz or less. When frequency-stabilized the Carcinotron linewidth is reduced to a few kHz. The stability-versus-time characteristic (better than ± 1 kHz within minutes) and the tunability (1–15 MHz) of the loop are excellent. The signal-to-noise ratio measured on the spectrum analyzer with 3-kHz bandwidth roughly obeys the approximate law $S/N(\text{dB}) \simeq 50 - 20 \log p$, where p is the integer for harmonic mixing between the Carcinotron frequency F_c and the klystron frequency F_k so that $p \simeq F_c/F_k$. This gives, for instance, the values $S/N \simeq 18$ dB at 460 GHz ($p = 38$) and $S/N \simeq 33$ dB at 80 GHz ($p =$

7) by comparison of the maximum signal with the sideband peaks at $\sim \pm 100$ kHz from the center (phase-noise rise just above the frequency limit of the loop feedback).

Phase locking can be performed in a way similar to frequency locking, reducing the linewidth to very low values (a few Hz). However, the long-term stability is not so good, and the phase-noise problem increases dramatically above 200 GHz.

4. *The Harmonic Generator Source*†

Carcinotrons can be very powerful (4 W at 80 GHz), and microwave leaks sometimes make it difficult to desaturate the most sensitive transitions for which power in the nW range is sufficient. Therefore instead of the Carcinotrons we have also used the harmonic mixer (Siemel, Micro Now HM–F) working as a frequency multiplier (top of Fig. 10) to generate millimeter waves. The harmonics $F = pF_f$ of the klystron frequency F_k are available at the millimeter waveguide output of the frequency multiplier as soon as their frequency is larger than the cut-off frequency of the waveguide. Starting with 100 mW from the klystron near 11.5 GHz, we have obtained the following optimized microwave powers: 0.1 mW at 80 GHz, 10 μW at 115 GHz, 1 μW at 230 GHz, and so on (Goy, 1982). The available power follows a linear decrease $P(\mathrm{dBm}) \simeq -2.1 \, p$. For a given harmonic p, reducing this microwave power is easily achieved by reducing the klystron power. Practically all single-photon transitions with $F \leq 340$ GHz ($P \geq 1$ nW) and most of the two-photon transitions with $F \leq 110$ GHz ($p \leq 10 \, \mu$W) can be obtained with this very simple arrangement.

5. *The Interaction Region*

In the interaction region, shown in Fig. 12, the microwave is propagated from the millimeter source through E-band (60–90 GHz) waveguides in a guided transverse electric mode valid from the frequency cutoff $\simeq 50$ GHz up to $\simeq 120$ GHz. For frequencies in the range 120–500 GHz the wave propagates without prohibitive attenuation in the oversized E-band waveguide.

Two setups are used for the coupling of the microwave to the atoms. In the horn configuration (Fig. 12a), the atoms are coupled to the progressive wave coming from a horn at the end of the waveguide. Section S of the interaction zone is of the order 3 cm². This very simple configuration permits the use of any frequency without any specific adjustment. In the cavity configuration (Fig. 12b), Rydberg atoms are coupled to a standing wave inside a Fabry–Perot semiconfocal cavity. This configuration involves precise tuning for each frequency but presents many advantages:

† See Goy (1982).

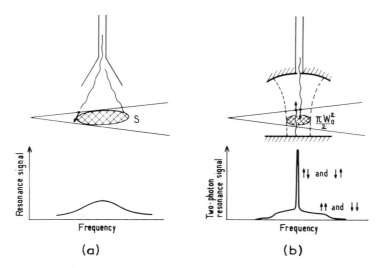

FIG. 12 The interaction region (a) in a progressive wave after a horn and (b) in a standing wave inside a cavity.

(i) The wave vector of the microwave is along the axis of the cavity and can be chosen perpendicular to the average beam velocity so that the Doppler effect is reduced.

(ii) For a two-photon resonance the standing wave permits a Doppler-free profile of the resonance to be obtained (Fig. 12b, bottom). The photon directions can be combined in four configurations, two of which are such that the Doppler effect of the first absorbed photon, propagating from down to up, is exactly compensated by the Doppler effect of the second photon, propagating from up to down.

(iii) The waist w_0 of the Gaussian electromagnetic field mode is given by

$$w_0 = \sqrt{r\lambda/2\pi}, \tag{20}$$

where r is the radius of curvature of the spherical mirror. The surface S in which the electromagnetic field is concentrated is proportional to the square of w_0,

$$S = \pi w_0^2/2 = r\lambda/4. \tag{21}$$

When S is smaller than 3cm^2, the value obtained with the horn configuration, a smaller microwave power coming from the millimeter source is enough to saturate the transition. Moreover, the cavity resonance makes the power flux ϕ enhanced by a factor proportional to the cavity finesse f. In an antinode position of the standing wave,

$$\phi_{\text{cav max}} = (4/\pi)f\phi_{\text{free space}}. \tag{22}$$

The cavity finesse is simply related to the intensity reflection coefficient R by

$$f = \pi\sqrt{R}/(1 - R) \sim \pi/(1 - R), \tag{23}$$

and the quality factor Q is defined as

$$Q = v/\Delta v, \tag{24}$$

Δv being the total frequency width where the microwave transmitted through the cavity, tuned at the frequency v, drops by -3 dB. Thus

$$f = Q/q, \tag{25}$$

where q is the integer giving the mode number, approximately the number of half wavelengths between the mirrors,

$$q \simeq 2L/\lambda \simeq r/\lambda, \tag{26}$$

where L is the distance between mirrors; close to the focal length $L \simeq r/2$.

   (iv)   Fabry–Perot-type cavities are very convenient for large transverse access. The semiconfocal case is well-suited to millimeter microwaves, because it permits one to reduce the diffraction losses with a transverse size $\simeq 10\lambda$. Moreover, it does not need any careful parallel alignment.

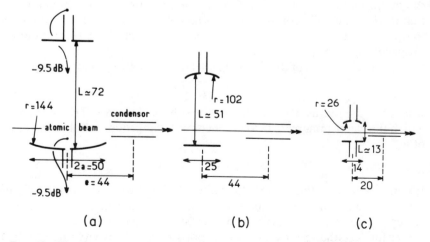

(a)   (b)   (c)

FIG. 13 The three Fabry–Perot semiconfocal cavities used in these experiments. All dimensions are in mm.

Three cavities (Fig. 13) have been used. Around the focal length $L \simeq r/2$ (Eq. 26) the resonance condition is

$$L = (\lambda/8)(4q + 1 + m + n), \qquad (27)$$

where q, m, and n are integers.

The transverse diameter $2a$ of the mirrors is chosen to increase the diffraction losses of the transverse modes $m + n \neq 0$ without damping the longitudinal q mode. The practical value is

$$a^2/r\lambda \simeq 1. \qquad (28)$$

A coupling hole is drilled at the center of each mirror of the cavity. The critical diameter of the hole is around 0.3 λ. A greater diameter reduces the Q cavity; a smaller diameter increases the loss T of the microwave transmitted through the cavity. Cooling the mirrors (Fig. 14) enhances the quality factor Q of the cavity. Table III summarizes all properties of the cavities we have used. Notice that these cavities have also been used for all the Rydberg maser studies (see Section IV).

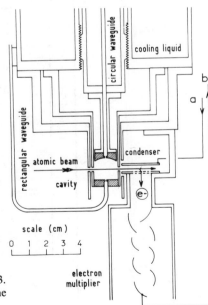

FIG. 14 Detailed drawing of cavity number 3. The electric ramp voltage is applied at a. The ionization current is collected at b.

TABLE III

THE THREE FABRY–PEROT SEMICONFOCAL CAVITIES USED IN OUR EXPERIMENTS

	General features						At the working frequency							
Cavity number	Optimal frequency band (GHz)	Material	Temperature (K)	Focal length $L \simeq r/2$ (mm)	Transverse diameter $2a$ (mm)	Coupling holes diameter (mm)	ν (GHz)	Q	q	f	Losses $T/2$ (dB)	Waist w_0 (mm)	Surface $\pi/2w_0^2$ (mm²)	Gain G compared with the horn (Fig. 12a) (dB)
1	50–120	Brass	Room	72	50	1.3	108	6,100	52	120	9.5	8	100	14
2	170–340	Copper	Room	51	25	0.3	256	110,000	84	1300	25	4	29	20
3	125–250	Electro-polished copper	77 K	13	14	0.5	145	20,000	13	1600	18	3	13	25

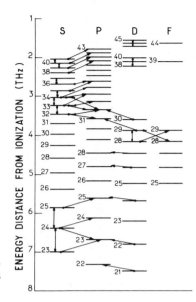

FIG. 15 Microwave transitions induced in sodium. The observed fine structure of the P levels is not represented.

IV. Millimeter Spectroscopy in Rydberg States of Sodium† and Cesium‡

A. SODIUM EXPERIMENTS

1. Introduction

The transitions observed in Na are summarized in Fig. 15. All were obtained using microwaves from the Carcinotrons of Table 2. At the beginning (Fabre *et al.*, 1978) the spectroscopy of sodium was performed with a frequency determination obtained by measuring the wavelength $\lambda = c/\nu$ of the Carcinotron with a limited accuracy $\Delta\lambda/\lambda \simeq 10^{-4}$. No variation of the quantum defects was thus detected. Then we used the digitized determination of the Carcinotron frequency ν (see Section III.C.1). According to Eq. (8) we have at resonance

$$\nu = R_X\{[n - \varepsilon_l(n)]^{-2} - [n' - \varepsilon_{l'}(n')]^{-2}\}. \tag{29}$$

2. Two-Photon Resonances§

The two-photon $nS \rightarrow (n + 1)S$ transitions performed in a cavity (as described in Section II.D.2) yielded a frequency determination in the 10^{-8}

† See Fabre *et al.* (1978), Goy *et al.* (1980), Fabre *et al.* (1980).
‡ See Goy *et al.* (1982).
§ See Goy *et al.* (1980).

TABLE IV

Quantum Defect Coefficients Obtained from Our High-Resolution Spectroscopy Results [†]

R_x(MHz)	Sodium		Cesium	
	3 289 763 408 (20)		3 289 828 299 (20)	
	$\epsilon_{ij}(\infty)$	a_{ij}	$\epsilon_{ij}(\infty)$	a_{ij}
$nS_{1/2}$	1.3479692 (4)	0.06137 (10)	4.049325 (15)	0.2462 (50)
$nP_{1/2}$	0.855424 (6)	0.1222 (2)	3.591556 (30)	0.3714 (40)
$nP_{3/2}$	0.854608 (3)	0.1220 (2)	3.559058 (30)	0.3740 (40)
$nD_{3/2}$	0.0149013	−0.042472	2.475365 (20)	0.0554 (60)
$nD_{5/2}$			2.466210 (15)	0.0670 (50)
$nF_{5/2}$	0.001629	−0.006850	0.033392 (30)	−0.191 (30)
$nF_{7/2}$			0.033537 (25)	−0.191 (20)

[†] The D and F coefficients in sodium are from Martin (1980). Otherwise see Goy et al. (1980) and Fabre et al. (1980) for sodium, Goy et al. (1982) for cesium.

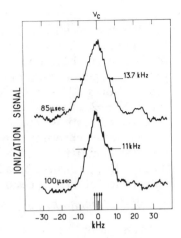

FIG. 16 Experimental traces of the two-photon resonance $39S \rightarrow 40S = 2 \times (59{,}260{,}222 \pm 2)$ kHz in sodium. Increasing the interaction time (given at left) permits a proportionnal decrease in the linewidth, showing that the sharpness is transit-time limited. The vertical arrows indicate the hyperfine structure, unresolved.

range and very precise values of $\varepsilon_s(n)$, exhibiting the variation of this parameter with n (see Table IV). The saturation power obeys the laws given by Eqs. (16), (17), and (18) with the experimental value $A \simeq 5 \times 10^{-4}$ W/cm^2 MHz, in relatively good agreement with the calculated value $A \simeq 10^{-4}$ W/cm^2 MHz (absolute power measurements are difficult).

Figure 16 gives an example of a two-photon resonance. Observe that the experimental width is inversely proportional to the interaction time. The lower trace represents the narrowest electron resonance ever observed in alkalis. The quality factor $v/\Delta v \simeq 5 \times 10^6$ is not too far from the maximum observable value $10^7 - 10^8$ (Eq. 5).

3. Single-Photon Resonances†

Observation of one-photon transitions $nS \rightarrow n'P_{1/2,3/2}$ yielded the determination of $\varepsilon_{p_{1/2}}(n)$ and $\varepsilon_{p_{3/2}}(n)$ variations from the previously determined value of $\varepsilon_s(n)$ (see Table IV).

Measurement of the saturating power on single-photon transitions induced by Carcinotrons is extremely difficult. Without a waveguide between the Carcinotron and the interaction region, and even with a closed waveguide at the output of the Carcinotron, microwave leaks are sufficient to saturate most of the transitions! The order of magnitude of the extremely low saturation power given by Eq. (14) is confirmed.

B. CESIUM EXPERIMENTS‡

1. Cesium

In cavity mounting No. 1 (Fig. 13), previously used for sodium, the

---

† See Fabre *et al.* (1980).
‡ See Goy *et al.* (1982).

distance between the interaction region (inside the waist of the cavity) and the detection region (at the middle of the condensor plates) is rather large ($e = 44$ mm). With the oven at around 350°C, the average thermal velocity of Na atoms in the beam is of the order of 600 m/sec, and the time necessary to cross a length $e = 44$ mm is of the order of 70 μs. Cesium atoms have a heavier mass ($M = 133$) than sodium, so their thermal velocity is $\sqrt{133/23} \simeq 2.4$ times smaller, of the order of 250 m/sec, and the time of flight over 44 mm is 180 μsec, far longer than the excited lifetime. For this reason the first experiments on cesium have been performed in a progressive wave (Section IIIB5a) with a typical interaction time around 5 μsec. Notice that the Doppler-effect broadening, of the order of 350 kHz in sodium, is reduced to 150 kHz in cesium due to the decreased beam velocity.

2. Single-Photon Resonances

As an example, consider the transition $23S \rightarrow 23P$ (Fig. 17a) in cesium.

It is split into two fine-structure components $23S \rightarrow 23P_{1/2}$ and $23S \rightarrow 23P_{3/2}$ (Fig. 17b) that are very far from each other due to the strong relativistic effects in the heavy cesium. (The distance $AB = 23P_{1/2} - 23P_{3/2} = 29.172$ GHz in cesium can be compared with $23P_{1/2} - 23P_{3/2} = 494$ MHz in sodium).

Also, the $23S_{1/2}$ level is divided into two hyperfine-structure components $F = 3$ and $F = 4$ (Fig. 17c), and the observed resonances A and B are clearly split into distinct components $C-D$ and $E-F$, with $CD = EF = (23S_{1/2}, F = 3 \rightarrow 23S_{1/2}, F = 4) \simeq 9$ MHz (see Fig. 18). This is the first experimental evidence of hyperfine structure in an excited level. The ground-state hyperfine structure ($6S_{1/2}, F = 3 \rightarrow 6S_{1/2}, F = 4$) = 9193 MHz is well-known for atomic-clock applications. The two values 9193 and 9 MHz verify the predicted n^{*-3} dependence, as in Eq. (2).

One can at least observe the splitting $G-H$ of the peak C and $I-J$ of the peak D. These structures have about the same amplitude as the noise but are observerable on many experimental recordings such as in Fig. 18 and also

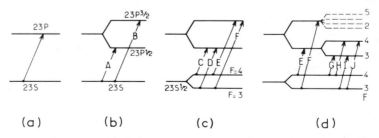

(a)   (b)   (c)   (d)

FIG. 17 (a) The $23S \rightarrow 23P$ transition in cesium, (b) fine structure of the $23P$ level, (c) hyperfine structure of the $23S_{1/2}$ level, and (d) hyperfine structure of the $23P_{1/2}$ level.

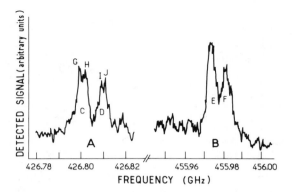

FIG. 18 Experimental trace of the $23S \rightarrow 23P$ resonance in cesium. The letters correspond to the arrows in Fig. 17.

on other transitions (Fig. 19). They correspond to the hyperfine structure of the $nP_{1/2}$ levels. In the case of the $23P_{1/2}$ level, we observe on Figure 18 that $GH = IJ = (23P_{1/2}, F = 3 \rightarrow 23P_{1/2}, F = 4) = 2.2$ MHz. This hyperfine structure, compared with the established value $(7P_{1/2}, F = 3 \rightarrow 7P_{1/2}, F = 4) = 398$ MHz, also verifies the n^{*-3} dependence. The hyperfine structure of the more complex $23P_{3/2}$ level is not experimentally visible.

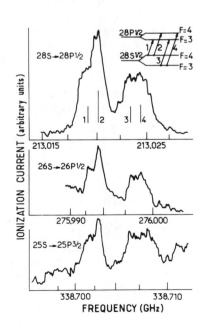

FIG. 19 Experimental traces of $nS \rightarrow nP$ resonances in cesium induced by the harmonics of the X-band klystron. The multiplication order from top to bottom is 18, 23, 28.

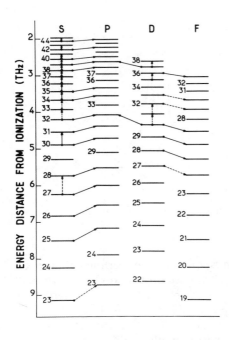

FIG. 20 Transitions observed in cesium (Goy *et al.*, 1982). The fine and hyperfine structures are not represented. Dashed arrows are for Carcinotron-induced transitions. Solid arrows are transitions induced by the harmonics of the *X*-band klystron generated by the frequency multiplier (Goy, 1982).

The transition $23S \rightarrow 23P$ just described occurs at high frequency and has been observed by using the submillimeter waves supplied by a Carcinotron. Transitions such as $25S \rightarrow 25P$ and $26S \rightarrow 26P$ (see Figs. 19 and 20) occurring at a slightly lower frequency have been observed simply by using the harmonic generator (Goy, 1982) fed by an *X*-band klystron. The availability of this device greatly simplified the experimental work; among all transitions observed in cesium only a small number needed the powerful (and expensive) Carcinotrons.

The measured saturation power of one-photon transitions corresponds fairly well to the value predicted by Eq. (14).

3. Two-Photon Resonances

The two-photon transitions $nS \rightarrow (n + 1)S$ need a saturation power acording to Eqs. (17) and (18) with an experimental constant $A \simeq 2 \times 10^{-2}$ W/cm² MHz, two orders of magnitude greater than the calculated value $A \simeq 10^{-4}$ W/cm² MHz. However, the power necessary to saturate a two-photon $nS \rightarrow (n + 1)S$ transition in cesium can be very small because $\delta\nu$ is very small for $20 < n \leq 50$. The weakest power flux 2×10^{-10} W/cm² is observed for the $42S \rightarrow 43S$ transition for which $\delta\nu \simeq 8$ MHz only.

Observation of the $nD \rightarrow (n + 1)D$ transitions needs a large microwave power, varying from 10^{-6} W/cm² for $37D \rightarrow 38D$ to 10^{-2} W/cm² for $30D \rightarrow$

$31D$. This power variation does not obey the predicted law (Eq. 17). [The relay levels $(n - 2)F_{5/2}$ and $(n - 2)F_{7/2}$ are such that δv is 9 and 16 GHz, respectively.]

C. SUMMARY

All spectroscopic values determined in Na and Cs by millimeter and submillimeter spectroscopy are summarized in Table IV. Note the following points:

(1) For any level $n > 20$, a given alkali presents a very large set of transition frequencies regularly spaced and easily predictable from the few constants given in Table 4, using Eqs. (29), (10), and (8).

(2) Different alkalis have different sets of frequencies.

(3) Due to its long lifetime, each level can be precisely defined, and high-resolution spectroscopy is observable (note possible metrological applications).

(4) The sensitivity of transition is extremely high.

(5) Each transition can be made widely tunable by applying a small electric field (strong Stark effect).

V. Blackbody Radiation Detection

A. FREE-SPACE DETECTION

The spectral density of blackbody radiation obeys Planck's law,

$$\frac{d\phi}{dv} = \frac{8h}{c^2} \frac{v^3}{(\exp hv/kT) - 1}. \tag{30}$$

At the typical temperatures $T = 300$ K, 77 K, and 4 K the thermal spectral density $d\phi/dv$ maximum is radiated at the respective frequencies 18, 4.5, and 0.23 THz. For our wavelength domain and at our temperatures we have, for instance, $kT/hv \simeq 60$ at 108 GHz and 300 K; $kT/hv \simeq 12$ at 135 GHz and 77 K. We can use the Rayleigh–Jeans approximation

$$\frac{d\phi}{dv} = \frac{8\pi k}{c^2} v^2 T. \tag{31}$$

Equations (31) and (14) yield the expression for the blackbody-induced transition probability $nS \to nP$, where $v \simeq R_X n^{*-3}$ (Eq. 3) and $G^2(\Delta n^* \simeq 0.5) \simeq 0.5$ (Fig. 6), which is

$$W_{nS \to nP_{1/2}} = 4.24 \times 10^3 \, n_c^{*-2} T, \tag{32}$$

and $W_{nS \to nP_{3/2}}$ is two times greater. Because $W_{nS \to nP} = W_{nS \to nP_{1/2}} + W_{nS \to nP_{3/2}}$

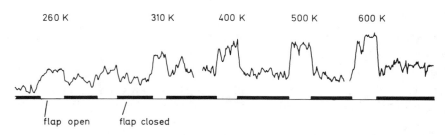

FIG. 21   Detection of blackbody radiation (Figger *et al.*, 1980).

we have, for instance, $W_{30S \to 30P} = 4.6 \times 10^3/\text{sec}$ in Na at 300 K. In an interaction time of the order of 20 μsec (smaller than the excited-state lifetime), about 9% of the atoms are transferred to the upper level by absorbing a resonant (134 GHz) thermal photon of blackbody radiation at 300 K.

Such extreme sensitivity of the excited states has been used in an experiment in which controlled blackbody radiation at variable temperatures interacts with Rydberg atoms (Figger *et al.*, 1980). Their detection sensitivity can be calibrated (Fig. 21), because they are surrounded by a shield cooled to a low temperature. With nonstabilized CW lasers used for the Rydberg atoms preparation, the obtained noise equivalent power is NEP = 10^{-17} W/Hz$^{1/2}$. It is predicted that with stabilized CW lasers a NEP $\simeq 10^{-19}$ W/Hz$^{1/2}$ could be achieved.

The main problem with absorbing blackbody radiation in free space is that many transitions, on different frequencies, are possible. It is fairly difficult to be sure that the observed signal is due entirely to a single transition. The noise equivalent power has also been measured in free space but with a coherent source (Ducas *et al.*, 1979) (a submillimeter laser around 600 GHz), giving NEP = 5×10^{-15} W/Hz$^{1/2}$ for the transition $26S \to 27P$ in Na, the exact tuning between the laser frequency and the transition frequency v being obtained by adjusting v by the Stark effect.

In the two following subsections (B and C) we describe different approaches in which well-defined modes of the blackbody field on a single Rydberg transition are selected.

B.   HIGH-PASS FILTER DETECTION†

Between parallel conducting planes there are families of modes for electromagnetic field propagation. Let us define $\rho(v)$ as the mode density for

† Vaidyanathan *et al.* (1981).

photons at the frequency v [in free space $\rho(v) = 4\pi v^2/c^3$]. Between the conducting planes the families of modes for the electric field parallel and perpendicular to a normal to the planes are $\rho_\parallel$ and $\rho_\perp$, respectively. If b is the distance between the planes the wave with $\lambda = c/v > 2b$ cannot propagate in the mode $\rho_\perp$ (i.e., with the electric field parallel to the planes). The frequency cutoff is $v_c = c/2b$, and the densities of modes in the interval $0 < v < 2_c$ are given by

$$\rho_\parallel = 4\pi v_c v/c^3, \tag{33}$$

$$\rho_\perp = \begin{cases} 4\pi v_c v/c^3, & v > v_c, \\ 0, & v < v_c. \end{cases} \tag{34}$$

The sodium atoms excited to the level $29D$ between parallel plates with $b = 3.3$ mm can absorb only thermal photons of the modes $\rho_\parallel$ in the transition $29D \rightarrow 30P \simeq 42.8$ GHz ($\lambda \simeq 7$ mm). If an increasing electric field is applied to the atoms (voltage on the two plates that constitute a condensor), the Stark effect shifts the levels so that the distance $29D \rightarrow 30P$ increases, crossing the cut-off frequency. The absorption of all polarization photons becomes possible at the high voltage part of the sweep (Fig. 22), resulting in an increase of the population transfer rate.

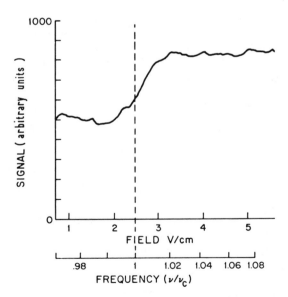

FIG. 22 Blackbody radiation detection through a high-pass filter. The experimental trace shows a step in the absorption. The $29D \rightarrow 30P$ transition in sodium is tuned by the Stark effect through the cut-off frequency of the parallel conducting plates. From (Vaidyanathan *et al.*, 1981).

C. BAND-PASS FILTER

In the nitrogen-cooled cavity (Fig. 14) the atoms are excited at the center in the 30S level and detected after a delay of 20 μsec at a distance $e = 20$ mm (Fig. 13, No. 3).

1. Cavity Untuned

When the cavity is untuned, no frequency mode is available in it for a transition. If we consider the up-transition $30S \rightarrow 30P \simeq 134$ GHz, there is nevertheless a transfer from the state 30S to the state 30P, due to absorption of the thermal photons at 300 K that the atoms can see after they have left the cavity, and even inside the cavity because there are large windows for laser and atomic-beam coupling.

Equation (14) gives the transition probability ($30S \rightarrow 30P_{1/2}$ and $P_{3/2}$) $W_{30S \rightarrow 30P} = 2.1 \times 10^{17} \, d\phi/dv$, and Eq. (31) gives at 300 K $d\phi/dv = 2.1 \, 10^{-14}$ W/m^{2}Hz, so that $W_{30S \rightarrow 30P} = 4.5 \times 10^3$/sec. In the interaction time 20 μsec we can deduce a population ratio $N(30P)/N(30S) = 9\%$. The observed ratio measured from the surfaces of the ionization peaks in Fig. 23 gives

$$|N(30P)/N(30S)|_{U.M.} = 14\%, \tag{35}$$

where U.M. is untuned measured.

In fact we have to take into account the finite lifetimes (Gounand, 1979) of the levels $\tau(30S) \simeq 30 \, \mu$sec and $\tau(30P) \simeq 200 \, \mu$sec according to Eq. (4). After a delay of 20 μsec we detect 90% of the 30P but only 51% of the 30S states. The predicted ratio due to blackbody absorption then becomes

$$|N(30P)/N(30S)|_{U.P.} = 16\%, \tag{36}$$

where U.P. is untuned predicted, and the equation is in very good agreement with Eq. (35).

2. Cavity Tuned

When the cavity is tuned at the frequency $v_{30S \rightarrow 30P_{1/2}} = 134.286$ GHz, it has a measured quality factor $Q = 9000$ (finesse $f = Q/q = 740$) partly limited by diffraction so that the 300-K blackbody radiation can be coupled into it although the mirrors are cooled to nitrogen temperature.

If the number of excited atoms is large, [$N(30S) > 4000$] no effect is visible, and the observed relative transfer remains as

$$|N(30P)/N(30S)|_{T.M._{N(30S)>4000}} = 14\%, \tag{37}$$

where T.M. is tuned measured.

If the number of excited atoms is reduced to a few hundreds, an increase of the relative transfer is visible due to a cavity-induced increase of the final

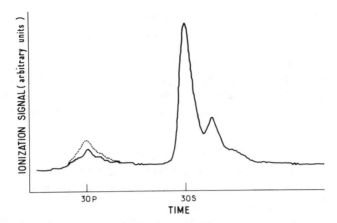

FIG. 23 Blackbody radiation detection by 200 atoms prepared in the $30S$ state in a cavity. The continuous trace (cavity untuned) shows about 28 atoms transferred to the upper state $30P$. The dotted line shows an increase by about 14 atoms of the blackbody radiation absorption when the cavity is tuned on the $30S \rightarrow 30P_{1/2}$ transition. Experimental traces obtained are averaged over 200 laser pulses. From Raimond et al. (1982).

$30P_{1/2}$ population. Figure 23 (dotted line) shows a 50% increase in the detected $30P$ population when the number of $30S$ states at detection is $N(30S) \simeq 200$:

$$|N(30P)/N(30S)|_{\text{T.M.}\cdot N(30S) \simeq 200} = 20\%. \qquad (38)$$

The number of absorbed (abs) photons at 134.286 GHz in the cavity mode during the time of flight of 3 μsec in the waist $W_0 = 3$ mm (Eq. 20) is

$$n_\phi = N(30P_{1/2})_{\text{abs}} = 200 \times 0.06 = 12. \qquad (39)$$

If we analyze in detail (Raimond et al., 1982) the variations of the number n_ϕ, we observe that it first increases linearly with $N(30S)$ and then saturates at the value $n_\phi^{\text{sat}} \simeq 94$, which makes the tuned/untuned contrast more and more difficult to observe with large $N(30S)$ values.

The observed limit corresponds in fact to the absorption of exactly $2kT/h\nu$ photons in the cavity. This limit is the equalization of the atomic excitation number with the cavity excitation in the two σ^+ and σ^- independent modes interacting with the atoms and can be interpreted as a thermalization of the atomic system behaving as a simple quantum object in the cavity. The dipoles belonging to different atoms are driven in phase by the spatially coherent field sustained by the cavity mode. Due to this phase correlation, the atomic system remains in a highly symmetric state with respect to atom exchange and behaves basically as a Bose gas, absorbing the thermal photons. The equilibrium energy of this Bose gas is precisely $2kT/hv$,

$$n_\phi = 2kT/hv, \tag{40}$$

the factor 2 being due to the σ^+ and σ^- independent polarization modes for the degenerate transitions $|30S_{1/2}, M_J = \pm\frac{1}{2}> \rightarrow |30P_{1/2}, M_J = \mp\frac{1}{2}>$.

By splitting the degeneracy with a magnetic field (Zeeman effect) we shift one of the two σ^+ and σ^- transitions out of resonance and then observe the single-mode limit,

$$n_\phi = kT/hv = 47. \tag{41}$$

On the other hand, a hot source at 900 K coupled to the cavity in a zero magnetic field corresponds to a larger number of absorbed photon at equilibrium,

$$n_\phi = 2kT/hv = 280. \tag{42}$$

VI.   The Rydberg Maser†

A.   THE PRINCIPLE

Let us now consider the situation in which the Rydberg atoms amplify the blackbody radiation field in the cavity instead of absorbing it. An nS excited level, for instance, represents a population inversion for $(n - 1)P$, $(n - 2)P$, ... levels. A resonant radiative transfer of atoms from the nS level to $n'P$ levels with $n' < n$, can occur even without any externally applied millimeter wave field. If a large enough number of atoms are prepared in the upper level of a transition that is resonant with the millimeter wave cavity, they undergo a fast decay towards the lower level and emit a millimeter-wave radiation burst at the frequency v of this transition. This is a transient maser effect that can be observed by directly detecting the emitted microwave (signal 1 in Fig. 24) or by monitoring the atomic population (signal 2 in Fig. 24).

Tuning the cavity resonance by adjusting the distance between mirrors permits us to select among the maser transitions that are possible from a given excited state. Figure 25 shows the many resonant positions (dots) observed when a sample of about 30,000 Cs atoms is excited in the $32D_{5/2}$ state in cavity number 3 of Fig. 13. The positions of the mirrors permit us to deduce the wavelength tuning from Eq. (27).

For practical reasons the resonance is generally monitored by the field ionization signal (signal 2 of Fig. 24) as is shown in Fig. 26.

Many maser transitions have been observed in cesium (Fig. 27) and

† See Gross et al. (1979); Moi et al. (1983); Goy et al. (1983).

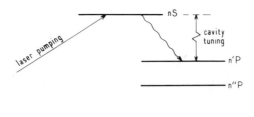

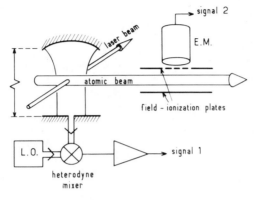

FIG. 24 General scheme of the Rydberg atoms maser. Signal 1 corresponds to the direct detection of the microwave emitted by the atoms. Signal 2 is an indirect detection of the microwave by monitoring the atomic populations by the field-ionization technique. The insert shows energy levels relevant for maser emission. The final $n'P$ state is selected by cavity tuning. From (Moi *et al.*, 1983).

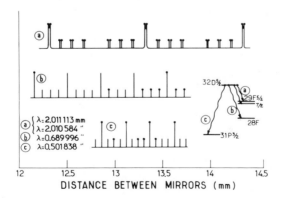

FIG. 25 Sweeping the cavity in which cesium atoms are prepared in the excited state $32D_{5/2}$ permits us to obtain a great number of positions where maser action occurs (black dots). From the spatial periodicity of the tuning, one can deduce the wavelength, i.e., identify the maser transitions given in the insert. The cavity modes obey Eq. (27) except for the $32D \rightarrow 29F$ transitions which show an unexplained $\lambda/16$ periodicity in (a).

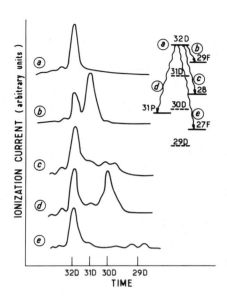

FIG. 26 Ionization signals obtained for masers starting from the excited level $32D_{5/2}$ in cesium. The nD positions in the time scale permit us to identify the levels, because the nD, $(n + 1)P$, and $(n - 2)F$ levels are ionized at about the same field. Besides, the $(n + 1)P$ levels have longer lifetimes and give greater ionization signals than the $(n - 2)$ levels. The differences in size and shape permit us to distinguish between P and F final states.

sodium (Fig. 28). The corresponding frequencies (and wavelengths) range from 69 GHz ($\lambda = 4.3$ mm) to 1.44 THz ($\lambda = 0.208$ mm). Because the submillimeter radiations belong to the light domain (infrared) and the millimeter radiations to the microwave domain, these masers can also be called lasers for $\lambda < 1$ mm.

Notice that the precise calibration of the distance between mirrors permits us to obtain a very simple way of performing spectroscopy between

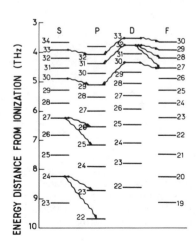

FIG. 27 All maser transitions observed in cesium with cavity number 3, from 135 GHz to 1.44 THz. See Table V.

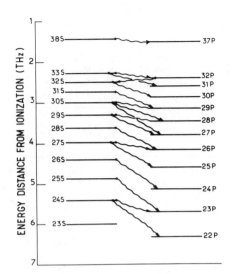

FIG. 28 Maser transitions observed in sodium in cavities number 1, 2, and 3.

Rydberg levels without an external microwave source, the atoms themselves creating the microwave inducing the transition from the upper to the lower level. See Table V in which the spectroscopy of cesium is performed by this means, to the relatively good accuracy of $\Delta\nu/\nu \simeq 10^{-4}$.

The fine-structure components are not always easily resolved by this method. Nevertheless they can produce different lineshapes in the ionization signal (Fig. 29), so identification is much easier.

B. MASER CALCULATIONS†

1. The Radiative Damping Rate

There is a collective radiation damping of the N Rydberg atoms system in the tuned cavity of quality factor Q. The electric dipoles are driven in phase by the small millimeter-wave field and coupled to their images in the cavity walls. The radiative damping rate is

$$T_R^{-1} = 2d^2QN/\hbar\varepsilon_0 V, \tag{43}$$

where d is the electric matrix element $d = eR_{nl}^{n'l'}$ with $R_{nl}^{n'l'}$ given by Eq. (13) and V is the effective volume of the mode in the cavity. (We assume here that all the atoms are efficiently coupled to the cavity mode in an antinode position.)

$$V = (\pi/4)LW_0^2, \tag{44}$$

† Moi et al. (1983).

TABLE V

Comparison of the Calculated Frequencies of Down-Transitions Predicted
from Eqs. (10) and (29) and from the Coefficients Given in Table IV,
with the Measured Values Deduced from the Tuning Positions (Eq. 27) in
Maser Action. The Relative Accuracy, Around 10^{-4}, is the Accuracy in
Determining the Distance Between Mirrors

Transition $nlJ \rightarrow n'l'J'$ cesium	Calculated frequency F (GHz)	Measured by cavity F' (GHz)	Accuracy $(F' - F)/F \ (10^{-4})$
$33D_{5/2} \rightarrow 30F_{5/2}$	134.771	134.78	0.7
$\rightarrow 30F_{7/2}$	134.806	134.82	1
$33S_{1/2} \rightarrow 32P_{3/2}$	142.010	141.99	-1.4
$32D_{5/2} \rightarrow 29F_{5/2}$	149.068	149.07	0.1
$\rightarrow 29F_{7/2}$	149.107	149.12	0.9
$30D_{5/2} \rightarrow 27F_{5/2}$	184.348	184.40	2.8
$\rightarrow 27F_{7/2}$	184.396	184.46	3.5
$30S_{1/2} \rightarrow 29P_{1/2}$	210.821	210.81	-0.5
$27S_{1/2} \rightarrow 26P_{3/2}$	287.136	287.08	-0.2
$\rightarrow 26P_{1/2}$	306.098	306.09	-0.2
$32D_{5/2} \rightarrow 28F_{5/2}$	434.463	434.46	-0.5
$\rightarrow 28F_{7/2}$	434.506		
$24S_{1/2} \rightarrow 23P_{3/2}$	439.480	439.47	-0.2
$\rightarrow 23P_{1/2}$	468.656	468.65	0
$30D_{5/2} \rightarrow 25F_{5/2}$	539.485	539.35	-3
$\rightarrow 25F_{7/2}$	539.539		
$32D_{5/2} \rightarrow 31P_{3/2}$	597.389	597.31	-1.3
$32D_{5/2} \rightarrow 27F_{5/2}$	752.191	752.24	0
$\rightarrow 27F_{7/2}$	752.240		
$33D_{5/2} \rightarrow 31P_{3/2}$	840.396	839.94	-5.4
$27S_{1/2} \rightarrow 25P_{3/2}$	910.822	910.67	-1.7
$\rightarrow 25P_{1/2}$	932.565	932.85	3
$24S_{1/2} \rightarrow 22P_{3/2}$	1409.373	1408.88	-3
$\rightarrow 22P_{1/2}$	1443.563	1443.82	1.7

where the effective surface for energy in the waist is $(\pi/2)W_0^2$ and the effective distance between mirrors is $L/2$ because of the standing wave.

The S and P quantum defects are such that $\varepsilon_s - \varepsilon_p \simeq 0.5$. For $nS \rightarrow (n-1)P$ transitions, the transition frequency v can be approximated by half the value given in Eq. (3), in MKSA units,

$$v_{nS \rightarrow (n-1)P} \simeq R_H n_c^{*-3} \simeq 3.29 10^{15} n_c^{*-3}. \tag{45}$$

Around the semiconfocal geometry, Eq. (43) can be simplified as

$$T_R^{-1}[nS \rightarrow (n-1)P] \simeq K n_c^{*-2} r^{-1} fN, \tag{46}$$

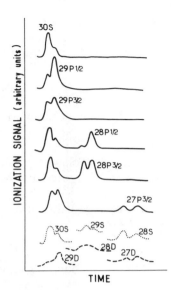

FIG. 29 Ionization signals obtained for masers starting from the excited $30S$ level in sodium. The $(n-1)P$
level is ionized at about the same field than the $(n-1)D$
level, just after the nS level. The $nP_{1/2}$ and $nP_{3/2}$ levels
appear at the same field, but the more split shape of the
$nP_{3/2}$ level permits us to distinguish between them.

where $K = 70$ for $nS \rightarrow (n-1)P_{1/2}$ and $K = 140$ for $nS \rightarrow (n-1)P_{3/2}$
transitions. For instance, with $N = 1000$ atoms involved in the maser
transition $30S \rightarrow 29P_{1/2}$ inside an antinode position of cavity number 3 ($r =$
26 mm and $f = 750$), we obtain $T_R = 376$ nsec.

In a similar way we have, for the $nS \rightarrow (n-2)P$ transitions,

$$v_{nS \rightarrow (n-2)P} \simeq 3R_H n_c^{*-3} \simeq 9.87 \times 10^{15} \, n_c^{*-3}, \qquad (47)$$

in MKSA units, and

$$T_R^{-1}[nS \rightarrow (n-2)P] \simeq K' n_c^{*-2} r^{-1} fN, \qquad (48)$$

where $K' = 13.4$ (MKSA units) for $P_{1/2}$ and $K' = 26.8$ for $P_{3/2}$. For instance,
with $N = 30,000$ atoms involved in the maser transition $30S \rightarrow 28P_{1/2}$ with
the cavity finesse slightly reduced ($f \simeq 500$) we obtain $T_R = 100$ nsec.

2. The Bloch Vector Representation†

The maser action between two Rydberg levels can be simply described
with the following model: with each atom having been initially prepared in
the up position of the two-level system, analogous to a spin-$\frac{1}{2}$ particle, the
whole atomic system of N atoms is similar to a $J = N/2$ angular momentum
called the Bloch vector. This Bloch vector evolves from the initial up

† Moi *et al.* (1983).

position (all atoms in the upper nS level of Fig. 24) to the final down position (all atoms in the $n'P$ level). The Bloch angle θ between the up position and the position at limit describes the maser evolution and obeys the equation

$$\frac{d^2\theta}{dt^2} + \frac{1}{2T_{cav}} \frac{d\theta}{dt} - \frac{1}{4T_{cav}T_R} \sin\theta = 0, \tag{49}$$

where T_{cav} is the characteristic damping time of the cavity,

$$T_{cav} = Q/2\pi v. \tag{50}$$

The angle θ obeys a pendulum-like equation (Eq. 49). At time $t = 0$ this pendulum, prepared in the up position, is triggered by the blackbody radiation field in the cavity which gives to the system a small tipping angle θ_i, the average value of which is (following statistical values) valid only if $N \gg kT/hv$,

$$\bar{\theta}_i = 2\sqrt{(1 + kT/hv)/N}. \tag{51}$$

If the cavity damping contribution in Eq. (49) is relatively small ($T_{cav} \gg T_R$ in the case of very large Q or N), there is an oscillatory regime of $\theta(t)$; and the atoms, which are very strongly coupled to the cavity, oscillate between the up and down levels, undergoing a self-nutation process induced by their own radiation field.

In our experimental conditions ($Q \simeq 8000$ at 134 GHz, for instance, so that $T_{cav} = 9$ nsec) the cavity time is small compared with T_R ($T_R > 20$ nsec for $N < 20,000$). Equation (49) can thus be approximated by

$$\frac{d\theta}{dt} = \frac{1}{2T_R} \sin\theta, \tag{52}$$

which corresponds to a monotonic increase of θ from 0 to π, the atomic system being irreversibly damped by its own radiation field within the time of a few T_R.

With the initial conditions of Eq. (51) one can obtain from Eq. (52) the variation $(d\theta/dt)^2$ of the microwave power radiated by the atoms,

$$\left(\frac{d\theta}{dt}\right)^2 = \left[\cosh\left(\frac{t - t_D}{T_R}\right)\right]^{-2}, \tag{53}$$

where t_D is the typical delay for the pulse emission

$$t_D = 2T_R \ln\bar{\theta}_i \simeq T_R \ln\frac{N}{1 + kT/hv}. \tag{54}$$

Equation (53) describes a so-called hyperbolic secant pulse that reaches its maximum after an average delay t_D depending upon, by Eqs. (43) and (54), the number of atoms N, the cavity Q, and the thermal background temperature T. The width of the pulse is of the order of $3.5\ T_R$.

3.   The Maser Threshold†

The condition for observing a self-oscillating Rydberg maser is simply that the typical delay of emission t_D (Eq. 54) be shorter than the time of flight of the atoms in the waist of the cavity T_t;

$$T_t = \frac{W_0}{\sqrt{2}\ \overline{V}},\tag{55}$$

where $\overline{V}$ is the average beam velocity and $W_0/\sqrt{2}$ the radius of the photon storage active region.

$$\overline{V} = 1.88\sqrt{kT/m},\qquad (\overline{V} = 890 \text{ m/sec in Na}).\tag{56}$$

The maser threshold condition

$$t_D < T_t\tag{57}$$

becomes [Eqs. (46), (48), (54), and (55)]

$$N \Big/ \ln\left(\frac{N}{1 + kT/h\nu}\right) > B\overline{V}f^{-1}\sqrt{rn_c^*},\tag{58}$$

where the constant $B = 157$ (MKSA units) for an $nS \rightarrow (n-1)P_{1/2}$ transition and 1480 for an $nS \rightarrow (n-2)P_{1/2}$ transition. Transitions $nS \rightarrow n'P_{3/2}$ give B two times smaller.

We have experimentally verified Eq. (58). The measured threshold N_0 for the transition $33S \rightarrow 32P_{3/2}$ with the Na beam crossing cavity number 1 is as predicted, $N_0 = 5600 \pm 2000$.

For the maser transition $30S \rightarrow 29P_{3/2}$ performed in cavity number 3 with its best finesse being $f = 1600$, the second term $B\overline{V}f^{-1}\sqrt{rn_c^*} = 38$ is of the order of the number of photons per mode $n_\phi = kT/h\nu = 46$. In these conditions the threshold is extremely low and the statistical function of Eq. (51) is no longer valid. We have experimentally observed $N_0 = 100 \pm 50$ atoms (Fig. 30). This is the smallest self-oscillating maser ever observed although the experimental conditions are not yet very severe. In particular, the quality factor $Q \simeq 20,000$ can certainly be increased by using cooled mirrors. Helium cooling is also necessary to reduce the transitions induced by blackbody photons. Experiments are in progress in our laboratory in an effort to obtain the "single atom maser," a single atom interacting resonantly with the high Q cavity which may enhance its spontaneous emission.

† Moi *et al.* (1983).

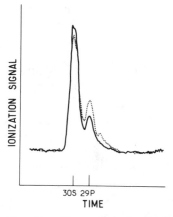

IONIZATION SIGNAL

30S 29P
TIME

FIG. 30 Evidence of the low threshold maser. Only 100 Na atoms are prepared in the $30S$ state. The cavity tuned on the transition frequency $30S \rightarrow 29P_{3/2} = 145{,}593$ MHz permits us to observe the beginning of the maser action. Solid line, cavity untuned; dotted line, cavity tuned.

C. THE RYDBERG ATOM MASER AS A MILLIMETER-WAVE SOURCE OR DETECTOR†

1. *The Source*

The field-ionization detector technique (signal 2 in Fig. 24) does not permit direct observation of the time evolution of the maser. On the contrary, the emitted microwave power (Moi *et al.*, 1983; Goy *et al.*, 1983) (signal 1 in Fig. 24) indirectly reflects this time evolution: after a time delay that fluctuates around t_D, given by Eq. (54), the microwave power is emitted in a bell-shaped peak according to Eq. (53) with a width of 3.5 T_R. The characteristic evolution time of the maser T_R is given by Eqs. (43), (46), and (48). The emitted peak power is given by

$$P_{\max} \simeq Nh\nu/3.5T_R. \tag{59}$$

For instance, for the transition $33S \rightarrow 32P_{1/2}$ in sodium at 107.892 GHz with 3×10^5 atoms in cavity number 1 (Fig. 13, Table III), one has $T_R \simeq 50$ nsec and $t_D \simeq 450$ nsec. The emitted peak power (Eq. 59) will be of the order of $P_{\max} \simeq 10^{-10}$ W. Because of the coupling loss, -9 dB leaving the cavity, the power available at the detector will be around 10 pW during a pulse time 3.5 $T_R \simeq 180$ nsec.

To detect such a small power we use a heterodyne receiver (see Fig. 31). Its bandwidth, 14 MHz, is large enough so that the fast microwave pulses are not broadened too much. The local oscillator L.O. (see Fig. 31) is a Carcinotron stabilized on a quartz-controlled X-band klystron. The frequency ν_c of this L.O. is 130 MHz (intermediate frequency) apart from the frequency of the maser ν_m.

† Goy *et al.* (1983).

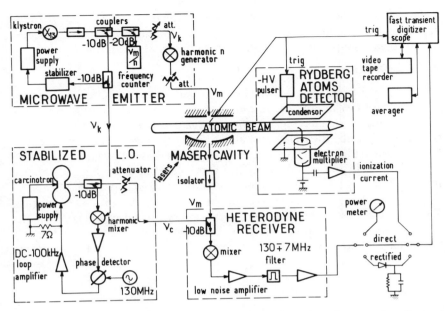

FIG. 31 General diagram of the microwave setup used for maser microwave detection and triggering. From (Goy *et al.*, 1983).

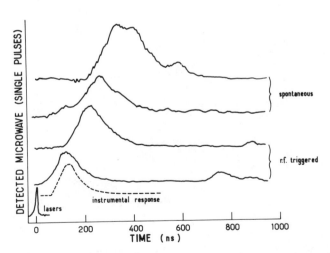

FIG. 32 Detection of single pulses emitted by the Rydberg maser on the transition $33S \rightarrow 32P_{1/2} = 107{,}892$ MHz.

Two spontaneous pulses are detected from the maser $33S \rightarrow 32P_{1/2}$ in the conditions described above (see Fig. 32, top). One can observe the predicted power, delay, shape, and fluctuation from one peak to another.

2.   The Detector

Instead of being triggered by blackbody photons the maser can be triggered by an external applied resonant microwave (Fig. 32, bottom). If we look at averaged signals (Fig. 33), we observe that the effect of triggering the maser is to reduce the delay and also to increase the detected power by a factor of around 2, which can be easily explained as follows.

The atoms in the tuned cavity are a two-level system (Fig. 34) of two independent classes of atoms. One class (about 50%) is prepared in the $|n, S_{1/2}, m_J = -\frac{1}{2} >$ state and evolves down to the $|n', P_{1/2}, m_J = +\frac{1}{2} >$ state. The other class, prepared in the $|n, S_{1/2}, m_J = +\frac{1}{2} >$ state, even evolves to the $|n', P_{1/2}, m_J = -\frac{1}{2} >$ state. The two classes give two σ^+ and σ^- circularly polarized microwaves. In the "spontaneous" maser the two components are

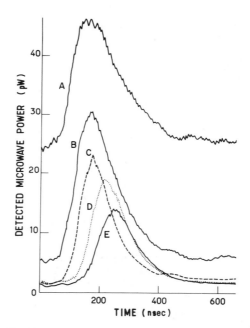

FIG. 33   Same as Fig. 32 but with signals averaged over 100 pulses for different amounts of impinging triggering microwaves, decreasing from A to E. In A the detected background power is 25 pW, although inside the cavity the power is 250 pW (there is a 10-dB loss from inside the cavity to the detector). In E no coherent microwave is sent to the atoms. The maser is then triggered by only the blackbody radiation.

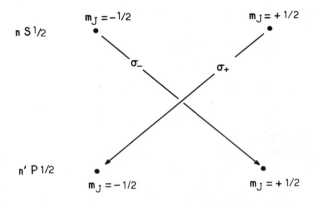

FIG. 34   Two-level scheme of the maser $nS \rightarrow n'P_{1/2}$.

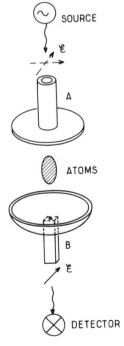

FIG. 35   Principle of triggering the maser with different polarizations. From Goy *et al.* (1983).

radiated independently with uncorrelated phases and small random delay differences. The resulting field then has generally an elliptical polarization —almost linear for most of the evolution time—with a large axis rotating randomly around the cavity axis from pulse to pulse. The average power detected through a rectangular waveguide, working as an analyzer (Fig. 35, bottom) is only one half of the average emitted power.

Otherwise, if a linearly polarized field impinges on the atoms, the two counter-rotating polarization components are locked in phase, and the emitted linear polarization is parallel to the impinging polarization. The circular input waveguide (Fig. 35, top) permits the propagation of any linear polarization, especially the one parallel to the small side of the output rectangular waveguide; then 100% of the maser pulses are detected. For the one perpendicular to that direction, the maser pulses are 0% detected (see Fig. 36).

To summarize, by triggering the maser with an applied microwave we have been able to observe a shortening of the maser emission delay, a locking of the signal polarization and phase. Figure 37 gives an example of delay shortening observed with very small triggering fields.

In the presence of coherent photons impinging on the atoms, the average delay of emission t_D is shortened according to the formula (Eq. 54 modified)

$$t_D = T_R \ln \frac{N}{1 + (kT/h\nu) + n_{\text{imp}}^{\text{trig}}}, \tag{60}$$

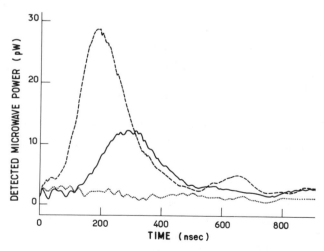

FIG. 36 Experimental traces with no triggering microwave (solid line), with parallel (dashed line) and perpendicular (dotted line) polarizations. From Goy *et al.* (1983).

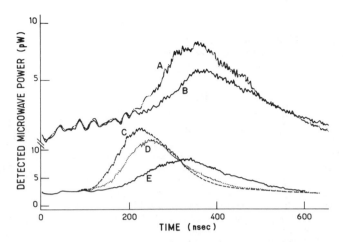

FIG. 37 The great sensitivity of the maser with small impinging microwave powers, $P_{imp} = $ 0.1 pW, 0 pW, 2 pW, 1 pW, and 0 pW for the traces A, B, C, D, and E, respectively. From Goy *et al.* (1983).

where n_{imp}^{trig} is the number of photons impinging into the cavity during the characteristic evolution time T_R.

$$n_{imp}^{trig} = (T_R/T_{cav})n_{cav}, \tag{61}$$

where T_R and T_{cav} are given by Eqs. (43) and (50) and n_{cav} is the number of microwave photons stored in the cavity, related to the impinging power P_{imp} and to the quality factor Q of the cavity.

$$n_{cav} = \frac{P_{imp}Q}{2\pi h\nu^2}. \tag{62}$$

The limit in sensitivity of the triggered maser, i.e., of the maser used as a transient amplifier, is obtained when the number n_{imp}^{trig} of coherent photons impinging into the cavity during the typical evolution time T_R is of the order of, or not far below, the number of blackbody photons $kT/h\nu$. In cavity number 3 (see Table III, $T_R/T_{cav} = 10$) we have experimentally observed a shortening of the delay with no more than one average photon stored in the cavity $n_{cav} = 1$, $n_{imp}^{trig} = 10$. This corresponds to (Eq. 62) $P_{imp} = 10^{-14}$ W and, in the bandwidth 10^7 Hz $= T_R^{-1}$, to the detectivity 3×10^{-18} W/Hz$^{1/2}$.

Helium cooling will permit us to reach much higher sensitivity.

These experiments clearly indicate that Rydberg atoms could serve as very good detectors and amplifiers of millimeter and submillimeter radiations, a frequency range where efficient receptors are in high demand.

ACKNOWLEDGMENTS

This work was supported in part by the Délégation Générale à la Recherche Scientifique et Technique, Ministère de la Recherche et de l'Industrie, under grant numbers 79.7.0195 and 80.7.0590.

REFERENCES

Ducas, T. W., Spencer, W. P., and Kleppner, D. (1979). *Appl. Phys. Lett.* **35**, 382.
Fabre, C. (1982). *Ann. Phys. Fr.* **7**, 5, and **4**, 247.
Fabre, C. Haroche, S., and Goy, P. (1978). *Phys. Rev. A* **18**, 229.
Fabre, C. Haroche, S., and Goy, P. (1980). *Phys. Rev. A* **22**, 778.
Figger, H., Leuchs, G., Straubinger, R., and Walther, H. (1980). *Opt. Commun.* **33**, 37.
Gounand, J. F. (1979). *J. Phys. Paris* **40**, 457.
Goy, P. (1982). *Int. J. Infrared Millimeter Waves* **3**, 221.
Goy, P., Fabre, C., Gross, M., and Haroche, S. (1980). *J. Phys. B* **13**, L83.
Goy, P., Raimond, J. M., Vitrant, G., and Haroche, S. (1982). *Phys. Rev. A* **26**, 2733.
Goy, P. *et al.* (1983). *Phys. Rev. A* **27**, 2065.
Gross, M. *et al.* (1979). *Phys. Rev. Lett.* **43**, 343.
Haroche, S. (1978). *La Recherche* **9**, 733.
Martin, W. C. (1980). *J. Opt. Soc. Am.* **70**, 784.
Moi, L. *et al.* (1980). *Opt. Commun.* **33**, 47.
Moi, L. *et al.* (1983). *Phys. Rev. A* **27**, 2043.
Raimond, J. M., Goy, P., Gross, M., Fabre, C., and Haroche, S. (1982). *Phys. Rev. Lett.* **49**, 117.
Vaidyanathan, A. G., Spencer, W. P., Kleppner, D. (1981). *Phys. Rev. Lett.* **47**, 1592.

CHAPTER 9

Far-Infrared Spectroscopy of InAs–GaSb Layered Structures

J. C. Maan

Max-Planck-Institut für Festkörperforschung
Hochfeld Magnetlabor
Grenoble, France

I. Introduction

The study of electronic systems in layered structures, interfaces, and surfaces of semiconductors is a very active research area. This activity is to a large extent inspired by the development of the molecular beam epitaxial technique (MBE), which permits the growth of very high-quality semiconducting thin layers with sharp interfaces. With this technique, layer thicknesses as small as a few atomic planes have been realized, and this possibility allows the creation of new electronic systems with promising potential applications. The interest from a more fundamental point of view lies in the possibility of the study of new physical properties due to the interplay between the physical dimensions of the layers with natural lengths such as the de Broglie wavelength, the cyclotron radius, and the screening length, because systems in which these lengths are of the same order of magnitude can be realized. In this review these latter aspects will be the *Leitmotiv* in the discussion of the far-infrared (FIR) properties of InAs–GaSb periodic layered structures.

In general, the confinement of carriers in a small layer by potential barriers created by space charge or compositional variation leads to the formation of two-dimensional subbands, whose energy separation depends on the thickness of the layers and on the height of the potential (the quantum size effect). In addition to the quantum size effect, there is also a coupling between successive wells in a superlattice (a system consisting of thin periodically arranged layers of different semiconducting materials). Therefore superlattices have properties that are substantially different from those of the constituent materials and that depend on the layer periodicity. For superlattices made from alternating layers of InAs and GaSb, or type II superlattices as they are sometimes called, a transition from semiconducting to semimetallic behavior is observed as the layer periodicity increases. The reason for this transition lies in the fact that the bulk InAs–CB is at lower energy than the bulk GaSb–VB and that if the layer thickness increases, both electrons and holes become less confined, which causes hole-like and electron-like subbands eventually to cross; thus the system becomes semimetallic because of the transfer of electrons from the GaSb into the InAs layers. For very thick layers of InAs and GaSb this electron transfer leads to a dipole field across the interface, which creates a band bending, and the electrons and holes are confined in a small region around the interface by this space-charge potential. In this region of thicknesses (~ 1000 Å) the individual InAs–GaSb heterojunction can be studied.

The electronic properties of InAs–GaSb superlattices and interfaces have been studied with far-infrared (FIR) magnetooptical experiments. Generally the FIR transmission at fixed frequency is studied as a function of the magnetic field. This technique is very useful for the study of these systems, because the subband spacing is in the range of FIR energies, and in addition the Landau level separation in a magnetic field for the magnetic fields used in the experiments is in the same range. In Section II of this chapter the interface between InAs and GaSb and the superlattice band structure will be discussed. The system is compared with a different type of superlattice consisting of GaAs and $Ga_{1-x}Al_xAs$ layers. The experimental results on superlattices are reviewed in Section III. Finally, in Section IV the experiments and the subband structure of the InAs heterojunction are discussed.

II.   InAs–GaSb Superlattices

A.   SUPERLATTICE PROPERTIES IN GENERAL

The term *superlattice,* a structure consisting of alternating periodically arranged thin layers of different semiconducting materials, was introduced for the first time by Esaki and Tsu (1970). They proposed that it would be possible to create a new type of crystal with new electronic properties that would be markedly different from the constituent materials by considering the effect on the band structure of a periodic spatially varying potential superimposed on the atomic potentials. This potential could be created either by doping variation or by growing thin layers of different materials having a close lattice match. This new periodicity leads to a division of the original Brillouin zone into subzones with zone boundaries at π/d, with d the superlattice periodicity, and to the formation of subbands separated by small gaps. If the de Broglie wavelength of the electrons in a potential well, $[2m^*(E - V)/\hbar^2]^{-1/2}$, where E is the electron energy, V the potential, and m^* the electron effective mass, is of the order of the barrier thicknesses or longer, the coupling between successive potential wells leads to subbands with a substantial width. This width expresses the degree of three- or two-dimensional character of the superlattice. For nearly flat subbands the effective mass in the direction of the superlattice is very high, which implies that the carriers are localized in the potential well (quantum size effect regime). If the subbands show a dispersion in k space, transport of carriers from one well to the next is possible. Due to symmetry, the subbands are flat in the center and at the edge of the mini Brillouin zone; and because they have a certain width, their curvature will be opposite and the effective mass will have an opposite sign at these extremal points. This effect leads to potentially interesting transport properties in the superlattice direction. For transport through the layers a negative differential conductivity could occur, and if the scattering time for transport through the layers is sufficiently long, this should result in oscillating behavior of the electrons, giving rise to emission of radiation (the so-called Bloch oscillator). Up to now there exists no clear experimental evidence for these latter two implications of subband formation, although some experimental indication for negative differential conductivity has been reported for GaAs–Al$_x$Ga$_{1-x}$As superlattices (Esaki and Chang, 1974). On the other hand the existence of a broadening due to the interaction between successive wells (Dingle *et al.*, 1975) and the existence of a subband with a finite width (Chang *et al.*, 1977; Chang, 1980; Maan *et al.*, 1981; and Stormer *et al.*, 1981) have been convincingly demonstrated.

Two types of semiconductor superlattices based on compositional variation, GaAs–Ga$_{1-x}$Al$_x$As and InAs–GaSb, have been studied (reviews of the earlier experimental work on these systems have been given by Dingle, 1975; Esaki, 1980; and Chang and Esaki, 1980). Both these systems have a nearly perfect lattice match, which of course is necessary to avoid localized interface states. The systems can be grown with the MBE technique, and because this method allows a slow (1 to 3Å/sec) crystal growth at relatively low temperatures (450 to 600°C), the superlattices have regular layer thicknesses and sharp boundaries (see for instance Esaki and Chang, 1976). Recently, the first experimental results showing evidence for subband formation in a new type of superlattice (based on the other method of creating a superlattice potential, i.e., by spatially varying the doping level in the material) have been reported by Döhler et al. (1981). However, the idea of doping modulation itself has been shown to be even more useful in single GaAs–Ga$_{1-x}$Al$_x$As heterojunctions in which very high electron mobilities have been obtained by the spatial separation of the dopant and the free electrons (Störmer, 1980).

B.   THE INTERFACE BETWEEN InAs AND GaSb

It was mentioned in the introduction that InAs and GaSb form a peculiar combination of III–V materials because of their specific band lineup. In Fig. 1 the band structure of InAs and GaSb is shown schematically at the Γ point of the Brillouin zone where both materials have their fundamental gap. In the figure the vacuum energy is the common reference for the energy scale of both materials. One sees that within this framework the InAs conduction band (CB) is 0.1 eV lower in energy than the GaSb valence band (VB). (The band-structure parameters were taken from Neuberger, 1971.) This band lineup is remarkably different from that of GaAs and Ga$_{1-x}$Al$_x$As. In that case the GaAs band gap is entirely contained within the wider Ga$_{1-x}$Al$_x$As band gap at the Γ point, and the band-gap difference is 15% in the VB and 85% in the CB (Dingle, 1975). It has been discussed in a number of theoretical papers that the lineup at the InAs–GaSb interface cannot in principle be obtained by simply subtracting the electron affinities. Frensley and Kroemer (1977) have argued that the difference in electronegativity of GaSb and InAs leads to a dipole field at the interface that raises the InAs–CB 0.1 eV above the GaSb valence band and that, therefore, the CB–VB overlap will not be preserved in the interface. In addition, in contrast with the GaAs–Ga$_{1-x}$Al$_x$As interface, that between InAs and GaSb could show deviations from the ideal crystallographic structure despite the very close match of the lattice parameters in the bulk. In the case of the GaAs–Ga$_{1-x}$Al$_x$As interface, Ga–As and Al–As bonds exist at the interface as well as in the bulk and are equally long. At the GaSb–InAs interface, however, there can exist either a Ga–As bond or an In–Sb bond. The

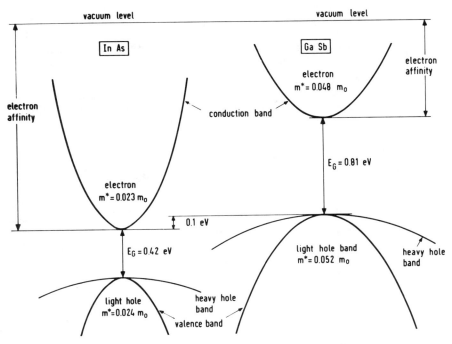

FIG. 1 Schematic band structure and band lineup of InAs and GaSb at the Γ point. The vacuum level is taken to be the same for both energy scales.

Ga–As bond is 7% shorter and the bulk In–Sb bond length is about 7% longer than that of Ga–Sb and In–As. Therefore interface reconstruction could take place, which would influence the band lineup. Saris *et al.* (1979) have given experimental evidence from Rutherford backscattering and channeling experiments for the existence of interface reconstruction.

This crystallographic aspect and especially its consequences on the electronic structure have attracted substantial theoretical interest. Ihm *et al.* (1979) have shown theoretically, on the basis of a self-consistent pseudopotential calculation applied to InAs–GaSb superlattices, that the charge density only a few lattice parameters away from the interface is similar to that in bulk InAs and GaSb, indicating that the interface reconstruction has a short range. Dandekar *et al.* (1980) have made a systematic theoretical study of the electronic structure of the InAs–GaSb interface using a three-dimensional tight-binding calculation. They have varied the bond lengths at the interface and found that for reasonable values around the In–Sb and Ga–As bulk bond lengths, the interface reconstruction does not give rise to localized interface states in the energy region around the fundamental energy gaps and that a finite density of states is found for energy regions throughout the gap of InAs and GaSb. It is clear, therefore, that there is a considerable theoretical interest in the InAs–GaSb interface. The study of superlattices and heterojunctions in InAs–GaSb has contributed substantially to elucidation of these matters. For instance, it has been unambiguously established that the GaSb–VB is higher in energy than the InAs–CB and that apparently no space-charge effects exist that change the lineup significantly from that obtained by the usual electron-affinity rule.

C. Superlattice Band Structure

In Fig. 2 the spatial band-edge variation and the subbands at the Γ point for InAs–GaSb and GaAs–Ga$_{1-x}$Al$_x$As are shown. In the following only the band structure in the neighborhood of the Γ point will be discussed. Bands of other points of the BZ have been considered by Mukherji and Nag (1975), but they will not be discussed here. Figure 2 clearly shows the differences

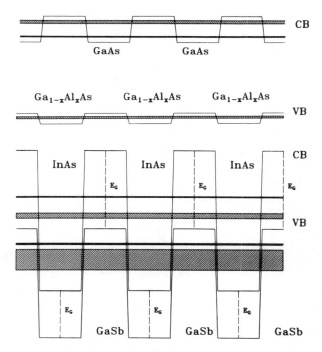

Fig. 2   Spatial band-edge variation and subbands in GaAs–GaAlAs superlattices (top) and InAs–GaSb superlattices (bottom). The hole subband in the lower drawing is indicated by a black line.

between the two types of superlattices, which are sometimes referred to as type I and type II superlattices, respectively. For the GaAs–Ga$_{1-x}$Al$_x$As system there exists a square wave modulated potential for both the conduction band and the valence band, and the band-edge modulation is small compared with the band gap. This energy-level structure is very similar to the Kronig–Penney model which has been succesfully used to calculate the subband structure. In this textbook model (for instance in Smith, 1961) the wave functions and their derivatives are matched at the material interfaces, and a propagating solution through the whole crystal is sought using periodic boundary conditions. In the case of GaAs–Ga$_{1-x}$Al$_x$As superlattices, this procedure can be applied to the CB and the VB separately. The model has recently been critized on formal grounds (Bastard, 1981; White and Sham, 1981) in that the effective-mass jumps at the boundary between GaAs and Ga$_{1-x}$Al$_x$As should be included in the boundary conditions; but nevertheless it has been able to explain several experimental results. This simple model cannot be used for the InAs–GaSb case, because the CB and VB cannot be treated separately. Using plane waves for the CB and VB and matching the wave functions at the materials boundaries were shown to lead to erroneous solutions (Sai-Halasz *et al.* 1977). The reason for this failure is clear, because such an approach completely neglects the orthogonality of CB and VB wave functions. Because these wave functions are orthogonal at the Γ point in the bulk materials, they are assumed also to be so between different materials. Therefore Sai-Halasz *et al.* (1977) used the full Bloch wave functions, including the orthogonality in the cell periodic part of the wave functions, and matched these and their derivatives at the interface. Because both InAs and GaSb are relatively narrow-gap materials, a correct description of the dispersion relations and the wave functions of both bands must include the CB–VB interaction in the bulk as can be described by the k.p method in the two-band Kane model (Kane, 1957). However, the assumption for the boundary conditions, i.e., the continuity of the wave functions and their derivatives at the interface, has been criticized. Although this choice is intuitively appealing, it is not correct if CB and VB wave functions with completely different symmetry and masses are considered. Bastard (1981) recently derived an explicit expression for the boundary conditions at the interface for the envelope wave functions, thereby focusing his attention on the slow spatial variation of these functions and neglecting all rapidly varying phenomena on the atomic scale. The calculation was performed using a more elaborate four-band Kane model (Kane, 1957) to describe the envelope wave functions and the dispersion relations in the constituent materials. Similar calculations by White and Sham (1981) have recently been reported. Bastard emphasized the specific importance of the CB–VB interaction in the bulk for the subband formation in the InAs–

GaSb superlattice, i.e., due to the CB–VB interaction in the bulk materials the InAs–CB is mainly S-type symmetry but also contains a P-type admixture. Similarly, the light hole GaSb is mainly P-type but contains an S-type admixture. This coupling is essential in the formation of the subbands in the InAs–GaSb superlattice, because it permits CB and VB wave functions to match. This effect is of minor importance for GaAs–Ga$_{1-x}$Al$_x$As superlattices because of the band lineup in that system and because the CB–VB interaction is less important in the bulk.

The calculational procedures described above have in common that they all seek to describe wave functions and dispersion relations of the bulk materials in an effective-mass type of formalism and achieve a coupling between these descriptions in a superlattice. Another approach starts from a more fundamental level and considers the superlattice primitive cell as a whole as consisting of Ga, Sb, In, and As ions. A full three-dimensional calculation has been made by Nucho and Madhukar (1978) and Madhukar and Nucho (1979) for superlattices with short periodicity. In this calculation the method of linear combination of atomic orbitals (LCAO) has been used. The description is simplified by tuning the parameters entering into the calculation in such a way that, in a simple nearest-neighbor approximation, the bulk band parameters are reproduced if a layer of only one type of material is considered. A similar approach has been adopted by Sai-Halasz et al. (1978) who considered only the atomic bonds in the superlattice direction. In this way it was possible to calculate the superlattice band structure for arbitrary layer thickness. Ihm et al. (1979) have performed a calculation using only the ionic pseudopotentials of In, As, Ga, and Sb without fitting parameters to the known band structure. They obtain the superlattice band structure and the microscopic electronic charge distribution of the valence electrons. Due to computational limitations, they could not include spin–orbit interaction, and therefore their results do not agree in detail with the results obtained by the other methods. For instance, they find the InAs–CB edge higher in energy than the GaSb–VB. The merit of their approach, however, is that it shows qualitatively the formation of a superlattice band structure and the range of the influence of the interface on the total charge distribution in the layers in a very first-principle-like way.

In Fig. 3 the band structure of InAs–GaSb superlattices with layers of equal thicknesses of each material is shown as a function of the superlattice periodicity. These results were actually calculated with the LCAO method (Sai-Halasz et al. 1978) but are in very close quantitative agreement with those calculated by the Bloch wave function matching scheme (Sai-Halasz et al., 1977 and Maan, 1979) and the matching of envelope wave functions (Bastard, 1981). In the calculations the InAs–CB is taken to be 0.15 eV

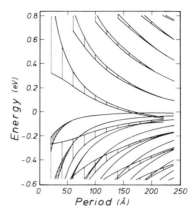

FIG. 3 Subband structure of InAs–GaSb superlattices with equal layer thicknesses $d/2$ as a function
of the periodicity d. The zero of the energy scale is
taken to be the top of the GaSb–VB. The subbands
whose energy increases with d are hole subbands;
those whose energy decreases are electron subbands.

below the GaSb–VB, because this value is found to give the best agreement
between the calculated superlattice band structure and the experimental
data. In Fig. 3 the evolution of three sets of subbands with the superlattice
periodicity can be distinguished; one set, at energies above zero on the
energy scale (taken to be at the level of the bulk GaSb–VB), converging to
-0.15 eV for increasing periodicities, and two sets of subbands increasing in
energy with increasing periodicity and approaching zero on the energy scale
(the energy of the bulk GaSb–VB). Of the latter two sets, one is fairly broad
and the other is nearly flat. At a periodicity of about 185 Å the two sets cross.

In a simplified manner, these results can be understood qualitatively by
considering the superlattice to be composed of two interwoven periodic
structures; InAs separated by GaSb barriers and GaSb separated by InAs
barriers. Thus the set of subbands that decrease in energy with increasing
periodicity is interpreted as due to the formation of electron subbands in
InAs confined by the GaSb layers, and the two sets that increase in energy as
hole subbands in GaSb confined by the InAs. For short periodicities the
InAs layers are thin, and due to the confining effect of GaSb they move to
higher energies. The InAs and GaSb have equal layer thickness, so the
coupling between successive InAs layers will be stronger, because for short
periodicities the GaSb barriers also are thin. This coupling is reflected by the
subband widths, which are broad for short periodicities and more narrow for
thicker layers. The same applies for the GaSb holes; but as can be seen from
Fig. 1, there are two sets of hole bands that have to be considered, one light
hole band (0.046 m_0) and one heavy hole band (0.33 m_0), which at the Γ
point are degenerate in the bulk. Due to the difference in mass, the light hole
bands (the broader set of subbands, which increases in energy for thicker
layers) move to lower energies than the heavy hole subbands (the narrower

set of bands). On the other hand, the light hole bands couple more efficiently through the InAs barriers (due to their longer de Broglie wavelength) than the heavy hole bands; therefore they are broader. The different behavior of the light and heavy hole bands lifts the degeneracy of these two bands at the center of the Brillouin zone, and two sets of hole subbands are formed. Figure 3 shows that the band gap in the superlattice is periodicity dependent. For periodicities below 150 Å this interesting fact has been observed experimentally by Sai-Halasz *et al.* (1978) with infrared measurements. The measured band gaps for superlattices with different thicknesses were fitted to calculated values using the energy difference between the InAs–CB and the GaSb–VB as a fitting parameter. The best agreement was found with a value of 0.15 eV, and this value has been used since.

For periodicities longer than 185 Å (see Fig. 3), the hole-like and electron-like subbands cross; and beyond 185 Å a hole subband exists at higher energy than the lowest electron subband, and the superlattice behaves as a semimetal. This transition can easily be understood in a simple double-confinement model; for thicker layers, electrons (and holes) become less confined and the lowest (highest) subband approaches the bulk band edge. Consequently, at some point the subbands cross, showing a semiconductor–semimetal transition. However, it is not an ordinary semimetal, because electrons and holes are mainly localized in different regions of space.

In the preceding section no account has been taken of space-charge effects due to free carriers. In the semiconducting region of thicknesses, this is certainly correct, as long as there is no selective doping of the layers. However, in the semimetallic region the hole subband, mostly confined in the GaSb, transfers electrons to the lower-lying electron subband, mostly confined in the InAs. Therefore an accumulation layer will exist of positively charged holes in the GaSb and negatively charged electrons in the InAs. This charge distribution creates an electric field across the interface, leading to band bending. If this is the case, the space-charge potential must be included in the band-structure calculation, because for thick layers the accumulation of charge at the interface will shield the band discontinuity, and the structure will ressemble that of several parallel, weakly interacting InAs–GaSb heterojunctions. The layer thicknesses for which this effect becomes important depends on the screening lengths for the electrons and holes in both materials, which will be discussed in section IV.

For clarity it has been somewhat emphasized that the InAs–GaSb superlattice can be considered as InAs confined between GaSb for electrons, and GaSb confined between InAs for holes. This approach to the superlattice band structure neglects one essential feature of this type of superlattice, namely, that a coupling exists between the wave functions of the InAs–CB and the GaSb–VB. This coupling between the wave functions of InAs–CB and GaSb–VB in a superlattice is possible because in the bulk materials the CB and VB interact. This interaction leads to a mixture of P-type and S-type symmetry for the CB and the VB wave functions; or otherwise stated, in an effective-mass approach both the light and the electron hole dispersion relations are nonparabolic and their wave functions at $k \neq 0$ are not orthogonal. This coupling, together with the light effective masses, makes the subbands much larger for a given layer thickness than is the case for $GaAs–Ga_{1-x}Al_xAs$ superlattices. The finite width of a the superlattice was, in fact, one of the important motivations for the study of this sort of material. Properly speaking, the term *superlattice,* a regular arranged periodic structure, is appropriate only if the subbands have a substantial width. If the bands are virtually flat the term *quantum well* is more suitable. Another consequence of the CB–VB interaction between the different layers is that in contrast with the usual Kronig–Penney model of electrons confined between barriers, as is the case for $GaAs–Ga_{1-x}Al_xAs$, in a InAs–GaSb superlattice the lowest subband of the electrons and the highest of the

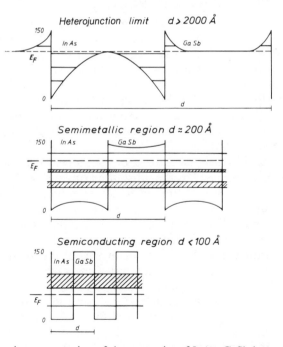

FIG. 4 Schematic representation of the properties of InAs–GaSb layered structures for different layer thicknesses.

holes have the largest width, whereas subbands that are further from the respective band edges are smaller. For GaAs–Ga$_{1-x}$Al$_x$As superlattices, on the other hand, the electrons at higher energy feel a smaller potential, giving rise to broader subbands, i.e., they are less localized.

The essential features of InAs–GaSb superlattices for different layer thicknesses are summarized in Fig. 4. For periodicities less than 185 Å they behave like a semiconductor, and for 185 Å $< d <$ 400 Å as a semimetal. For layers thicker than 1000 Å the space charge at the interface dominates, and the system is essentially a set of noninteracting heterojunctions.

III. Experimental Results of InAs–GaSb Superlattices

A. The Dispersion Relation in the Plane of the Layers

In experiments the FIR transmission at fixed frequency is measured as a function of the magnetic field. The magnetic field is generally perpendicular to the plane of the layers, and the electrons (holes) trace their orbits in the plane. In this case the level structure in a magnetic field is determined by the dispersion relation in the plane. The band structure shown in Fig. 3 has been calculated for the lowest kinetic energy in the plane of the layers. It may be assumed that as long as the wave functions of the electrons are localized mainly in the InAs layers, and those for the holes in the GaSb, the dispersion relations in the plane will be given by their respective bulk formula. Bastard (1982) has performed a full calculation for arbitrary transverse energy for semiconducting superlattices and has shown that this assumption is correct for layer thicknesses greater than 70 Å but is increasingly wrong for progressively thinner superlattices. This result is a direct consequence of the increasing three-dimensional nature of the superlattice with decreasing periodicities, as is expressed by the substantial subband width in that regime. His calculation, however, is limited to semiconducting superlattices. For semimetallic superlattices the closeness of hole and electron subbands and the crossing (or more likely, anticrossing) of their dispersion relations in the plane (see Fig. 5) require an even more elaborate calculation, which has not yet been performed. For the layer thicknesses of the samples used in the experiments, the assumption that the transverse dispersion relation can be described by the bulk is adequate.

In Fig. 5 the schematic three-dimensional band structure of a superlattice with a periodicity of 200 Å is shown with, in the superlattice direction (z-direction), one hole and three electron subbands. The effect of the superlattice band structure is that electron (hole) subbands are formed at higher (lower) energies with respect to the InAs–CB (GaSb–VB) edge (the

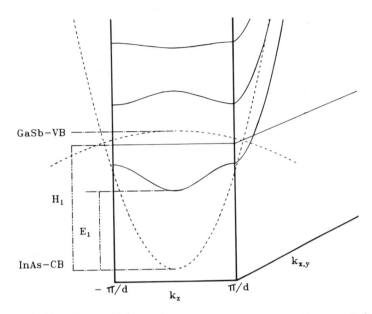

FIG. 5 Schematic three-dimensional band structure of a semimetallic InAs–GaSb super-lattice. In the superlattice direction (z-direction) the subbands are shown. In the plane the dispersion relations are assumed to be given by those of the InAs–CB and a heavy hole for the VB.

dotted line). Because the InAs–CB is nonparabolic, this will have an effect on the transverse dispersion relation. The InAs–CB in the two-band Kane model adequate for the range of energies of interest) is given by

$$\frac{\hbar^2}{2m_0^*}(k_x^2 + k_y^2 + k_z^2) = E\left(1 + \frac{E}{E_G}\right), \tag{1}$$

where m_0^* is the bottom of the band effective mass; k_x, k_y, and k_z are the wave vectors; and E is the energy measured from the InAs–CB edge. According to the analysis of Bluyssen et $al.$ (1979) the bottom of the band in a superlattice is given by Eq. (1) for $k_x = 0$ and $k_y = 0$, whereas $\hbar^2 k_z^2/2m_0^*$ has the lowest energy allowed in the superlattice, E_1, i.e., the shift in the CB edge. For a magnetic field perpendicular to the plane of the layers, as corresponds to the experimental situation, k_x and k_y can be replaced by

$$(N + \tfrac{1}{2})\,\hbar eB/m_0^*, \tag{2}$$

where N is the Landau level quantum number, B is the magnetic field, and

m_0^* is the effective mass at the bottom of the bulk InAs–CB. The Landau levels are then given by

$$\frac{\hbar eB}{m_0^*}(N + \tfrac{1}{2}) = E_N\left(1 + \frac{E_N}{E_G}\right) - E_1\left(1 + \frac{E_1}{E_G}\right), \tag{3}$$

and the cyclotron effective mass, the quantity observed in the experiments, is

$$m^*(E_1,B) = m_0^*\left[1 + \frac{E_{N+1}(E_1,B) + E_N(E_1,B)}{E_G}\right]. \tag{4}$$

B. EXPERIMENTAL RESULTS OF SEMICONDUCTING SUPERLATTICES

The equations derived in the previous section were used (Bluyssen *et al.*, 1979) to interpret the first magnetooptical data on InAs–GaSb superlattices. The properties of the sample used in their experiments are listed in Table I. Because the sample is in the semiconducting region of layer thicknesses, it is intentionally *n*-type doped in order to observe free-carrier absorption. The Fermi energy corresponding to the two-dimensional carrier density can be calculated by integrating the density of states. Neglecting, at present, the subband curvature and including the nonparabolicity of the effective mass, E_F can be obtained from

$$\int_{E_1}^{E_F} g(E)\,dE = \int_{E_1}^{E_F} \frac{m_0^*}{\pi\hbar^2}\left(1 + \frac{2E}{E_G}\right) = n_s. \tag{5}$$

The value of 40 meV above the lowest subband (Table I) was obtained this way. In this calculation the theoretical band structure value for E_1 has been used. Because the Fermi energy is in the band, the sample is a degenerate semiconductor.

The results of the FIR transmission experiments are shown in Fig. 6. The

TABLE I

PARAMETERS OF THE SEMICONDUCTING SAMPLE S_1

Parameter	Value
InAs layer thickness	65 Å
GaSb layer thickness	80 Å
Number of periods	125
Carrier concentration (Hall effect)	4.8×10^{17} cm^{-3}
Carrier mobility (Hall effect)	7300 cm^2/Vsec
Calculated subband energy E (measured)	195 meV (145 meV)
Calculated Fermi energy (above first subband)	40 meV

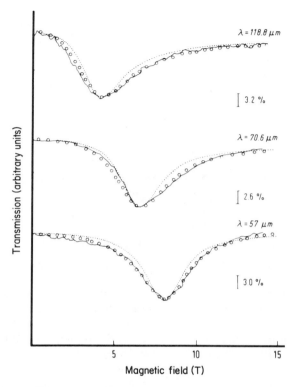

FIG. 6 Measured FIR transmission as a function of the magnetic field for the semiconducting superlattice S_1. The dots and the dashed curves represent the calculated transmission for different values of the momentum scattering time, as described in the text.

magnetic field in these experiments was perpendicular to the plane of the layers. A fairly broad absorption maximum is observed, and its resonance position depends linearly on the FIR energy, as is shown in Fig. 7. Interpreting the results as cyclotron resonance, one derives a mass of 0.043 m_0 from the slope in Fig. 7. This effective mass is much higher than that of bulk InAs with the same carrier density. The observed effective-mass enhancement is a direct consequence of the shift in the InAs subband edge by the superlattice formation. The Landau levels in this superlattice were calculated with the use of Eq. (3) in Fig. 8, and the observed transitions are shown by the arrows. Transitions take place below and above the Fermi energy. For $\lambda = 118.8\ \mu m$ a transition between Landau levels with $N = 3$ and 4 is observed, and for $\lambda = 70.6\ \mu m$ and $\lambda = 57\ \mu m$ a transition between $N = 1$ and 2. The effective mass calculated from Eq. (4) at an energy corresponding to these transitions

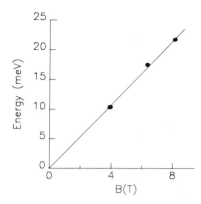

FIG. 7 Resonance position as a function of the radiation energy for sample S_1. The straight line represents the resonance position of an effective mass of 0.043 m_0.

using the calculated value of E_1 is 0.047 m_0, in fair agreement with the experiment.

A more rigorous analysis of these data has been presented by Bluyssen *et al.* (1982). In this paper, the transmission of the superlattice is calculated as a function of the magnetic field, taking into account plasma effects and line-shape distortion due to interference in the plane substrate. The procedure is to calculate the response of a two-dimensional sheet of carriers with density n_s, an effective mass m_0^*, and a momentum scattering time τ as a

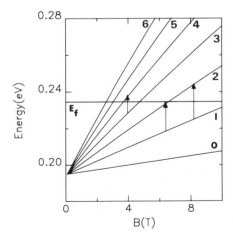

FIG. 8 Landau levels for sample S_1 using Eq. (3) with the calculated band-structure parameters. The observed transitions are denoted by the arrows.

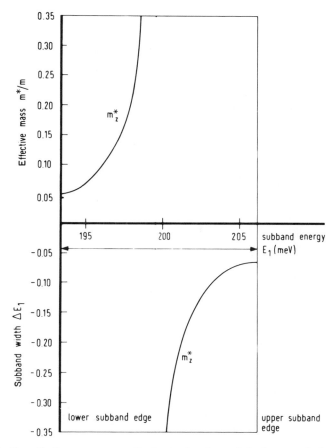

FIG. 9   The effective mass perpendicular to the plane of the layers, calculated from the subband curvature for sample S_1 as a function of the energy.

function of the magnetic field. The mass is related through Eq. (4) to the magnetic field, the Landau level quantum number N, and the shift in the subband E_1. The subband energy E_1, a superlattice band-structure parameter, can then be deduced from a cyclotron resonance measurement, using the energy dependence of the effective mass. From the fit of the calculated line shape to the experimental data, two parameters can be determined: the shift in the subband E_1 and the scattering time. The dots in Fig. 6 represent the result with fit parameters $E = 145$ meV and $\tau = 1.4 \times 10^{-13}$ sec. This scattering time is 75% shorter than that obtained from the measured Hall mobility. For comparison, the fit with the same E_1 but with $\tau = 1.8 \times 10^{-13}$ sec equal to the Hall value is shown as the dotted line in Fig. 6. It is a

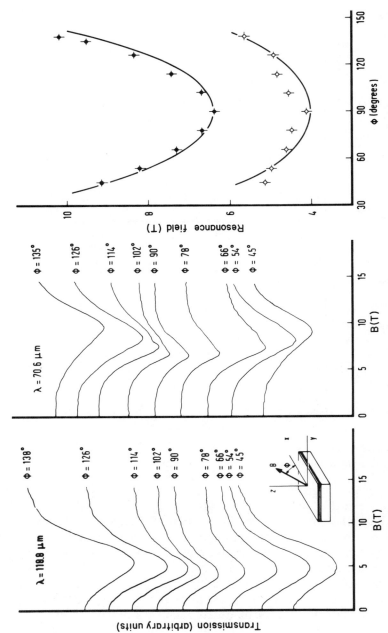

FIG. 10 Evolution of the transmission spectra of sample S_1 as a function of the angle between the plane of the layers and the magnetic field.

well known fact from extensive studies of the cyclotron resonance linewidth in two-dimensional systems (Ando, 1975; Abstreiter, 1976) that the observed width is related to the dc scattering time τ_0 by

$$\frac{1}{\omega_c\tau} \approx 0.65 \left(\frac{1}{\omega_c\tau_0}\right)^{1/2}. \tag{6}$$

In the present case this leads to a calculated τ of 1.4×10^{-13} sec at 8. T, in fairly good agreement with the experimental results. The obtained value of E_1 of 145 meV is somewhat lower than that calculated from the superlattice band structure. This deviation has been attributed to the shortcomings of the model that describes the parallel dispersion relation by that of InAs at a higher energy.

The effective mass in the direction of the superlattice has been calculated from the subband curvature, and the results are shown in Fig. 9. The effective mass is fairly low in the center and at the edge of the Brillouin zone and very high between these points. Therefore an attempt has been made to observe this longitudinal effective mass by measuring the transmission for different angles between the plane of the layers and the magnetic field. In Fig. 10 the angular dependence of the CR spectra and the position of the transmission minimum as a function of the angle φ is shown. The resonance position is compared with the relation $B_{res} = B_0 \cos \varphi$ which describes the angular dependence of an entirely two-dimensional electron gas and neglects, therefore, the regions of a finite transverse electron mass, as shown in Fig. 9. It was found that in general the two-dimensional relation is well obeyed. This behavior has been explained as being caused by the relatively large broadening of the subband due to scattering. With the measured scattering time from the CR, a subband broadening due to scattering of the order of 5 MeV was derived. As can be seen from Fig. 9, the energy regions of the subband in which the transverse mass is small are much smaller than this broadening. Therefore the broadening masks the three-dimensional aspect of the superlattice.

C. EXPERIMENTS ON SEMIMETALLIC SUPERLATTICES

The first magnetooptical experiments of semimetallic superlattices were reported by Guldner *et al.* (1980) with magnetic fields up to 5 T. The same samples were later measured using high magnetic fields up to 20 T (Maan *et al.*, 1981). The samples studied were S_2 and S_3, and their electronic properties are summarized in Table II. In these samples the free carriers come essentially from the transfer of electrons from the GaSb hole subband to the electron subband, mainly localized in InAs. The Fermi energy is determined by the condition that the hole and electron densities must be equal. For the layer thicknesses in these samples, band bending due to the accumulated

TABLE II

PARAMETERS OF THE SEMIMETALLIC SAMPLES S_2 AND S_3

Parameter	S_1	S_2
InAs layer thickness	120 Å	200 Å
GaSb layer thickness	80 Å	100 Å
Calculated subband energy E_1 (measured)	87.5(100 ± 15) meV	64.5(70 ± 10) meV
Calculated subband energy H_1 (measured)	129.5(139 ± 15) meV	132.5(128 ± 10) meV
Calculated subband width ΔE (measured)	21(23 ± 1) meV	16(19 ± 1) meV

charge at the interface is of minor importance, and the subband positions are given essentially by the results of the LCAO calculation.

The experimental results for the magnetotransmission are shown in Fig. 11. The transmission shows several absorption minima superimposed on an increasing background. In Fig. 12 the position of these resonance peaks as a

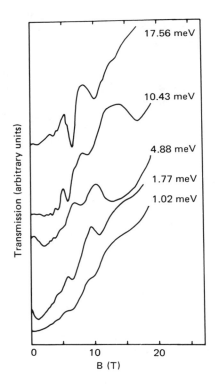

FIG. 11   FIR transmission of the semimetallic superlattice sample S_2 as a function of the magnetic field for different radiation energies.

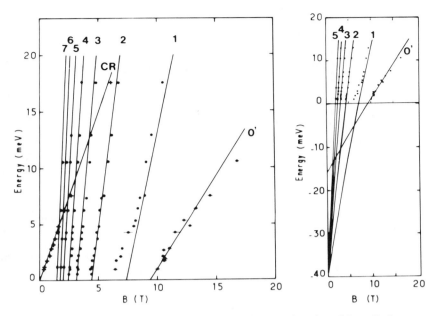

FIG. 12 Magnetic field position of resonant transitions as a function of the radiation energy for sample S_2. The lines drawn correspond to theoretical fits as described in the text.

function of FIR energy is plotted. Three sets of transitions are observed. One set extrapolates to zero energy for zero magnetic field, one set of several different transitions extrapolates to a negative energy, and one set extrapolates to a somewhat less negative energy. These sets of transitions were interpreted, respectively, as the CR of the electrons (a), the interband transition between the hole subband and the electron subband at the center (b), and at the edge at $k_z = \pi/d$ (c) of the superlattice Brillouin zone (indicated schematically in Fig. 13).

As before, the cyclotron resonance transitions take place between Landau levels below and above the Fermi energy; similarly, the interband transitions take place between the filled hole subbands and the empty electron subbands. The cyclotron resonance data are interpreted as for the semiconducting sample using Eq. (3), and the interband transitions are described in a similar fashion. The Landau levels for the electrons is given by Eq. (3); the Landau levels for the holes are taken to be those of a parabolic band and are given by

$$H_1^N = H_1 - \frac{\hbar eB}{m_{HH}^*}, \qquad (7)$$

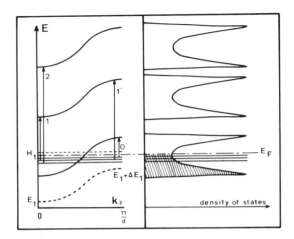

FIG. 13 Subband structure in a magnetic field with possible transitions (left) and associated density of states of a semimetallic sample (right).

where m^*_{HH} is the heavy hole mass and H_1 the heavy hole subband energy. In principle, the valence-band structure is much more complicated than that of the conduction band, especially in the semimetallic samples in which the valence and conduction bands are very close in energy. However, it has been found experimentally that using the heavy hole mass in Eq. (7) gives satisfactory agreement between the experimental and theoretical values for the superlattice band-structure parameters. Assuming a selection rule $\Delta k_z = 0$, one then determines the interband transitions by

$$\hbar \omega = -\frac{E_G}{2} + \frac{E_G}{2}\left[1 + 4\left(N + \frac{1}{2}\right)\frac{\hbar eB}{m^*_0} + 4E_s\left(1 + \frac{E_s}{E_G}\right)\right]^{1/2}$$
$$-H_1 + \left(N + \frac{1}{2}\right)\frac{\hbar eB}{m^*_{HH}}, \tag{8}$$

with $E_s = E_1$ for the transitions at $k_z = 0$ and $E_s = E_1 + \Delta E_1$ for those at $k_z = \pi/d$. These equations were used to fit the experimental data. The only fitting parameters are E_1, ΔE, and H_1 which are thus determined directly from the experiment. The resulting values, given in Table II, compare favorably with the calculated values. These results unambiguously show the semimetallic nature of these samples by the observation of a negative energy gap.

Furthermore, the observation of a negative energy gap proves that the bulk InAs–CB edge must be lower in energy than the bulk GaSb–VB. Without reference to a model the difference between E_1 and H_1 can be derived directly from experiment as the intersection of the observed inter-band transitions with the energy axis in Fig. 12. The values thus obtained are 40 and 58 meV for S_2 and S_3, respectively. These values are more accurate than the energy values given in Table II, because the principal error is due to the fitted value for E_1. These values show that the overlap of the CB and the VB must be much more than 58 meV. A value of 150 meV was used in the band-structure calculation, and with this value a good agreement between calculated band structure and experiments is obtained. This value was based on a best fit of the calculated band structure and the observed periodicity dependent band gap. Madhukar and Nucho (1979) obtained a best agreement with these same experimental data using a different model with a value of 60 meV for the band overlap. However, the experimental results show a measured difference between H_1 and E_1 of 58 meV, meaning that the CB–VB overlap must be considerably larger.

Another important result from these experiments is that the subband width in the superlattice is measured in a very direct way, and as such it is clear evidence for the pseudo three-dimensional character of these systems. The width obtained is quite substantial for the layer thicknesses involved and is, in fact, much larger than would have been obtained with similar thicknesses in the usual GaAs–$Ga_{1-x}Al_xAs$ system. The agreement between the experimental value and the calculation shows a specific feature of these superlattices, i.e., the importance of the coupling between the GaSb–VB and InAs–CB wave functions that is responsible for the substantial subband width.

D. HELICON WAVE PROPAGATION

The electromagnetic properties of a different type of sample have been studied by Maan et al. (1982). This sample was highly doped with Sn, leading to a carrier concentration of 3.4×10^{18} cm^{-3} which corresponds to a calculated Fermi energy of 340 MeV above the lowest subband. It has been established that Sn acts as a donor in InAs and as an acceptor in GaSb. Therefore the free-carrier density is almost entirely confined in the InAs layers; so the sample can be considered as consisting of several layers with a high carrier density separated by isolating barriers. The high carrier density implies that the plasma frequency is much higher than the FIR frequency; therefore mainly plasma effects can be studied in this sample. Recently there has been a theoretical interest in the electromagnetic properties of this type of multicomponent plasma (Belyantsev and Ignatov, 1977; Das Sarma and Madhukar, 1981; and Das Sarma and Quinn, 1982). In addition, from the band-structure point of view the interest in the experiments on this sample is that up to three subbands are occupied and that the calculated Fermi energy is within the GaSb band gap, whereas with the previous samples the Fermi energy was in or close to the overlap region (the parameters of the helicon wave sample are summarized in Table III).

The experimental results (Fig. 14) show a drastic increase in transmission with a pronounced transmission maximum. The transmission at the maximum was found to be up to 60 times as high as the transmission at zero magnetic field. At the higher radiation energies the transmission maximum is preceded by a shoulder. The positions of both the maximum and the shoulder depend linearly on the radiation energy and the magnetic field, as shown in Fig. 15. These results were interpreted as helicon wave propagation (Maan *et al.*, 1982).

To explain their results we briefly review the essential properties of a plama in a magnetic field (the subject is treated extensively by Brazis *et al.*, 1979). The conductivity tensor of a three-dimensional electron gas with a magnetic field pointing in the positive z direction is given by

$$\sigma = \begin{pmatrix} \sigma_+ & 0 & 0 \\ 0 & \sigma_- & 0 \\ 0 & 0 & \sigma_z \end{pmatrix}, \tag{9}$$

where the tensor components σ_+, σ_-, and σ_z in a rotating coordinate frame are given by

$$\sigma_\pm = \frac{ne^2}{m^*}\left[\frac{\tau}{1 - i(\omega \pm \omega_c)\tau}\right],$$
$$\sigma_z = \frac{ne^2}{m^*}\left[\frac{\tau}{1 - i\omega\tau}\right], \tag{10}$$

with m^* the effective mass, ω the signal frequency, τ the momentum

TABLE III

PARAMETERS OF THE HEAVILY DOPED HELICON WAVE SAMPLE S_4

Parameter	Value
InAs layer thickness	150 Å
GaSb layer thickness	150 Å
Number of layers	67
Carrier density (Hall effect)	3.4×10^{18} cm^{-3}
Mobility	14000 cm^2/Vsec
Electron subband energies E_1	58 meV
E_2	178 meV
E_3	308 meV
Fermi energy (above lowest subband)	286 meV

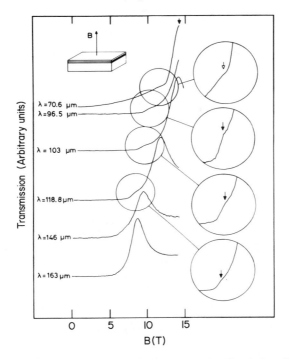

FIG. 14   FIR transmission as a function of the magnetic field for the heavily doped sample S_4.

scattering time, ω_c the cyclotron frequency, and n the three-dimensional electron concentration. The dynamic conductivity tensor is related to the complex dielectric tensor by

$$\varepsilon = \varepsilon_0 \left(\varepsilon_2 \mathbf{I} + i\, \frac{\sigma}{\omega \varepsilon_0} \right),$$ (11)

where ε_2 is the static dielectric constant, $\mathbf{I}$ the unit matrix, and ε_0 the permittivity of free space. Under the conditions that the cyclotron frequency is of the order of the signal frequency, much smaller than the plasma frequency ($\omega_p^2 = ne^2/m^*\varepsilon_0\varepsilon_2$) and much higher than the collision frequency τ, the dielectric tensor reduces to

$$\varepsilon = \varepsilon_2 \left[1 - \frac{\omega_p^2}{\omega(\omega \pm \omega_c)} \right].$$ (12)

The $+$ sign corresponds to the cyclotron inactive and the $-$ sign to the

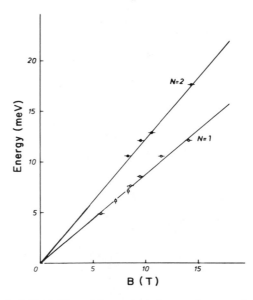

FIG. 15   Magnetic field position of the transmission maximum and the shoulder at lower magnetic fields (indicated by the arrows in Fig. 14) as a function of the radiation energy. The lines drawn represent the theoretical fit as described in the text.

cyclotron resonance active (CRA) polarization mode. In this case one sees that the dielectric constant is a function of the magnetic field. Wave propagation can take place only when $\varepsilon > 0$, and this occurs for $\omega_c > \omega$ for the CRA propagation mode. These propagating modes are called helicon waves.

Because the dielectric constant is a function of the magnetic field, interference can take place as half of the wavelength in the medium $(\lambda/\sqrt{\varepsilon})$ is equal to an integer N times the total sample thickness d; $\frac{1}{2}\lambda/\sqrt{\varepsilon} = d/N$. Neglecting at present the layered nature of the superlattice and considering it simply as a material with a carrier density n and a total thickness d, one can write the expression for the positions of the transmission maxima as

$$\frac{B}{\omega} = \frac{m^*}{e}\left(1 + \frac{ne^2d^2}{m^*\varepsilon_0 N^2\pi^2c^2}\right). \tag{13}$$

This equation can be fitted to the experimental results with n and m^* as fitting parameters. The results of the fit are shown in Fig. 15 for $n = 3.4 \times 10^{18}$ cm^{-3} and $m^* = 0.0082\ m_0$. The free-carrier density is very close to the average carrier density as obtained with the Hall measurement. It should be recalled that this result is derived from the standard three-dimensional formula; therefore the three-dimensional carrier density, and not the density of the InAs layers only, is obtained. This may come as a surprise because the InAs layers contain twice the average carrier density whereas the GaSb layers have essentially no free carriers. An intuitive explanation is that the wavelength in the medium is twice (once) the superlattice total thickness for $N = 1$ ($N = 2$) and is in any case much longer than the individual layer thickness. Therefore the wave sees only an average carrier density. In fact it can be shown in a more rigorous way that this should indeed be the case. The mass derived from the experiments, $(0.082 + 0.005)m_0$, is substantially higher than that which one gets from Eq. (4) using the calculated band-structure parameters, $0.063\,m_0$.

E. THE EFFECTIVE MASS

It is instructive to compare the effective masses obtained from several experiments with those obtained for bulk InAs. In the comparison, for bulk InAs the energy E is the Fermi energy either for highly doped samples or for experiments using high magnetic fields. For the superlattice $E = E_F + E_{pot}$, the latter being the potential energy of the subbands in the InAs well. The result (Fig. 16) shows a fair agreement for energies below 200 meV but

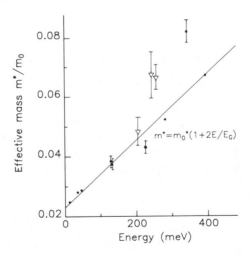

FIG. 16 A comparison of the InAs effective mass as a function of energy with the effective mass in a InAs–GaSb superlattice. The straight line represents the usual energy dependence of the InAs effective mass, and the closed data points are literature values. (After Palik and Stevenson, 1963, and Litton et al., 1969, from cyclotron resonance; and Thomas and Woolley, 1971, from plasma reflection.) The closed data points with the error bars are the superlattice results from magnetooptical measurement, and the triangles are from Shubnikov–de Haas measurements (Sakaki et al., 1978) at an energy $E = E_F + E_{pot}$.

shows discrepancies at higher energies. In the figure the results of the effective-mass determination of other superlattices using the temperature dependence of the Shubnikov – de Haas effect are also included (Sakaki *et al.* 1978). At these higher energies a systematic higher effective mass is observed than one would expect from Eq. (4). At present no satisfying explanation exists for this discrepancy; either the assumption that the superlattice effective mass is described by that of InAs at higher energies is in error, or the calculated bandstructure is wrong at energies above 200 meV. The observed descrepancy is not easy to explain because one would expect that a confinement picture is more correct for higher energies, as is suggested by Fig. 3 which shows the decreasing subband width under these conditions. Especially at the energies within the GaSb band gap, a model that assumes carriers to be localized in the InAs would be expected to be fairly correct. In addition, it should be noted that a mass enhancement due to many-body interactions, as has been observed in a purely two-dimensional system like the MOS field transistor (Wilson *et al.* 1980, 1981), is not very likely for InAs – GaSb heterojunctions. Many-body effects become important whenever the kinetic energy of the carriers becomes small compared with the potential energy due to their mutual Coulomb interaction. Because the electrons in the InAs layer possess a high kinetic energy these effects will be unimportant.

IV. The InAs-GaSb Heterojunction

A. THE HETEROJUNCTION SUBBAND STRUCTURE

For very large layer thicknesses, because a hole subband is formed at higher energy than the electron subband, electrons will flood from the GaSb layer into the InAs, leaving holes behind; thus an accumulation layer of holes in the GaSb and of electrons in the InAs is built up. The band structure in this case will be determined by the electrostatic potential due to the space-charge layer and not by the band interaction as for the superlattices. This band bending will start to play a role in the semimetallic region; however, as long as the layer thicknesses remain small compared with the width of the accumulation layer, as was the case for the semimetallic samples used in the previously described experiments, the effect can be neglected. Space-charge layers have been extensively studied in several systems, such as inversion and accumulation layers, in several semiconducting materials. (A complete review is given by Ando *et al.,* 1982; see also Kotthaus, 1978 and Ando, 1978.) In these cases charge is accumulated at an interface due to an externally applied electric field (Stern and Howard,

1967), and extensive calculations of the band structure of these systems exist. These calculations show the formation of two-dimensional (flat) subbands due to the confinement of the carrier motion along the interface. The InAs–GaSb interface is very similar to these systems except for the important difference that the calculation has to be performed self-consistently in the sense that at the interface the total number of electrons and holes must be equal. A rather simplified model has been used (Guldner *et al.*, 1982) to calculate the band structure. This model turns out to be sufficiently accurate to describe the experimental results and will be described here.

Figure 17 shows schematically the band structure of the interface between GaSb and InAs. At the interface the band edges are at their bulk positions, and away from the interface the bands are bent, due to the dipole field caused by the accumulated charge. If GaSb is lightly *p*-type doped and InAs lightly *n*-type (as was found to be the case for unintentionally doped GaSb and InAs layers), the electron (hole) bands are bent up (down) to the Fermi energy located at the VB in GaSb and at the CB in InAs. The position of the Fermi energy at the interface is dictated by the condition of charge neutrality. A simplified expression for band bending has been derived by Keyes (1976) using the Thomas–Fermi approximation. At low temperatures the electron density at any position z is

$$n(z) = \int_0^{E_b(z)} N(E) \, dE, \tag{14}$$

where E_b is the the energy measured from the band edge, which depends, of course, on z (see Fig. 17). For the density of states $N(E)$ in a parabolic band

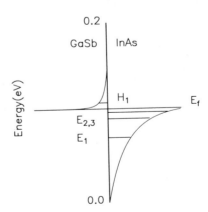

FIG. 17 Electronic structure of the InAs–GaSb interface. Potential wells are created by space charge for the holes in GaSb and the electrons in InAs. One hole and three electron subbands are found in the wells.

one has

$$N(E) = 4\pi \left(\frac{2m^*}{h^2}\right)^{3/2} E_b^{1/2}. \tag{15}$$

The potential φ, which equals E_b/e, must satisfy the Poisson equation

$$\frac{d^2\varphi}{dz^2} = \frac{n(z)e}{\varepsilon} = \frac{e}{\varepsilon}\frac{8\pi}{3h^3}(2me\varphi)^{3/2}. \tag{16}$$

The solution is given by

$$E_b = E_b^0 / \left(1 + \frac{z}{z_0}\right)^4, \tag{17}$$

with

$$z_0 = \left(\frac{(4\pi)^2 225\,\hbar^6\pi^2\varepsilon^2}{8m^{*3}e^4E_b^0}\right)^{1/4}, \tag{18}$$

and E_b^0 the band bending at the interface ($z = 0$).

These equations describe the band bending of the VB and the CB when the appropriate masses for these bands are substituted. The Fermi energy E_F should be determined next. First the subbands are calculated using the classical Bohr–Sommerfeld quantization rule

$$\int_0^{z_t} k_z dz = \pi(\phi + l), \tag{19}$$

where l is an integer and ϕ is a phase factor that generally depends on the shape of the potential well. For a triangular shape this phase factor can be shown to be exactly equal to 0.75, which is the value used here as an approximation (Landau and Lifschitz, 1958). The classical turning point is z_t, i.e., the value of z for which E_b is zero; as before, E_b is the energy measured from the band edge.

Antcliffe et al. (1971) used this approach to calculate the subbands in nonparabolic materials, and it was also applied by Guldner et al. (1982) to the InAs–GaSb heterojunction. The nonparabolic band for InAs has been given by Eq. (1), and from this equation k_z can be obtained:

$$\int_0^{z_t} \left\{\frac{2m_0^*}{\hbar^2}\left[E - E_b^0\left(1 - \left(1 + \frac{z}{z_0}\right)^{-4}\right)\right]\right.$$
$$\left. \times \left[1 + \frac{E}{E_G} - \frac{E_b^0}{E_G}\left(1 - \left(1 + \frac{z}{z_0}\right)^{-4}\right)\right] - k_\parallel^2\right\}^{1/2} = \pi\left(\frac{3}{4} + l\right) \tag{20}$$

This equation can be solved numerically for a certain E, and the different subbands follow for different values of N. For the holes a similar approach can be used, neglecting nonparabolicity and using the energy distance at the interface minus E_b^0 for the band bending. The electron and hole density at the interface can be calculated for the subbands thus obtained. If the densities are not equal, the calculation is repeated for other trial values of E_b^0 until charge neutrality is obtained. This value of E_b^0 is the Fermi energy at the interface. Table IV summarizes the subband energies and the Fermi energy for the InAs–GaSb heterojunction. One heavy hole subband is found above and three electron subbands are found below the Fermi energy. The characteristic length of the band bending z_0 is 550 Å for the electrons and 110 Å for the holes with an electron and hole density at the heterojunction of 4.6×10^{11} cm^{-2}.

B. EXPERIMENTS

Guldner *et al.* (1982) have measured the FIR magnetotransmission of an InAs–GaSb sample with 1000 Å layer thicknesses. Because the characteristic length for the band bending is 550 Å and 110 Å in InAs and GaSb, respectively, they interpreted their results as being typical for a single heterojunction. The transmission data and the plot of the observed transmission minima as a function of the radiation energy (Figs. 18 and 19) are very similar to the results obtained with the semimetallic samples. This is not too surprising because in both cases a hole subband is formed at an energy higher than the lowest electron subbands and "a negative energy gap" is observed. The qualitative difference with the superlattice results is the observation of a split CR peak that is clearly observed in the spectra with the highest radiation energy. This doubling of the CR peak is explained as being due to CR of different subbands that have different masses. This effect is typical for CR in accumulation layers of nonparabolic materials with multiple occupied subbands, and it is directly related to the formation of a potential well by band bending. In a superlattice the mass observed at the Fermi energy is equal for all subbands, because at the Fermi energy all levels

TABLE IV

CALCULATED BAND STRUCTURE PARAMETERS OF THE InAs–GaSb
HETEROJUNCTION AT THE INTERFACE

Parameter	Value (meV)
Electron subbands in InAs E_1	72
E_2	95
E_3	102.5
Hole Subband H_1	107.5
Fermi energy	104

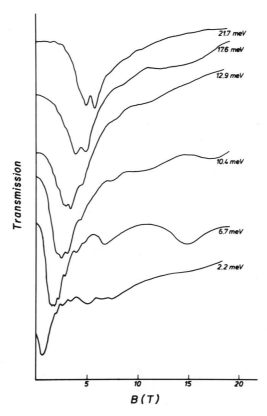

FIG. 18 Measured FIR transmission as a function of the magnetic field for different for energies of sample S_5.

from different subbands are equally far from the bulk band edge. In a potential well caused by band bending this is not the case. Here the mass that is associated with a certain subband depends on the average position of the electron in the well. This position is determined by the probability density, which is obtained from the amplitude of the wave function perpendicular to the plane. The Bohr–Sommerfeld approximation is very appropriate to describe this effect. The integral in Eq (19) automatically constructs the average of all possible electron states with a certain momentum parallel to the plane $(k_\parallel)$ that are compatible with the boundary conditions imposed by the potential well. Solving Eq (19) for every $k_\parallel$ gives directly the in-plane dispersion relation, and the Landau levels in a magnetic field can be directly calculated by substituting $\hbar^2 k_\parallel^2 / 2m^*$ by $(N + \frac{1}{2})\hbar\omega_c$. This way the Landau levels for the electrons can be calculated. The same technique can be applied

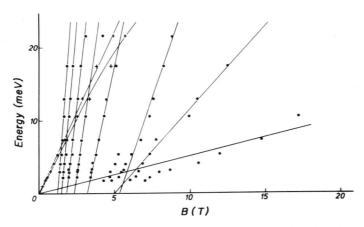

FIG. 19 Position of the observed resonant transitions (Fig. 18) as a function of the radiation energy.

for holes and electrons, and the cyclotron resonance and interband positions are obtained and can be compared with the experiment.

The lines drawn in Fig. 19 represent the calculation and show very good agreement with the experimental data. The two lines going through the origin represent the calculated double cyclotron resonance of the two lowest subbands. Also shown in Fig. 19 are the observed transitions between hole and electron subbands. The lines drawn are calculated results for these transitions. In this sample, also, a transition is observed at high magnetic fields that was the attributed CR of the heavy holes. The line with the smallest slope in Fig. 19 represents the position of the hole CR for a mass of 0.26 m_0, which is equal to the hole mass in the 100 direction. This fact must be considered coincidental because it can hardly be expected that the bulk heavy hole mass is obtained from the in-plane dispersion relation, and the dispersion relation is expected to be much more complicated. As with the semimetallic samples there will be interaction between the hole and electron bands as they approach each other in the transverse plane. As can be seen from Fig. 19, the resonance attributed to the holes has no simple linear magnetic field versus energy dependence, which is probably due to this effect. It is interesting to note that the difference between E_1 and H_1 obtained from the calculations and the experiments in the heterojunction is 37 meV, which is less than the 58 meV obtained for the semiconducting sample S_3. This is a direct demonstration of the influence of band bending, without which a much larger value would be found for a 1000 Å/1000 Å sample. The potential due to the accumulation layers, however, once more confines the electrons and holes as the confining due to layer thickness becomes less for thicker layers.

C. Hybrid Subband Formation

One of the usual techniques for verifying the dimensional character of a system is to measure the angular dependence of the Shubnikov–de Haas oscillations or of CR. In two-dimensional systems it is usually assumed that because carrier motion is in a plane, only the magnetic field component perpendicular to the interface determines the resonance position. Classically, one would intuitively expect that if the cyclotron radius fits in the dimensions of the well, the two-dimensional character is no longer observable. Another way of expressing this condition is that if the Landau level splitting is comparable to the splitting between electric subbands, mixing between electric and magnetic subbands will take place. The angular dependence of the sample described above was studied by Maan *et al.* (1982) to investigate this effect. In this experiment the magnetic field component

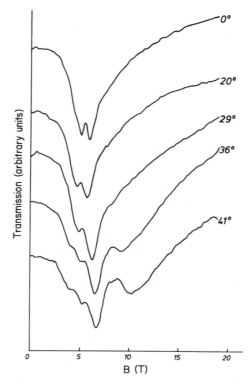

Fig. 20 FIR transmission of sample S_5 as a function of the magnetic field at $\lambda = 57 \ \mu m$ (E = 21.7 meV) for different angles between the magnetic field and the normal to the plane of the layers.

parallel to the interface is increased by turning the sample. The experiments of the angular dependence have shown pronounced deviations of a simple $\cos(\varphi)$ behavior (Fig. 20), as would be expected if only the magnetic field component perpendicular to the surface determines the resonance position. Even for the case of a magnetic field parallel to the layers, a resonance has been observed (Fig. 21). The position of this resonance is plotted in Fig. 22. At lower radiation energies no structure is observed because the resonance position does not extrapolate to zero energy for zero magnetic field. The observed transition is clearly different from that of simple CR in the InAs layer as is indicated by the dotted line in Fig. 22.

These results have been explained (Maan *et al.*, 1982) as being due to the mixing of electrostatic and magnetic subbands. The Hamiltonian of an electron in a potential well in a magnetic field perpendicular to the potential variation is given by

$$\mathcal{H} = \frac{\hbar^2}{2m^*}\left[\frac{\delta^2}{\delta z'^2} + \frac{z'^2}{l^4} + \frac{2m^*}{\hbar^2}\,V(z)\right]$$

$$z' = z + \frac{\hbar k_y}{eB}, \tag{21}$$

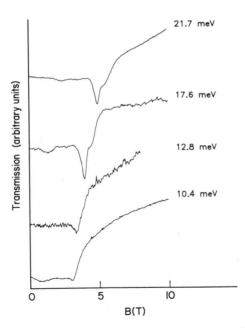

FIG. 21   FIR transmission as a function of the magnetic field parallel to the plane of the layers (sample S_5).

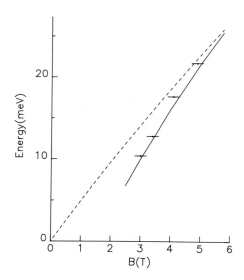

FIG. 22 Position of the transmission minimum (Fig. 21) as a function of the radiation energy. The solid line represents the calculated transition between hybrid subbands as described in the text. The dotted line is the position of CR in pure InAs.

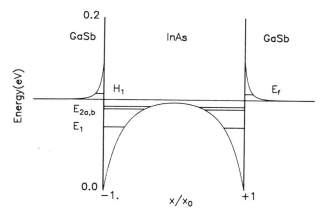

FIG. 23 Energy-level structure of the double GaSb–InAs–GaSb heterostructure. The characteristic length of the bond bending x_0 is half the layer thickness.

where l is the magnetic length, which equals the CR orbit radius for the lowest Landau level. It is clear from Eq. (21) that a simple $\cos(\varphi)$ behavior can be observed only if the potential term $V(x)$ dominates over the diamagnetic term z'^2/l^4. In many two-dimensional electronic systems, as in the Si MOSFET or the GaAs–Ga$_{1-x}$Al$_x$As heterojunction, this is the case for attainable magnetic fields, but for the InAs–GaSb heterojunction it is true only for the GaSb. For the InAs the characteristic length of the well is 5 times longer than for GaSb, and its mass is 14 times lighter. Equation (21) conveniently expresses the conditions under which the effect of a parallel magnetic field component can be neglected. This depends on the relative strength of the ratio z'^2/l^4, i.e., the well width compared to the cylotron radius to the potential term $(2m^*/\hbar^2\, V(x)$, which is the inverse of the square of the de Broglie wavelength in the well. If these two quantities are comparable, as is the case for the InAs layer, the effect of the presence of $V(x)$ in Eq. (21) is to lift the k_y degeneracy; otherwise stated, the energy depends on hk_y/eB, the position of the cyclotron orbit center. This result is expected classically because it expresses the fact that orbits that fit in the potential well and those that touch the boundaries of the well will have different energies. At 4 T the solution of Eq. 21 (see Fig. 24) shows a Landau level-like energy level for the orbit center in the middle of the InAs layer ($k_y = 0$) and a magnetic surface state-like level structure with a spacing of about $2\hbar\omega_c$ (see for instance Wanner *et al.*, 1975) for the orbit center at the interface $(\hbar k_y/eBx_0) = 1$ or -1. At 1 T (Fig. 25) the level structure is much more complicated. In both cases the level structure shows a remarkable magnetic field dependence and is very different from the zero field result shown in Fig.

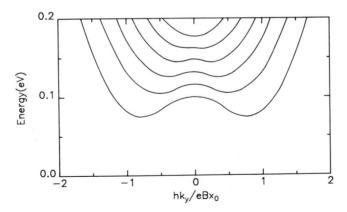

FIG. 24 Energy-level structure of sample S$_5$ with a magnetic field of 4 T parallel to the plane of the layers (z direction) as a function of $\hbar k_y/eBx_0$, the cyclotron orbit center coordinate normalized to the characteristic length of the band bending x_0, which is half the layer thickness.

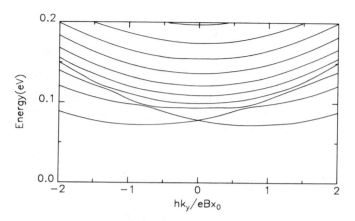

FIG. 25   Same as Fig. 24 but at 1 T.

25. This behavior is caused by the mixing of electrostatic subbands with Landau levels into a set of so-called hybrid subbands. At 4 T the magnetic quantization starts to dominate, and at 1 T the subband splitting of both types is of comparable magnitude, which causes the complex level structure.

The results for $B_\parallel$ (Fig. 21) can be interpreted as transitions between the hybrid subbands at $k_y = 0$. The comparison between calculated and measured data is shown in Fig. 22, and the agreement is quite satisfying. Qualitatively the angular dependence can also be understood (see Fig. 20); for instance, it is clear that the double cyclotron resonance peak due to CR of different electric subbands disappears when $B_\parallel$ increases, because in this case the distinction between electric and magnetic subbands is meaningless, the two being completely mixed in one set of hybrid subbands. Also, the interband transitions between H_1 and $E_{1,2}$ broaden as the tilt angle increases and disappear for $B_\parallel$. This is a direct consequence of the different influence of a $B_\parallel$ component on H_1 and $E_{1,2}$; H_1 remains virtually unaffected and $E_{1,2}$ depends strongly on the position of the orbit center. Consequently, transitions between the two bands will broaden and eventually disappear as the $B_\parallel$ component increases.

V.   Summary and Conclusions

The results previously discussed have demonstrated that InAs–GaSb layered structures can be considered as a new material with properties that are in some sense related to those of the bulk material but that also are different, as shown for instance, by the observation of the semimetallc character of some samples. It is clear that detailed FIR magnetooptical investigations are valuable in establishing the electronic properties of these crystals. This conclusion is not too surprising because the same technique has already shown its merit in establishing the properties of bulk semiconductors.

The description of the results and their interpretation has not been rigorous; instead, simple models have been used to explain the essential phenomena and were generally adequate to describe the experimental results. It has been emphasized that InAs–GaSb systems are interesting because of the delicate interplay of natural lengths such as the de Broglie wavelength, the screening length, the magnetic length, and the wavelength with sample dimensions and layer thickness. This interplay is demonstrated by the experimental results showing the finite width of the subbands, the angular dependence of the heterojunction transmission, the description of the helicon wave transmission, the semiconductor-to-semimetal transition, and the observation of the heterojunction properties for thick layers. These results show that InAs–GaSb layered structures cannot be considered as purely two-dimensional systems, because they have some three-dimensional character.

ACKNOWLEDGMENTS

The work reviewed above is the result of the effort of several people. Special credit should be given to Leroy Chang and Leo Esaki who conceived and realized these systems; to a large extent the foundation of the understanding of their properties has relied on their work. In addition, I am pleased to acknowledge the very pleasant and fruitful collaboration with Yves Guldner, Paul Voisin, Michel Voos, and Peter Wyder. Finally, I would like to thank Prof. G. Landwehr for his interest in this work.

REFERENCES

Abstreiter, G., Kotthaus, J. P., Koch, J. F., Dorda, G. (1976). *Phys. Rev. B* **14**, 2480.
Ando, T. (1975). *J. Phys. Soc. Japan* **38**, 989.
Ando, T. (1978). *Surface Sci.* **73**, 1.
Ando, T., Fowler, A. B., and Stern, F. (1982). *Rev. Mod. Phys.* **54**, 437.
Antcliffe, G. A., Bate, R. T., and Reynolds, R. A. (1971). *In* "Proceedings Conference on the Physics of Semimetals and Narrow Gap Semiconductors" (D. L. Carter and R. T. Bate, eds.), pp. 499. Pergamon, Oxford.
Bastard, G. (1981). *Phys. Rev. B* **24**, 5693.
Bastard, G. (1982). *Phys. Rev. B* unpublished.
Belyantsev, A. M., Ignatov, A. A. (1977). *Solid State Commun.* **24**, 817.
Bluyssen, H., Maan, J. C., Wyder, P., Chang, L. L., Esaki, L. (1979). *Solid State Commun.* **31**, 35.
Bluyssen, H., Maan, J. C., Wyder, P., Chang, L. L., Esaki, L. (1982). *Phys. Rev. B* **25**, 5364.
Brazis, R. S., Furdyna, J. K., Pozela, J. K. (1979). *Phys. Status Solid A* **53**, 11; **54**, 11.
Chang, L. L., Sakaki, H., Chang, C. A., and Esaki, L. (1977). *Phys. Rev. Lett.* **38**, 1489.
Chang, L. L. (1980). *J. Phys. Soc. Japan A* **49**, 997.
Chang, L. L., and Easki, L. (1980). *Surface Sci.* **98**, 70.

Dandekar, M. V., Madhukar, A., Lowy, D. N. (1980). *Phys. Rev. B* **21**, 5687.

Das Sarma, S., and Madhukar, A. (1981). *Phys. Rev. B* **23**, 805.

Das Sarma, S., and Quinn, J. J. (1982). Unpublished.

Dingle, R., Gossard, A. C., and Wiegman, W. (1975). *Phys. Rev. Lett.* **34**, 1327.

Dingle, R. (1975). *In* "Festkörperprobleme" (Vol. 15), p. 21. Pergamon, Oxford.

Döhler, G. H., Kunzel, H., Olego, D., Ploog, K., Ruden, P., Stolz, H. J. (1981). *Phys. Rev. Lett.* **47**, 864.

Esaki, L., Tsu, R. (1970). *IBM J. Res. Dev.* **14**, 61.

Esaki, L., and Chang, L. L. (1974). *Phys. Rev. Lett.* **33**, 495.

Esaki, L., and Chang, L. L. (1976). *Thin Solid Films* **36**, 285.

Esaki, L. (1980) *In* "Narrow Gap Semiconductors, Physics and Applications," Vol. 133 of "Lecture Notes in Physics" (W. Zawadzki, ed.) Springer Verlag, Berlin, p. 302.

Frensley, W. R., and Kroemer, H. (1977). *Phys. Rev. B* **16**, 2642.

Guldner, Y. *et al.* (1980). *Phys. Rev. Lett.* **45**, 1719.

Guldner, Y., Vieren, J. P., Voisin, P., Voos, M., Maan, J. C., Chang, L. L., Esaki, L. (1982). *Solid State Commun.* **41**, 755.

Ihm, J., Lam, Pui K., and Cohen, M. L. (1979). *Phys. Rev. B* **20**, 4120.

Kane, E. O. (1957). *J. Phys. Chem. Solids* **1**, 249.

Keyes, R. W. (1976). *Comment Solid State Phys.* **7**, 55.

Kotthaus, J. P. (1978). *Surf. Sci.* **73**, 472.

Landau, L. D., and Lifschitz, E. M. (1958). Ch. 7 In "Quantum Mechanics." Addison-Wesley, Reading, Massachusettes.

Litton, C. W., Dennis, R. B., and Smith, S. D. (1969). *J. Phys. C* **2**, 2146.

Maan, J. C. (1979). Ph.D thesis, Univ. of Nijmegen. Unpublished.

Maan, J. C. *et al.* (1981). *Solid State Commun.* **39**, 683.

Maan, J. C. *et al.* (1982). *Surface Sci.* **113**, 347.

Maan, J. C., Uihlein, Chr., Chang, L. L. and Esaki, L. (1982). *Solid State Commun.* **44**, 653.

Madhukar, A., and Nucho, R. N. (1979). *Solid State Commun.* **32**, 331.

Mukherji, D., and Nag, B. R. (1975). *Phys. Rev. B* **12**, 4338.

Neuberger, M. (1971). "Handbook of Electronic Materials," Vol. 2, Plenum, New York.

Nucho, R. N., and Madhukar, A. (1978). *J. Vac. Sci. Technol.* **15**, 1530.

Palik, E. D., and Stevenson, J. R. (1963). *Phys. Rev.* **130**, 1344.

Sai-Halasz, G. A., Tsu, R., and Esaki, L. (1977). *Appl. Phys. Lett.* **30**, 651.

Sai-Halasz, G. A., Chang, L. L., Welter, J. M., Chang, C. A., Esaki, L. (1978). *Solid State Commun.* **27**, 935.

Sai-Halasz, G. A., Esaki, L., and Harrison, W. A. (1978). *Phys. Rev. B* **18**, 2812.

Sakaki, H., Chang, L. L., Sai-Halasz, G. A., Chang, C. A., Esaki, L. (1978). *Solid State Commun.* **26**, 589.

Saris, F. W., Chang, C. A., Chu, W. K., and Esaki, L. (1979). *J. Vac. Sci. Technol.* **16**, 1506.

Smith, R. A. (1961). "Wave Mechanics of Crystalline Solids." Wiley, New York.

Stern, F., and Howard, W. E. (1967). *Phys. Rev.* **163**, 816.

Störmer, H. L. (1980). *J. Phys. Soc. Japan A* **49**, 1010.

Störmer, H. L., Gossard, A. C., and Wiegman, W. (1981). *Bull. Am. Phys. Soc.* **26**, 420.

Thomas, M. B., and Woolley, J. C. (1971). *Can. J. Phys.* **49**, 2052.

Wanner, M., Doezema, R. E., and Strom, U. (1975). *Phys. Rev. B* **12**, 2883.

White, S. R., and Sham, L. J. (1981). *Phys. Rev. Lett.* **47**, 879.

Wilson, B. A., Allen, S. J., and Tsui, D. C. (1980). *Phys. Rev. Lett.* **44**, 479.

Wilson, B. A., Allen, S. J., and Tsui, D. C. (1981). *Phys. Rev. B* **24**, 5887.

INDEX